Selected Titles in This Series

42 **Ding-Zhu Du and Frank K. Hwang, Editors,** Advances in Switching Networks

41 **David Aldous and James Propp, Editors,** Microsurveys in Discrete Probability

40 **Panos M. Pardalos and Dingzhu Du, Editors,** Network Design: Connectivity and Facilities Location

39 **Paul W. Beame and Samuel R Buss, Editors,** Proof Complexity and Feasible Arithmetics

38 **Rebecca N. Wright and Peter G. Neumann, Editors,** Network Threats

37 **Boris Mirkin, F. R. McMorris, Fred S. Roberts, and Andrey Rzhetsky, Editors,** Mathematical Hierarchies and Biology

36 **Joseph G. Rosenstein, Deborah S. Franzblau, and Fred S. Roberts, Editors,** Discrete Mathematics in the Schools

35 **Dingzhu Du, Jun Gu, and Panos M. Pardalos, Editors,** Satisfiability Problem: Theory and Applications

34 **Nathaniel Dean, Editor,** African Americans in Mathematics

33 **Ravi B. Boppana and James F. Lynch, Editors,** Logic and Random Structures

32 **Jean-Charles Grégoire, Gerard J. Holzmann, and Doron A. Peled, Editors,** The SPIN Verification System

31 **Neil Immerman and Phokion G. Kolaitis, Editors,** Descriptive Complexity and Finite Models

30 **Sandeep N. Bhatt, Editor,** Parallel Algorithms: Third DIMACS Implementation Challenge

29 **Doron A. Peled, Vaughan R. Pratt, and Gerard J. Holzmann, Editors,** Partial Order Methods in Verification

28 **Larry Finkelstein and William M. Kantor, Editors,** Groups and Computation II

27 **Richard J. Lipton and Eric B. Baum, Editors,** DNA Based Computers

26 **David S. Johnson and Michael A. Trick, Editors,** Cliques, Coloring, and Satisfiability: Second DIMACS Implementation Challenge

25 **Gilbert Baumslag, David Epstein, Robert Gilman, Hamish Short, and Charles Sims, Editors,** Geometric and Computational Perspectives on Infinite Groups

24 **Louis J. Billera, Curtis Greene, Rodica Simion, and Richard P. Stanley, Editors,** Formal Power Series and Algebraic Combinatorics/Séries Formelles et Combinatoire Algébrique, 1994

23 **Panos M. Pardalos, David I. Shalloway, and Guoliang Xue, Editors,** Global Minimization of Nonconvex Energy Functions: Molecular Conformation and Protein Folding

22 **Panos M. Pardalos, Mauricio G. C. Resende, and K. G. Ramakrishnan, Editors,** Parallel Processing of Discrete Optimization Problems

21 **D. Frank Hsu, Arnold L. Rosenberg, and Dominique Sotteau, Editors,** Interconnection Networks and Mapping and Scheduling Parallel Computations

20 **William Cook, László Lovász, and Paul Seymour, Editors,** Combinatorial Optimization

19 **Ingemar J. Cox, Pierre Hansen, and Bela Julesz, Editors,** Partitioning Data Sets

18 **Guy E. Blelloch, K. Mani Chandy, and Suresh Jagannathan, Editors,** Specification of Parallel Algorithms

17 **Eric Sven Ristad, Editor,** Language Computations

16 **Panos M. Pardalos and Henry Wolkowicz, Editors,** Quadratic Assignment and Related Problems

15 **Nathaniel Dean and Gregory E. Shannon, Editors,** Computational Support for Discrete Mathematics

14 **Robert Calderbank, G. David Forney, Jr., and Nader Moayeri, Editors,** Coding and Quantization: DIMACS/IEEE Workshop

(*Continued in the back of this publication*)

Advances in Switching Networks

DIMACS

Series in Discrete Mathematics
and Theoretical Computer Science

Volume 42

Advances in Switching Networks

DIMACS Workshop
July 7–9, 1997

Ding-Zhu Du
Frank K. Hwang
Editors

NSF Science and Technology Center
in Discrete Mathematics and Theoretical Computer Science
A consortium of Rutgers University, Princeton University,
AT&T Labs, Bell Labs, and Bellcore

American Mathematical Society

This DIMACS volume contains papers from the DIMACS Workshop on Network Switching, which was part of the DIMACS Special Year on Networks. The Workshop was held on July 7–9, 1997.

1991 *Mathematics Subject Classification.* Primary 05C40, 05C90.

Library of Congress Cataloging-in-Publication Data

Advances in switching networks : DIMACS workshop, July 7–9, 1997 / Ding-Zhu Du, Frank K. Hwang, editors.

p. cm. — (DIMACS series in discrete mathematics and theoretical computer science ; v. 42)

Includes bibliographical references (p.).

ISBN 0-8218-0831-1 (alk. paper)

1. Computer networks. 2. Switching theory. 3. Object-oriented programming (Computer science) 4. COMMON LISP (Computer program language) I. Du, Dingzhu. II. Hwang, Frank. III. DIMACS Workshop on Network Switching (1997 : Princeton University) IV. Series.

TK105.5.A34 1998

004.6′6–dc21

98-6284

CIP

∞ The paper used in this book is acid-free and falls within the guidelines
established to ensure permanence and durability.
Visit the AMS home page at URL: http://www.ams.org/

10 9 8 7 6 5 4 3 2 1   03 02 01 00 99 98

Contents

Foreword ix

Preface xi

Average-case bounds for the complexity of path-search
NICHOLAS PIPPENGER 1

Extended generalized shuffle networks: Sufficient conditions for strictly nonblocking operation
G. W. RICHARDS AND F. K. HWANG 15

Constructing nonblocking multicast switching networks with fanout reduction
CHUN-FAI CHAN AND CHIN-TAU LEA 49

Multirate broadcast switching networks nonblocking in a wide sense
DONGSOO S. KIM AND DING-ZHU DU 59

Repackable networks—The concept and applications
ANDRZEJ JAJSZCZYK AND MAREK KUBALE 75

Modeling the blocking behavior of Clos networks
YUANYUAN YANG AND NEIL H. KESSLER 85

Performance of a new decomposition algorithm for rearrangeable fault-tolerant Clos interconnection networks under sub-maximal and no-fault conditions
JOHN D. CARPINELLI AND CATHERINE B. WANG 103

Non-blocking multistage interconnection networks with limited depth
ITALO BUSI AND ACHILLE PATTAVINA 119

Isomorphism of classical rearrangeable networks
M. DOUGLAS MCILROY AND JOSEPH P. SAVICKI 147

Characterizing bit permutation networks
GERARD J. CHANG, FRANK K. HWANG, AND LI-DA TONG 157

Structure and density of sparse crossbar concentrators
EMRE GÜNDÜZHAN AND A. YAVUZ ORUÇ 169

Non-blocking routing properties of Clos networks
TONY T. LEE AND PHILIP P. TO 181

An algorithm for the construction of concentrators from 2×2 sorters
SHOU-YEN ROBERT LI, GAR MAN KOO, AND HUI LI 197

Strictly nonblocking conferencing meshes
YI DU AND GERALD M. MASSON 221

An $O(N^{1.695})$ permutation routing algorithm on augmented data manipulators
EHAB S. ELMALLAH AND CHIN-HUNG LAM 235

Preventing conflicts in input buffering baseline-based ATM switches
HASAN ÇAM 245

Routing strategy and performance evaluation of multiple-ring ShuffleNet topology for high speed wavelength-division multiplexed optical communications
DONGJIE HUANG AND KAMRAN KIASALEH 273

Multispace search for quorumcast routing
JUN GU, BIN DU, DANNY H. K. TSANG, AND WEI WANG 295

On 3-rate rearrangeability of Clos networks
GUO-HUI LIN, DING-ZHU DU, WEILI WU, AND KYEONGAH YOO 315

Foreword

The Workshop on Network Switching was held in July 1997 at Princeton University. We would like to express our appreciation to Ding-Zhu Du and Frank K. Hwang for their efforts to organize and plan this successful workshop.

The workshop was part of the broader Special Year on Networks. We extend our thanks to Stuart Haber, David Johnson and Mihalis Yannakakis for their work over many months as special year organizers.

The workshop focused on all aspects of network switching, including network environment, routing, network topology, switching components, nonblockingness and optimization.

DIMACS gratefully acknowledges the generous support that makes these programs possible. The National Science Foundation, through its Science and Technology Center program, the New Jersey Commission on Science and Technology, DIMACS partners at Rutgers, Princeton, AT&T Labs, Bell Labs and Bellcore generously supported the special year.

Fred S. Roberts
Director

Bernard Chazelle
Co-Director for Princeton

Preface

The articles collected in this book were presented in the DIMACS Workshop on Network Switching, held on July 7–9, 1997 at Computer Science Department, Princeton University.

There are many worthwhile and interesting issues about switching networks. This collection focuses on connectivity properties, such as nonblockingness and routing. By nonblockingness, we mean that an intended connection does not fail.

We wish to thank Professors C. T. Lea, G. M. Masson, A. Y. Oruc, P. M. Pardalos, N. Pippenger, C. S. Raghavendra, S. Sahni, J. S. Turner, and A. C. Yao for their advice on the organization of the workshop. We wish to thank all who contributed their articles to this book. We also wish to thank the National Science Foundation for their sponsorship.

Ding-Zhu Du and Frank K. Hwang

DIMACS Series in Discrete Mathematics
and Theoretical Computer Science
Volume 42, 1998

Average-Case Bounds for the Complexity of Path-Search

Nicholas Pippenger

ABSTRACT. A channel graph is the union of all paths between a given input and a given output in an interconnection network. At any moment in time, each vertex in such a graph is either idle or busy. The search problem we consider is to find a path (from the given input to the given output) consisting entirely of idle vertices, or to find a cut (separating the given input from the given output) consisting entirely of busy vertices. We shall also allow the search to fail to find either a path or a cut with some probability bounded by a parameter called the failure probability. This is to be accomplished by sequentially probing the idle-or-busy status of vertices, where the vertex chosen for each probe may depend on the outcome of previous probes. Thus a search algorithm may be modelled as a decision tree. For average-case analysis, we assume that each vertex is independently idle with some fixed probability, called the vacancy probability (and therefore busy with the complementary probability).

For one commonly studied type channel graph, the parallel graph, we show that the expected number of probes is at most proportional to the length of a path, irrespective of the vacancy probability, and even if the allowed failure probability is zero. Another type of channel graph we study is the spider-web graph, which is superior to the parallel graph as regard linking probability (the probability that an idle path, rather than a busy cut, exists). For this graph we give an algorithm for which, as the vacancy probability is varied while the positive failure probability is held fixed, the expected number of probes reaches its maximum near the critical vacancy probability (where the linking probability make a rapid transition from a very small value to a substantial value). This maximum expected number of probes is about the cube-root of the diversity (the number of paths between the input and output).

1. Introduction

A *channel graph* is an acyclic directed graph $G = (V, E)$ with *vertices* V and *edges* E in which there exists a *source* vertex $s \in V$ and a *target* vertex $t \in V$ such that every vertex lies on a directed path from s to t. (Such a source and target, if they exist, are clearly unique.) The vertices other than the source and target will be called *links*.

We shall deal in this paper with two particular families of channel graphs that have been proposed and analyzed for use in telephone-switching networks. The first

1991 *Mathematics Subject Classification*. Primary 68R10, 94C15; Secondary 60C05.

This research was supported by an NSERC Research Grant.

family, comprising the *fully parallel* channel graphs F_k (or simply *parallel graphs*), can be described as follows. Let T_k be a tree whose vertices are the $2^{k+1} - 1$ words of length at most k over the alphabet $\{0, 1\}$, with the empty word ε as the root, with edges directed from each word w of length at most $k - 1$ to each of the two words $w0$ and $w1$, and with the 2^k words of length k as leaves. Let T'_k be a tree like T_k, but with all vertices "primed" (to distinguish them from the corresponding vertices in T_k) and with all edges reversed. Then F_k is obtained from T_k and T'_k by joining each leaf w of T_k with the corresponding leaf w' of T'_k by an edge (w, w'). The source of F_k is the root of T_k, and the target is the root of T'_k. The second family, comprising the *spider-web* channel graphs G_k (or simply *spider-web graphs*), is formed like F_k, except that each leaf w of T_k is joined to the leaf $\varrho(w)'$ of T'_k by an edge $(w, \varrho(w)')$, where $\varrho(w) = \sigma_k \cdots \sigma_1$ denotes the *reversal* of the word $w = \sigma_1 \cdots \sigma_k$.

A *state* of a channel graph is an assignment of a *status* (*busy* or *idle*) to each link of the graph. We shall extend such an assignment to all vertices by agreeing that the source and target are always idle. We shall deal in this paper with a particular probability distribution on the states of a channel graph, which again has been proposed and analyzed in the context of telephone-switching networks. We choose a real number q, in the range $0 < q < 1$, which we call the *vacancy probability*; its complement $p = 1 - q$ is called the *occupancy probability*. We then define a random state of a channel graph to be one in which each link is independently idle with probability q or busy with probability p. This probability distribution on states, which was introduced independently by Lee [**L1**] and Le Gall [**L2**], is one most commonly used in the probabilistic analysis of interconnection networks.

We shall say that a channel graph is *linked* in a given state if there exists a directed path from the source to the target consisting entirely of idle links. We shall say that a channel graph is *blocked* in a given state if there exists a cut between the source and the target consisting entirely of busy links. (Clearly a channel graph in a given state is either linked or blocked, but not both.) If a channel graph H is in a random state with vacancy probability q, the *linking probability* will be denoted $Q(H, q)$, and the complementary *blocking probability* will be denoted $P(H, q) = 1 - Q(H, q)$. For the channel graphs we consider, the asymptotic dependence of the linking probability on q can be described fairly easily. For q strictly less than the critical probability $q_0 = 1/\sqrt{2}$, $Q(H, q)$ tends to zero as k tend to infinity for either $H = F_k$ or $H = G_k$, for the simple reason that the expected number of idle paths from the source to the target tends to zero. For $q > q_0$, the linking probability tends to a limit (as k tends to infinity). This limit (which is $1 - (1 - q)^4/q^4$ for G_k and the smaller $1 - (1 - q^2)^2/q^4$ for F_k) is strictly positive (for $q > q_0$), but also strictly less than unity (for $q < 1$). We shall refer to the range $0 < q \leq q_0$ as the "blocking regime", and the range $q_0 < q < 1$ as the "linking regime". See Ikeno [**I**] for analysis applicable to parallel graphs, and Pippenger [**P3**] for spider-web graphs.

A few words are in order here concerning the channel graphs and the probability distribution that we are using. Channel graphs arise form the study of telephone-switching networks, in which a number of inputs are connected to a number of outputs by means of a network comprising a number of switches interconnected by wires. We model the inputs, outputs and wires (which we call links) by vertices, and individual switches by edges. (There is an alternative convention used in modelling switching networks, whereby wires are modelled by edges and aggregations of switches (called crossbars) are modelled by vertices. This convention is only applicable to networks in which the switches aggregate into crossbars. This condition is satisfied by the networks considered in this paper, but we prefer the more flexible convention that we have adopted.) Both parallel graphs and spider-web graphs are among those considered by Ikeno [**I**]. Although spider-web graphs are not optimal (in the sense that they do not have the largest possible linking probability among graphs arising from networks with a given number of components; see Chung and Hwang [**CH**]), they have been shown by Pippenger [**P3**] to be asymptotically optimal. In the terminology of Pippenger [**P3**], the parallel graphs have rhyme scheme $12\cdots(k-1)k(k-1)\cdots 21$, and the spider-web graphs have rhyme scheme $12\cdots(k-1)k1\cdots(k-2)(k-1)$. The corresponding interconnection networks have also been studied for their combinatorial properties: the parallel graphs arise from a well-known network shown to be rearrangeable by Beneš [**B1**]; the question of whether the network corresponding to the spider-web graph is rearrangeable is the "shuffle-exchange conjecture" (see Beneš [**B2**]).

Consider now an algorithm that seeks to determine whether a known channel graph in an unknown state is linked or blocked. The algorithm gathers information about the state of the channel graph by sequentially probing the status of links until all the links of either an idle path or a busy cut have been probed. (The decision as to which link to probe at any step may depend upon the outcomes of all previous probes.)

Such an algorithm can be modelled as a decision tree. The elements of such a tree will be called *nodes* and *arcs* (to distinguish them from the vertices and edges of the channel graph). Each node is either a *probe* node, in which case it is labelled with the name of a link in the channel graph and has two outgoing arcs (one labelled "idle" and the other labelled "busy") leading to other nodes, or a *leaf* node, in which case it is labelled with one of the two possible outcomes ("linked" or "blocked") and has no outgoing arcs. There is a distinguished probe node, called the *root*, that has no incoming arcs; every other node has exactly one incoming arc. Execution of the algorithm begins at the root and proceeds in an obvious way, probing links and following the appropriate arcs, until it arrives at a leaf node that announces the final result. There is an obvious notion of such an algorithm being correct: every trajectory from the root to a "linked" leaf probes a set of links corresponding to a path in the channel graph and departs each of these nodes along the "idle" arc, and every trajectory from the root to a "blocked" leaf probes a set of links corresponding to a cut in the channel graph and departs each of these nodes along the "busy" arc. We shall denote by $\Omega(H, q)$ the minimum possible expected number of probes for any algorithm that correctly searches the channel graph H with vacancy probability q. Bounds to $\Omega(H, q)$ for series-parallel channel graphs (which include the case $H = F_k$ of parallel graphs, also considered below) have been give by Lin and Pippenger [**LP**].

We shall also want to consider search algorithms that occasionally fail to reach a decision as to whether the graph is linked or blocked. We shall model such algorithms by decision trees in which there are leaves of a third kind, "failure" leaves. We shall say that such an algorithm is *ε-correct* (for a given graph H and vacancy probability q) if the trajectory from the root reaches a failure leaf with probability at most ε, and if the algorithm reaches a correct leaf whenever it does not reach a failure leaf. We shall denote by $\Omega_\varepsilon(H, q)$ the minimum possible expected number of probes for any algorithm that ε-correctly searches the channel graph H

with vacancy probability q. We have of course

$$\Omega_\varepsilon(H,q) \le \Omega_0(H,q) = \Omega(H,q).$$

We shall only need to allow $\varepsilon > 0$ in one situation: that of spider-web graph with $q > 1/\sqrt{2}$, where the linking probability is bounded away from zero. But in this situation the blocking probability is also bounded away from zero, and it seems difficult to object to an algorithm that fails with at most a fixed proability that is small compared with the blocking probability.

Two final comments are in order. First, we have not considered randomized search algorithms: the probability space is that of the states of the channel graph. This entails no loss of generality: since the search process is a "game against nature" (rather than against an adversary), the payoff of a mixed strategy can always be matched or exceeded by that of a pure strategy. Second, we have taken the position that q is known and fixed while k tends to infinity (so that the graphs F_k and G_k become larger and larger). This of course does not correspond to the situation in practice where a specific network is subjected to varying traffic. It does accord well, however, with the fact that a network is usually designed so that the heaviest traffic it ordinarily receives leaves it with a vacancy probability q just slightly larger than the critical probability $q_0 = 1/\sqrt{2}$, which is independent of k; this allows the network to carry the largest amount of traffic consistent with a blocking probability bounded away from unity. (The provision of a blocking probability sufficiently close to zero to be satisfactory to subscribers is the task of additional stages of switching, outside the network considered here and beyond the scope of this paper.) Finally, we observe that in a situation in which the vacancy probability is not known, it is of course possible to estimate it closely and with high confidence by sampling.

2. Parallel Graphs

In this section we shall study the expected search cost $\Omega(F_k,q)$ of the fully parallel channel graph F_k. The following is our main result.

THEOREM 2.1. *For all $k \ge 1$ and all $0 < q < 1$, we have*

$$\Omega(F_k,q) \le 4k.$$

Observe that this bound is independent of q. Since there are just $2k$ ranks of links in F_k, on average at most 2 links in each rank need to be probed.

Let $\Omega'(F_k,q)$ denote the expected search cost of F_k when the source and target are, like the links, independently busy with probability p. (In this definition, say that F_k is blocked if either the source or target is busy.) We have the inequality

$$\Omega(F_k,q) \le 2\Omega'(F_{k-1},q), \tag{2.1}$$

since apart from its source and target F_k comprises two disjoint copies of F_{k-1}. Thus Theorem 2.1 will follow from (2.1) and the following proposition.

PROPOSITION 2.2. *For all $k \ge 0$ and all $0 < q < 1$, we have*

$$\Omega'(F_k,q) \le 2(k+1).$$

In the case $k = 0$, F_0 is a single edge, joining the source to the target. Thus we have

$$\Omega'(F_0,q) \le 2. \tag{2.2}$$

We also have the recurrence

$$\Omega'(F_k, q) \le 2 + q^2\big(1 + P(F_{k-1}, q)\big)\Omega'(F_{k-1}, q), \tag{2.3}$$

since we can search F_k by the following algorithm. First probe the source and target, at cost 2. Halt with the result "blocked" unless both the source and target are idle, an event which occurs with probability q^2. In this case, search one of the two disjoint copies of F_{k-1}, at expected cost $\Omega'(F_{k-1}, q)$. Halt with the result "linked" unless the searched copy is blocked, an event which occurs with probability $P(F_{k-1}, q)$. In this case, search the remaining copy of F_{k-1}. This algorithm establishes (2.3).

To use (2.3), we need an estimate for the blocking probability $P(F_k, q)$, which is provided by the following lemma.

LEMMA 2.3. *For all $k \ge 0$ and all $0 < q < 1$, we have*

$$P(F_k, q) \le q^{-2} - 1.$$

PROOF. We proceed by induction on k. For the basis, $k = 0$, we have

$$\begin{aligned} P(F_0, q) &= 1 - q^2 \\ &\le q^{-2} - 1, \end{aligned}$$

where the inequality, which is equivalent to $1 \le (q^2 + q^{-2})/2$, compares the geometric and arithmetic means of q^2 and q^{-2}.

For the inductive step, $k \ge 1$, we have

$$\begin{aligned} P(F_k, q) &= 1 - q^2\big(1 - P(F_{k-1}, q)^2\big) \\ &= 1 - q^2 + q^2 P(F_{k-1}, q)^2 \\ &\le 1 - q^2 + q^2\big(q^{-2} - 1\big)^2 \\ &= 1 - q^2 + q^2\big(q^{-4} - 2q^{-2} + 1\big) \\ &= q^{-2} - 1, \end{aligned}$$

where the inequality follows by inductive hypothesis.

Substituting this estimate into (2.3) yields the recurrence

$$\Omega'(F_k, q) \le 2 + \Omega'(F_{k-1}, q). \tag{2.4}$$

Proposition 2.2 now follows by induction, with (2.2) as the basis and (2.4) as the inductive step.

The foregoing algorithm and its analysis have a simple interpretation that will be useful later. Consider the problem of finding an idle path from the root to an idle leaf of T_k, supposing as usual that the root is idle and that every other vertex is independently idle with probability q. This problem can be solved by an obvious "depth-first search", and the analysis given above yields the following result.

COROLLARY 2.4. *If the root of T_k is idle, and every other vertex is independently idle with probability q, then depth-first search finds an idle path from the root to an idle leaf (or finds a busy cut showing that no such path exists) using at most $2k$ probes (and finds a cut rather than a path with probability at most $(1-q)^2/q^2$).*

If we fold the graph F_k back upon itself, we obtain a tree similar to T_k, which we shall denote T''_k, with two vertices of F_k corresponding to each vertex of T_k. Thus

Theorem 2.1 can be recovered from Corollary 2.4 by observing that each probe in T_k'' can be simulated by 2 probes in F_k (so that we have a bound of $4k$ rather than $2k$), and 2 vertices in F_k must be idle for the corresponding vertex in T_k'' to be idle (so that in the bound $(1-q)^2/q^2$, q must be replaced by q^2 to obtain $(1-q^2)^2/q^4$).

3. Spider-Web Graphs

In this section we shall begin our study of the expected search cost $\Omega(G_k, q)$ of the spider-web channel graph G_k. We shall present an algorithm that performs well in the blocking regime, and in the next section we shall combine this algorithm with other elements to obtain good performance in the linking regime as well.

We can search G_k using an algorithm comprising the following sequence of k phases. During the j-th phase, for $1 \leq j \leq k$, we shall probe each link w in ranks j and j' that meets the following condition: there exists a path from the source through w to the target on which every link that has been probed is idle.

This algorithm will always find either an idle path or a busy cut. If there is an idle path from the source to the target, all its links will be probed. (Thus this algorithm will probe the links on all idle paths, and not just one.) If a path from the source to the target has a busy link, then such a link will be probed in the j-th phase, where j is the smallest number such that the path has a busy link in rank j or j'. Thus if every path has a busy link, the algorithm will probe every link of some busy cut.

It remains to analyze the expected cost of this algorithm, which we shall denote $A(k, q)$. We may write

$$A(k,q) = \sum_{1\leq j\leq k} A_j(k,q),$$

where $A_j(k,q)$ denotes the expected number of probes in the j-th phase. The expected number of probes in the j-th phase may be expressed as the sum over links w in rank j or j' of the probability that w is probed. There are 2^j links in rank j and 2^j links in rank j', and thus there are $2 \cdot 2^j$ links i rank j or j'. If v and w are two links in rank j, then the probability that v is probed is the same as the probability that w is probed. To see this, observe that there is an automorphism of G_k taking the union of all paths from the source through v to the target to the union of all paths from the source through w to the target, and that the probability distribution on states is invariant under automorphisms of G_k. Furthermore, the probability that a link w is probed is the same as the probability that w' is probed. This can be seen by considering anti-automorphisms, which exchange source and target and reverse the directions of edges. Denoting this common probability of being probed by $P_j(k,q)$, we have

$$A(k,q) = 2 \sum_{1\leq j\leq k} 2^j P_j(k,q).$$

To estimate $P_j(k,q)$, consider a link w in rank j. There are $j-1$ links on the path in T_k from the source to w (not including w), and all of these links must be idle if w is to be probed; this event occurs with probability q^{j-1}. Thus $P_j(k,q) \leq q^{j-1}$, and we have

$$A(k,q) \leq 2 \sum_{1\leq j\leq k} 2^j q^{j-1}.$$

There are k terms in this summation; if $q \leq 1/2$, the largest of these is 2, which arises from $j = 1$. Thus we have (for $q \leq 1/2$)

$$A(k,q) \leq 4k. \tag{3.1}$$

To analyze this algorithm for $1/2 < q < 1$, we shall need a sharper estimate for $P_j(k,q)$. Consider a link w in rank j. There are 2^{k-j} paths from the source through w to the target. Denote by K the union of these paths. The portion of K between the source and w is a path containing $j-1$ links (not including w). These $j-1$ links have already been probed (in phases earlier than the j-th), and if w is to be probed they must all be idle; this event occurs with probability q^{j-1}. Now consider the portion of K joining w to the target. The links of this portion that have already been probed are those in ranks $(j-1)', \ldots, 2', 1'$. They form a tree (a subtree of T'_k) that is the union of at most 2^{k-j} paths each containing $j-1$ links, and if w is to be probed all the links on at least one such path must be idle; this event occurs with probability at most $2^{k-j}q^{j-1}$ (and of course with probability at most 1). Thus we have

$$P_j(k,q) \leq q^{j-1} \min\{1, 2^{k-j}q^{j-1}\},$$

and therefore

$$A(k,q) \leq 2 \sum_{1 \leq j \leq k} 2^j q^{j-1} \min\{1, 2^{k-j}q^{j-1}\}.$$

Since we are now assuming $q > 1/2$, we can write

$$A(k,q) \leq 8 \sum_{1 \leq j \leq k} 2^j q^j \min\{1, 2^{k-j}q^j\}.$$

To estimate this sum, we define $\alpha(q)$ by

$$\alpha(q) = \frac{\log 2}{\log(2/q)},$$

so that $1 < 2^{k-j}q^j$ if and only if $j < \alpha(q)k$. Thus we have

$$A(k,q) \leq 8 \sum_{1 \leq j < \alpha(q)k} 2^j q^j + 8 \sum_{\alpha(q)k \leq j \leq k} 2^k q^{2j}.$$

The two summations contain k terms between them, and each of these terms is at most the maximum of these terms. In the case of the first summation, this maximum is (since $2q > 1$) at most $(2q)^{\alpha(q)k} = 2^{\beta(q)k}$, where we define

$$\beta(q) = \frac{\log(2q)}{\log(2/q)}$$

for $q > 1/2$, and in the case of the second summation it is (since $q^2 < 1$) at most $2^k q^{2\alpha(q)k} = 2^{\beta(q)k}$. Thus we have (for $1/2 < q < 1$)

$$A(k,q) \leq 8k\, 2^{\beta(q)k}. \tag{3.2}$$

We combine the results of (3.1) and (3.2) in the following theorem.

THEOREM 3.1. *For all $k \geq 1$ and $0 < q < 1$, we have*

$$\Omega(G_k, q) \leq 8k\, 2^{\beta(q)k},$$

where $\beta(q) = 0$ for $0 < q \leq 1/2$ and

$$\beta(q) = \frac{\log(2q)}{\log(2/q)}$$

for $1/2 < q < 1$. This is achieved by an algorithm that finds all idle paths from the source to the target.

We shall now restate the results of this section in a form that simplifies the expressions involved and gives them some intuitive significance. We may regard the search for an idle path as taking place in a search-space containing 2^k paths. We shall think of the subset of paths actually probed as constituting a subspace of fractional dimension: if 2^{dk} paths are probed, then the dimension of this subspace is d $(0 \leq d \leq 1)$. This suggests that we transfer attention from $\Omega(G_k, q)$ to $k^{-1} \log \Omega(G_k, q)$. Furthermore, it will be convenient to make the substitution $q = 2^\lambda$ as well, so that $-\infty < \lambda < 0$. Thus we define

$$d(k, \lambda) = k^{-1} \log \Omega(G_k, 2^\lambda).$$

We can now express the analysis in this section by

$$d(k, \lambda) \leq b(\lambda) + O\left(\frac{\log k}{k}\right),$$

where we define $b(\lambda)$ by $b(\lambda) = 0$ for $-\infty < \lambda \leq -1$ and $b(\lambda) = (1+\lambda)/(1-\lambda)$ for $-1 < \lambda < 0$. We observe that $(1+\lambda)/(1-\lambda)$ increases from 0 to 1 as λ increases from -1 to 0, and that it reaches the value $1/3$ at $\lambda = -1/2$. In the next section we shall derive for $-1/2 < \lambda < 0$ a sharper upper bound to $d(k, \lambda)$ (more precisely, to an analogous quantity $d_\varepsilon(k, \lambda)$ that allows for positive, but arbitrarily small, probability ε of failure) that decreases from $1/3$ to 0 as λ increases from $-1/2$ to 0. Thus the value $1/3$ represents the maximum "search dimension" required by our algorithms, achieved for the value $\lambda = -1/2$ corresponding to the critical vacancy probability $q = q_0 = 1/\sqrt{2}$.

4. Spider-Web Graphs in the Linking Regime

In this section we shall present an algorithm that improves the performance in the linking regime of the algorithm presented in the preceding section. The latter algorithm is limited by the fact that it finds all idle paths, so that it cannot use fewer probes than the number of such paths. We might contemplate reducing the number of idle paths explored by constraining the initial and final segment of the path. If the length of these constrained segments is chosen appropriately, the number of idle paths will be reduced to a small number, just large enough to give a blocking probability near $P(G_k, q)$. The problem with this strategy is that the task of finding suitable initial and final segments also contributes a blocking probability near $P(G_k, q)$. The solution to this problem is to allow a number of alternatives (about k) for the initial and final segments. We shall present the resulting algorithm as a sequence of three steps, interspersed with their analysis.

Set

$$i = \lceil 4 \log(2k) \rceil.$$

STEP 1. Find the set I of all links v in rank i of T_k such that all links on the path from the source to v (including v itself) are idle. Similarly, find the set I' of all links v' in rank i' of T'_k such that all links on the path from v' to the target (including v' itself) are idle. If either I or I' is empty, there is a busy cut separating the source from the target. If neither I not I' is empty, but either I or I' contains fewer than $2k$ links, the algorithm fails. Otherwise, let $v_1, \ldots, v_{2k}$ be $2k$ distinct links in I, and let $v'_1, \ldots, v'_{2k}$ be $2k$ distinct links in I'.

The number of links probed in Step 1 is at most

$$2 + 2^2 + \cdots + 2^i \leq 64k^4.$$

To estimate the probability of failure in Step 1, we need to estimate the probability that I contains at least 1 but fewer than $2k$ links. The cardinality of I is the population Z_i in the i-th generation of a branching process that has been analyzed by Pippenger [**P3**]. By the analysis preceding Lemma 8.1 of that paper, we have

$$\Pr(1 \leq Z_i < 2k) = O\big((2k/R)^\alpha\big),$$

where $R = (2q)^i \geq (2q_0)^i \geq 4k^2$, and $\alpha = -\log\big(2(1-q)\big)/\log(2q) \geq -\log\big(2(1-q_0)\big)/\log(2q_0) > 1$. Thus we have

$$\Pr(1 \leq Z_i < 2k) = O(1/k).$$

The probability of failure in Step 1 is also $O(1/k)$, since it is at most $\Pr(1 \leq Z_i < 2k)$ plus an equal contribution from the primed side of the graph.

Set

$$j = \left\lfloor \frac{k\log(2q^2) - 4\log k}{2\log(2q)} \right\rfloor.$$

We shall assume that k is sufficiently large that $j > i$, which we may do since $\log(2q^2)/\log(2q) > 0$ for $q > q_0$. Furthermore, we shall assume that $j < (k-1)/2$, which we may do since $\log(2q^2)/\log(2q) < 1/2$ for $q < 1$.

STEP 2. For each link v_h $(1 \leq h \leq 2k)$ found in Step 1, find if possible using depth-first search a link w in rank j of T_k such that all links on the path from v_h to w (including w itself) are idle. If fewer than k of these $2k$ attempts are successful, the algorithm fails. Otherwise, let $w_1, \ldots, w_k$ be k such links, which then have the property that all the links on the path from the source to w_h, inclusive, are idle. Similarly, for each link v'_h $(1 \leq h \leq 2k)$ found in Step 1, find if possible using depth-first search a link w' in rank j' of T'_k such that all links on the path from w' (including w' itself) to v'_h are idle. If fewer than k of these $2k$ attempts are successful, the algorithm fails. Otherwise, let $w'_1, \ldots, w'_k$ be k such links, which then have the property that all the links on the path from w'_h, inclusive, to the target are idle.

According to Corollary 2.4, the expected number of probes required for each of the $2k+2k = 4k$ trials in Step 2 is at most $2(j-i) \leq 2k$. Thus the expected number of links probed in Step 2 is at most $8k^2$. To estimate the probability of failure in Step 2, we must estimate the probability of having fewer than k successes in $2k$ independent trials. According to Corollary 2.4, each trial succeeds with probability at least $1 - (1-q)^2/q^2 \geq 1 - (1-q_0)^2/q_0^2 > 3/4$. By Chebyshev's inequality, the probability $\Pr(Y < k)$ that the number Y of successes in $2k$ trials, which has expectation at least $(3/4)\,2k = 3k/2$ and variance at most $(1/4)\,2k = k/2$, deviates by at least $3k/2 - k = k/2$ from its expectation is at most $(k/2) \,/\, (k/2)^2 = 2/k$. The probability of failure in Step 2 is thus at most $4/k$, since it is $\Pr(Y < k)$ plus a equal contribution from the primed side of the graph.

For $1 \le h \le k$, denote by L_h the channel graph comprising all vertices of G_k lying on paths from w_h to w'_h, with w_h as source and w'_h as target. The graph L_h is a generalized spider-web graph of a type dealt with by Pippenger [**P3**]. Since G_k arises from a network with rhyme scheme $12\cdots(k-1)k12\cdots(k-2)(k-1)$, L_h arises from an network with rhyme scheme $(j+1)(j+2)\cdots(k-1)k12\cdots(k-2-j)(k-1-j)$. This is equivalent to the rhyme scheme $12\cdots(s+t-1)(s+t)12\cdots(s+2t-1)(s+2t)$, where $s = 2j+1$ and $t = k-1-2j$. The channel graph arising from a network with this rhyme scheme will be denoted $G_t \circ H_s$; it is obtained from G_t by substituting for each central edge $\big(u, \varrho(u)'\big)$ a copy of a channel graph H_s that has a unique path from the source u through s edges and $s-1$ intermediate links to the target $\varrho(u)'$. Thus each L_h is isomorphic to $G_t \circ H_s$. Furthermore, for $1 \le f < g \le k$, the graphs L_f and L_g are disjoint: the portions in T_k are disjoint because w_f and w_g are distinct, and the portions in T'_k are disjoint because w'_f and w'_g are distinct.

STEP 3. For $1 \le h \le k$, determine in the following way if there is an idle path in L_h from w_h to w'_h. To determine if $L_h \approx G_t \circ H_s$ is linked, apply the algorithm of Section 3 to G_t to find the set U of central edges of idle paths in G_t. Then, for each edge $\big(u, \varrho(u)'\big)$ in U, determine if all links in the copy of H_s from u to $\varrho(u)'$ are idle, so that the idle path through $\big(u, \varrho(u)'\big)$ in G_t extends to an idle path in L_h. If an idle path through some L_h is found, it extends to an idle path through G_k. If all L_h are blocked, the algorithm fails.

According to Theorem 3.1, each application of the algorithm of Section 3 to a copy of G_t uses an expected number of probes at most $8t\,2^{\beta(q)t} \le 8k\,2^{\beta(q)(k-2j)}$. This expression of course also bounds the number of idle paths found, and thus the expected cardinality of U. For each edge in U, the number of probes needed to search the corresponding copy of H_s is at most $s-1 \le k-1$. Thus the expected number of probes needed to search each L_h is at most $8k^2\,2^{\beta(q)(k-2j)}$. Since there are k such channel graphs L_h, the expected number of links probed in Step 3 is at most $8k^3\,2^{\beta(q)(k-2j)}$.

From the definition of j we have

$$\begin{aligned} k - 2j &\le k - \frac{k\log(2q^2) - 4\log k}{\log(2q)} + 2 \\ &= \frac{k\log(1/q) + 4\log k + 2\log(2q)}{\log(2q)}. \end{aligned}$$

Since $\beta(q) = \log(2q) \,/\, \log(2/q)$ for $q > q_0$, this yields

$$\beta(q)\,(k-2j) \le \frac{k\log(1/q) + 4\log k + 2\log(2q)}{\log(2/q)},$$

and since $\log(2q) \,/\, \log(2/q) < 1 \,/\, \log(2/q) < 1$ for $q_0 < q < 1$, we obtain

$$\beta(q)\,(k-2j) \le \gamma(q)\,k + 2 + 4\log k,$$

where we define

$$\gamma(q) = \frac{\log(1/q)}{\log(2/q)}$$

for $q_0 < q < 1$. Thus the expected number of links probed in Step 3 is at most

$$8k^3\, 2^{\beta(q)(k-2j)} \leq 32k^7\, 2^{\gamma(q)k}.$$

To estimate the probability of failure in Step 3, we need to estimate the probability of blocking in each L_h. To do this we shall need to estimate the expected number $J = 2^t q^{s+2t-1} = 2^{k-2j-1} q^{2k-2j-2} \geq 2^{k-2j} q^{2k-2j}$ of idle paths from w_h to w'_h in L_h. Since

$$2j \leq \frac{k\log(2q^2) - 4\log k}{\log(2q)},$$

we obtain $J \geq k^4$. According to Theorem 10.1 from Pippenger [**P3**], the blocking probability of each L_h is then $1-(1-\xi)^2 + O(J^{-\beta})$, where $\xi = (1-q)^2/q^2$ and $\beta = \alpha/2(1+\alpha) > 1/4$ (since $\alpha > 1$). Since $\xi = (1-q)^2/q^2 \leq (1-q_0)^2/q_0^2 < 1/4$, we have $1-(1-\xi)^2 < 1/2$. Thus the blocking probability in each L_h is at most $1/2 + O(1/k)$, and we may assume that k is sufficiently large that it is at most $2/3$. Since failure occurs in Step 3 only if blocking independently occurs in all k disjoint channel graphs L_h, the probability of failure in Step 3 is at most $(2/3)^k$.

Combining the expected number of probes from the three steps yields

$$64k^4 + 8k^2 + 32k^7\, 2^{\gamma(q)k} \leq 128k^7\, 2^{\gamma(q)k}.$$

Combining the estimates for probability of failure yields

$$O(1/k) + 4/k + (2/3)^k = O(1/k),$$

and thus for any fixed $\varepsilon > 0$ we may assume that k is large enough that the probability of failure for the entire algorithm is at most ε. Thus we have obtained the following result.

THEOREM 4.1. *For any fixed $q > q_0$ and $\varepsilon > 0$, and for all sufficiently large k, we have*

$$\Omega_\varepsilon(G_k, q) \leq 128k^7\, 2^{\gamma(q)k}.$$

Combining this result with Theorem 3.1, and extending the domain of definition of $\gamma(q)$, we have the following result.

THEOREM 4.2. *For any fixed $0 < q < 1$ and $\varepsilon > 0$, and for all sufficiently large k, we have*

$$\Omega_\varepsilon(G_k, q) \leq 128k^7\, 2^{\gamma(q)k},$$

where $\gamma(q) = 0$ for $0 < q \leq 1/2$,

$$\gamma(q) = \frac{\log(2q)}{\log(2/q)}$$

for $1/2 < q < q_0$ and

$$\gamma(q) = \frac{\log(1/q)}{\log(2/q)}$$

for $q_0 < q < 1$.

Finally, if we define

$$d_\varepsilon(k,\lambda) = k^{-1}\log \Omega_\varepsilon(G_k, 2^\lambda)$$

as the search dimension analogous to $d(k, \lambda)$, then can express the analysis in this section by

$$d_\varepsilon(k, \lambda) \leq c(\lambda) + O\left(\frac{\log k}{k}\right),$$

where we define $c(\lambda)$ by $c(\lambda) = 0$ for $-\infty < \lambda \leq -1$, $c(\lambda) = (1+\lambda)/(1-\lambda)$ for $-1 < \lambda < -1/2$ and $c(\lambda) = -\lambda/(1-\lambda)$ for $-1/2 < \lambda < 0$. As described before, the function $c(\lambda)$ assumes its maximum value 1/3 for the unique critical value $\lambda = -1/2$.

5. Conclusion

The most obvious problem left by this paper is to determine whether the algorithms presented for spider-web networks are asymptotically optimal. We can show that this is the case in the blocking regime, but in the linking regime the best lower bound we have been able to obtain falls below the upper bound. It is also an open question as to whether the upper bound we have presented in the linking regime can be obtained for $\varepsilon = 0$.

It would also be of interest to address some of the shortcomings of the models we have employed. The probability distribution on states introduced by Lee [**L1**] and Le Gall [**L2**] is the simplest to use for analysis, but is somewhat unrealistic. Some of its defects are rectified in a more complicated model introduced by Pippenger [**P1**] (and in a slightly different form by Koverninskiĭ [**K**]). This model has been used to analyze the linking probabilities of both series-parallel networks (Pippenger [**P1**]) and spider-web networks (Pippenger [**P2**]). The decision-tree model we have used gives a good account of time requirements (at least in so far as they are reflected by probes), but takes no account of space requirements. This shortcoming could be addressed by replacing the decision-tree with a decision-graph, and reckoning space as the logarithm of the number of nodes in the graph. It should then be possible to analyze the time-space trade-off for searching spider-web networks.

Finally, we should mention the possibility of extending the analysis to other families of channel graphs. The spider-web graphs are superior to parallel graphs as regards linking probability, but they require more time for path-search. This raises the problem of constructing a new family of channel graphs that might enjoy the advantages of both, and the disadvantages of neither. Since the linking probability is determined largely by events in the outermost stages of a network, while much of the search time is consumed by links deeper in the network, it seems reasonable to hope that a network that is spider-web-like in its outer stages and parallel-like within (say with the rhyme scheme $12\cdots j(j+1)\cdots(k-1)k(k-1)\cdots(j+1)1\cdots(j-1)j$, for some j growing very slowly with k) would accomplish these objectives.

6. References

[**B1**] V. E. Beneš, Optimal Rearrangeable Multistage Connecting Networks, *Bell System Tech. J.*, vol. 43, 1964, pp. 1641–1656.

[**B2**] V. E. Beneš, Proving the Rearrangeability of Connecting Networks by Group Calculations, *Bell System Tech. J.*, vol. 45, 1975, pp. 421–434.

[**CH**] F. R. K. Chung and F. K. Hwang, The Connection Patterns of Two Complete Binary Trees, *SIAM J. Algeb. and Discr. Meth.*, vol. 1, 1980, pp. 322–335.

[**I**] N. Ikeno, A Limit on Crosspoint Number, *IRE Trans. on Inform. Theory*, vol. 5, 1959, pp. 187–196.

[**K**] I. V. Koverninskiĭ, Estimation of the Blocking Probability for Switching Circuits by Means of Probability Graphs, *Problems of Info. Transm.*, vol. 11, 1975, pp. 63–71.

[**L1**] C. Y. Lee, Analysis of Switching Networks, *Bell Syst. Tech. J.*, vol. 34, 1955, pp. 1287–1315.

[**L2**] P. Le Gall, Étude du blocage dans les systèmes de commutation téléphoniques automatiques utilisant des commutateurs électroniques du type crossbar, *Ann. des Téécomm.*, vol. 11, 1956, pp. 159–171, 180–194.

[**LP**] G. Lin and N. Pippenger, Routing Algorithms for Switching Networks with Probabilistic Traffic, *Networks*, vol. 28, 1996, pp. 21–29.

[**P1**] N. Pippenger, On Crossbar Switching Networks, *IEEE Trans. on Comm.*, vol. 23, 1975, pp. 646–659.

[**P2**] N. Pippenger, The Blocking Probability of Spider-Web Networks, *Random Structures and Algorithms*, vol. 2, 1991, pp. 121–149.

[**P3**] N. Pippenger, The Asymptotic Optimality of Spider-Web Networks, *Discr. Appl. Math.*, vol. 37/38, 1992, pp. 437–450.

Department of Computer Science, University of British Columbia, 2366 Main Mall, Vancouver, BC V6T 1Z4, Canada

E-mail address: nicholas@cs.ubc.ca

DIMACS Series in Discrete Mathematics
and Theoretical Computer Science
Volume 42, 1998

Extended Generalized Shuffle Networks: Sufficient Conditions for Strictly Nonblocking Operation

G. W. Richards
F. K. Hwang

Abstract

Since Clos gave the first construction of a strictly nonblocking multistage interconnection network, only a few other constructions have been proposed in almost half a century. In this paper we introduce a constructive class of networks which utilizes crossbars of virtually any size, and for which the sizes can vary from stage to stage. The interconnection between stages is a generalized shuffle pattern. We derive sufficient conditions for strictly nonblocking operation and suggest the potential for wide application of these networks by providing several special case results.

1.0 Introduction

This work was initially motivated during the process of attempting to design efficient photonic nonblocking switching networks. Various photonic switching constraints suggested a need for flexible design capabilities. It became apparent that it would be useful to be able to design networks composed of various numbers of stages of diverse functionality switching modules of (nearly) arbitrary sizes and shapes.

This led to the development of a broad class of networks, which we call Extended Generalized Shuffle (EGS) networks, whose members demonstrate the diversity just mentioned while at the same time satisfy a simple inter-stage interconnection rule. (Formal definitions of these networks and other terminology used in this introduction follow in Section 2). The generality of EGS networks is such that they include all three-stage Clos networks[2] and all so called baseline networks[5] (Omega, Banyan, etc.). Also, although photonic switching constraints prompted this work, our results are technology independent and thus EGS networks should in no way be considered as being limited to photonic applications.

EGS networks can exhibit numerous interesting attributes including multi-stage modularity, fault tolerance, elegant path hunting and control algorithms, etc. In this paper we will primarily focus on sufficient conditions for which EGS networks are strictly nonblocking for point-to-point connections. As will be presently seen, this is a rather large undertaking in itself and will serve to introduce the reader to some fundamental properties of EGS networks.

1991 Mathematics Subject Classification Number---94C10

The results concerning nonblocking networks are theoretically interesting and in addition they serve as points of departure for the design of low probability blocking networks. This is important because strictly nonblocking networks are seldom implemented in practice due to cost and performance considerations. One interesting result to be presented is that the expressions that give the conditions for nonblocking operation are global as opposed to being specific to particular switching modules; i.e., nowhere in these expressions does there exist the ratio of the number of inlets to the number of outlets of any switching module in the network. The implication is that for the most part nonblocking networks can be designed with arbitrarily sized switching modules.

The nonblocking conditions to be derived will be given in general terms of number of inlets and outlets on the network, the number of inlets and the number of outlets on the switching modules in each stage of the network, and the number of stages in the network. For academic comparison we will show that a strictly nonblocking EGS network with N inlets and N outlets can be constructed with $O(N(\log N)^2)$ crosspoints. This is asymptotically as good as any construction we know of[1], but perhaps a more important consideration is whether practically sized EGS networks can be efficiently and conveniently constructed for low (but nonzero) probabilities of blocking while meeting the constraints of the technology being used. We think that the answer to this question is usually yes. In the next section we will formalize via definitions much of what has been casually mentioned in this introduction.

Some aspects of our approach may look similar to that of Lea and Shyy[3,4], but there are a number important fundamental differences. First of all, their construction applies only to k-extra-stage, $0 \le k \le \log_2 N - 1$, networks using 2x2 crossbars in which the first N stages are a Banyan network and the last k states are a mirror image of the first k stages (and other networks isomorphic to these). Our construction imposes no constraints on either the size of the crossbars or the number of stages. In addition, we employ the the same general interconnection definition for all stages. Consequently, the channel graphs of their networks have a special structure and are relatively easy to analyze. Our channel graphs are very general and thus demand a different and more comprehensive analysis.

2.0 Terminology and Definitions

As noted in the introduction, this work was motivated by constraints imposed by photonic technology on the design of switching networks. We begin by considering the switching modules from which a switching network is constructed, followed by the logical association and interconnection of these modules. We then discuss strictly non-blocking operation, including the concepts of paths and blocked paths.

2.1 Nonblocking Operation

We are interested in networks that are *strictly nonblocking* for *point-to-point connections*. A point-to-point connection is one that connects a single inlet and a single outlet. (This is distinguished, for example, from a multi-point connection in which a single inlet may be connected to one or more outlets simultaneously.) A network which allows only point-to-point connections is said to be strictly nonblocking if any idle inlet can be connected to any idle outlet, regardless of the other point-to-point connections existing in the network. Unless otherwise indicated, from this point on, we will refer to such networks simply as nonblocking networks.

2.2 Switching Modules

A switching module is defined as a device having two sets of terminals denoted inlets and outlets, plus a set of operational states such that for every inlet/outlet pair there exists at least one operational state in which that inlet/outlet pair are connected and at least one operational state in which that inlet/outlet pair are not connected. No operational states connect inlets to inlets nor outlets to outlets. Furthermore, an operational state does not necessarily isolate connected inlet/outlet pairs; i.e., a given inlet can be simultaneously connected to more than one outlet and at the same time a given outlet can be simultaneously connected to more than one inlet.

Thus, by our definition, a switching module is a device that has the capability to connect and disconnect all inlet/outlet pairs but not necessarily independently. In a subsequent paper we will consider various functional types of switching modules, corresponding to different connect/disconnect capability. However, in this paper we will limit our consideration to that of conventional "crossbar" type switching modules.

Consider an n by m switching module (n inlets and m outlets) represented equivalently by either an n by m crossbar or a complete bipartite graph having two vertex sets corresponding to the n inlets and m outlets, respectively, and the nm edges correspond to the nm crosspoints in the n by m crossbar. An existing edge in the bipartite graph corresponds to a closed crosspoint in the crossbar, and a removed edge in the bipartite graph corresponds to an open crosspoint in the crossbar. See Figure 1.

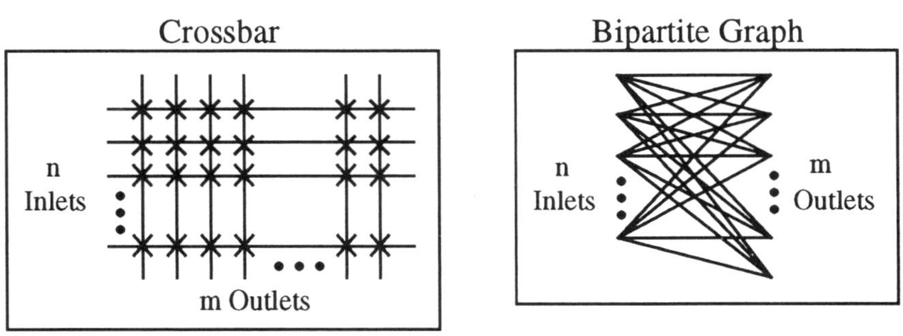

Figure 1. Switching Module Representations

In this type of switching module crosspoints may be opened or closed individually. This is the normal assumption for a crossbar switch and results in a nonblocking module since any inlet/outlet pair can be connected and at the same time be isolated from any other inlet or outlet by requiring that the only closed crosspoint in any row or column is the one at the intersection of the inlet row and outlet column of the inlet/outlet pair to be connected.

2.3 Multi-Stage Interconnection Network (MIN)

A multi-stage interconnection network (MIN) is an interconnection of *stages* of switching modules. A *stage* is a set of identical switching modules. Let S_i, $i = 1, \ldots, s$, denote the ith stage of an s-stage MIN, where S_i contains r_i modules, each having n_i inlets and m_i outlets. The $N = r_1 n_1$ inlets of the switching modules of S_1 are the N inlets of the MIN and the $M = r_s m_s$ outlets of the switching modules of S_s are the M outlets of the MIN.

For $i = 2, 3, \ldots, s$, the inlets of the switching modules of S_i are connected by links *only* to outlets of the switching modules of S_{i-1} and for, $i = 1, 2, \ldots, s-1$, the outlets of the switching modules of S_i are connected by links *only* to inlets of the switching modules of S_{i+1}. So that all of these inlets and outlets can be connected, we require that $r_i m_i = r_{i+1} n_{i+1}$, for $1 \le i \le s\text{-}1$.

2.4 Extended Generalized Shuffle (EGS) Network

An EGS network is simply a MIN with a particularly specified interconnection pattern. Formally, an EGS network is defined as a MIN in which, for $i = 1, 2, \ldots, s$, $S_i =$ the integer set $\{0, 1, \ldots, r_i - 1\}$, and, for $i = 1, 2, \ldots, s-1$, switching module $\alpha \in S_i$ is connected to switching module $\beta \in S_{i+1}$ if and only if $\beta \in \{ [\alpha m_i + o_i]_{\text{mod } r_{i+1}} : o_i \in \{0, 1, \ldots, m_i - 1\} \}$. (Our results will obviously apply to any isomorphic networks.)

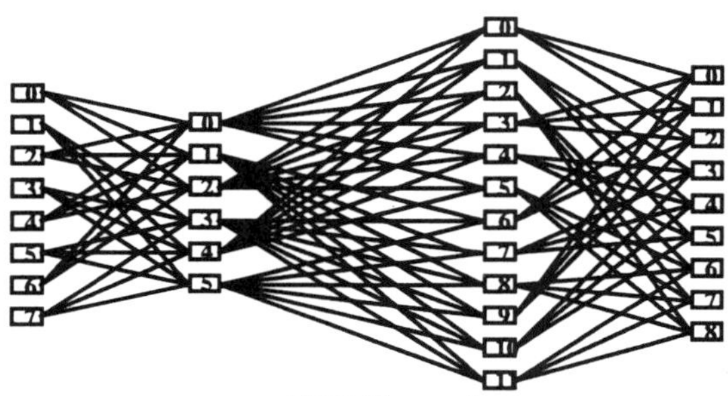

EGS Network η

Figure 2

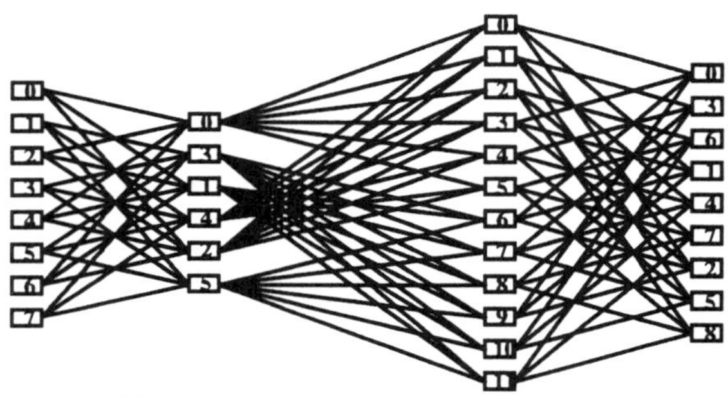

EGS Network Isomorphic to Network η

Figure 3

Figures 2 and 3 depict two isomorphic EGS networks. The only difference between the two is that in Figure 2 all switching modules in each stage are positioned in numerical order and in Figure 3 they are not.

2.5 Paths

Define L_k, $k = 1, 2, \ldots, s-1$, to be the set of links connecting the outlets of S_k to the inlets of S_{k+1}. We denote $\lambda \in L_k$ as a *stage-k link* and say that λ is *incident* to both the stage-k and k+1 switching modules that it connects. For $1 \le i < j \le s$, let $w \in S_i$ and $z \in S_j$. A set of j-i links, one from each L_k, $i \le k < j$, is said to satisfy the *w~z chain condition* if the stage i link is incident to w; for $k = i, i+1, \ldots, j-2$ the stage k and k+1 links are incident to the same stage k+1 switching module; and the stage j-1 link is incident to z. A *path* between w and z is defined as a set of j-i links that satisfies the *w~z chain condition,* plus the switching modules that these links connect.

Any two non-identical sets of j-i links (i.e. two sets differing by at least one link) that satisfy the w~z chain condition comprise two different paths between w and z. Thus, different paths between a given pair of switching modules may have some, but not all, links in common and, of course, they may also have no links in common. The same is true for different paths between different pairs of switching modules. The total number P of paths between w and z is equal to the number of j-i link sets satisfying the w~z chain condition, where no two of these sets are identical.

For network inlet x appearing on stage-1 switching module w and for network outlet y appearing on stage-s switching module z, we say that a set of s-1 links satisfying the w~z chain condition comprises a path between x and y as well as a path between w and z. The *channel graph* $L(x,y)$ of input x and output y is defined as the union of all paths between x and y.

2.6 Blocked Paths

A connection from inlet x to outlet y via some path p is established by operating the appropriate crosspoint in each of the s switching modules (one for each stage) that succcessively connects x, the s-1 links of p, and y. Any link so involved in a connection is said to be *busy*. A *blocked path* between a given input/output pair is one that contains at least one link that is busy from some other connection. A path is *idle* if it contains no such busy links.

3.0 The Fundamental Principle of Strictly Nonblocking Networks

In Section 2.1 we defined a strictly nonblocking network as one in which any idle inlet/outlet pair can be connected regardless of the other exisiting connections in the network. In terms of paths this means that there always exists at least one idle path between any idle input/output pair. Since the number of idle paths between an idle input/output pair is simply the total number of paths between that pair minus the number of blocked paths between that pair, we have the following condition which represents the fundamental principle of strictly nonblocking networks.

> Strictly Nonblocking <u>Necessary</u> Condition:
>
> The number of paths between any inlet/outlet pair must exceed the maximum number of paths which can be blocked between <u>that</u> pair.

If the numbers of paths between all inlet/outlet pairs differ by only a small amount and if the maximum numbers of paths which can be blocked between all inlet/outlet pairs also differ by only a small amount, then the following condition proves useful.

> Strictly Nonblocking <u>Sufficient</u> Condition:
>
> The minimum number of paths between any inlet/outlet pair exceeds the maximum number of paths which can be blocked between <u>any</u> inlet/outlet pair.

For the sake of brevity we define:

PATHS = the minimum number of paths between any inlet/outlet pair

BLOCKED PATHS = the maximum number of paths which can be blocked between any inlet/outlet pair

Which yields:

Strictly Nonblocking Sufficient Condition:

PATHS > BLOCKED PATHS

Thus, if we can count (determine expressions for) both PATHS and BLOCKED PATHS for EGS networks, we will have ascertained generalized sufficient conditions for these networks to be nonblocking.

The logic to determine PATHS is relatively straightforward but the logic to determine BLOCKED PATHS is multi-leveled and rather intricate. To aid the reader in following this process we are providing a logic map (Figure 4). We will continually be referring to this map throughout most of the rest of this paper. The various logic modules on the map are labeled identically with corresponding results from the text for ease of reference. The arrows indicate flows of logic progression. Thus, a given logic module can only be considered after providing results for the modules which direct arrows to the module in question. Our ultimate objective is to provide results for the row of modules second from the bottom; i.e., modules (4.2a), (9.1a), (9.1b), and (9.1c) because these are the components of PATHS > BLOCKED PATHS.

The results for module (2.3) and the two modules (2.4) have already been provided via the definitions of a MIN and an EGS network in Section 2. We next consider modules (4.1i) and (4.2a) to determine a lowerbound on the number of paths between any inlet/outlet pair.

4.0 A Lowerbound on PATHS

While developing an expression for a lowerbound on PATHS, we will find it useful to provide a more general result on the number of paths between switching modules in any two stages of an EGS network.

4.1 Numbers of Paths between Switching Modules in Different Stages

We will need the following preliminary results.

Definition 4.1a:

$\lfloor x \rfloor$ denotes the largest integer $\le x$ and is called the *floor function* of x. $\lceil x \rceil$ denotes the smallest integer $\ge x$ and is called the *ceiling function* of x.

Definition 4.1b:

$x_{\bmod m}$ denotes the smallest nonnegative remainder when dividing nonnegative integer x by positive integer m. The equivalent mathematical formulation is $x_{\bmod m} \equiv x - m\left\lfloor \frac{x}{m} \right\rfloor$.

Lemma 4.1c:

For nonnegative integers w and x and positive integers y and z, if z divides wy then

$$(w\,(x_{\bmod y}))_{\bmod z} = (wx)_{\bmod z} \quad .$$

Proof:

THE LOGIC MAP FOR STRICTLY NONBLOCKING EGS NETWORKS

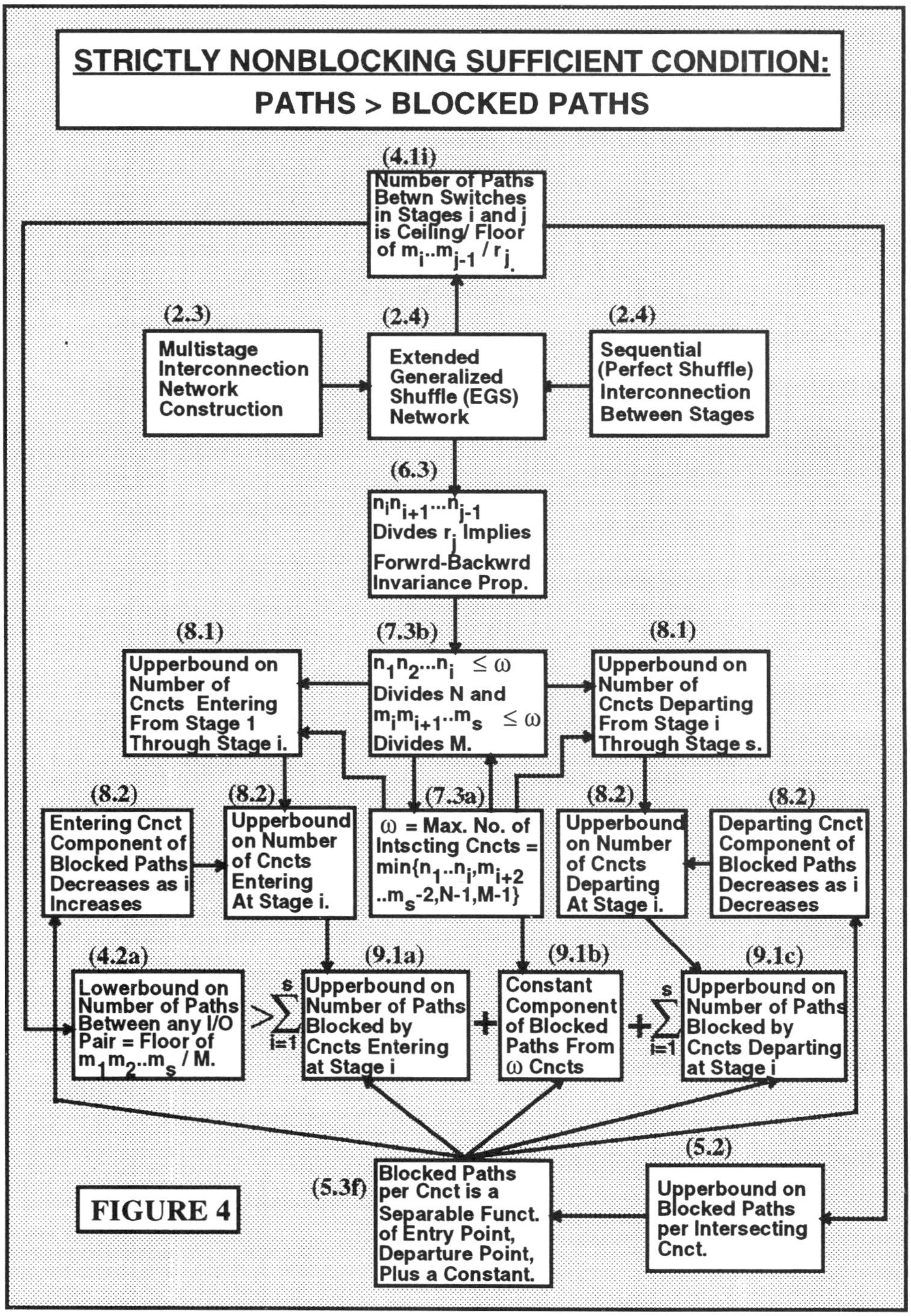

FIGURE 4

For integers a and b with $0 \le a$ and $0 \le b \le y$-1, write $x = ay + b$. Then, $(w(x_{\text{mod } y}))_{\text{mod } z} = (wb)_{\text{mod } z}$. Also, $(wx)_{\text{mod } z} = (way + wb)_{\text{mod } z} = (wb)_{\text{mod } z}$, because z divides wy.

Lemma 4.1d:

For any s consecutive integers and some positive integer m, if each integer d of the s integers is replaced with m consecutive integers, given by md, md+1, ..., md+m-1, the resulting ms integers are consecutive.

Proof:

Consider any two consecutive integers, d and d+1, in the original group of s. The smallest integer replacing d+1 is md+m which is consecutive with md+m-1, the largest integer replacing d.

Definition 4.1e:

Define $N_{i,j} \equiv \prod_{k=i}^{j} n_k$ and $M_{i,j} \equiv \prod_{k=i}^{j} m_k$.

Lemma 4.1f:

In a MIN, for $1 \le i < j \le s$, $r_i M_{i,j-1} = r_j N_{i+1,j}$.

Proof:

By induction on the stage number. According to the definition of a MIN, $r_i m_i = r_{i+1} n_{i+1}$ for $1 \le i \le s$-1 and thus the lemma is true for $j = i$+1. We now show that if it is true for $j = i$+t it is true for $j = i$+t+1.

We assume that $r_i M_{i,i+t-1} = r_{i+t} N_{i+1,i+t}$. Again, by the definition of a MIN we have that $r_{i+t} m_{i+t} = r_{i+t+1} n_{i+t+1}$ or $m_{i+t} = \dfrac{r_{i+t+1} n_{i+t+1}}{r_{i+t}}$. Thus we have

$$r_i M_{i,i+t-1} m_{i+t} = r_i M_{i,i+t} = \frac{r_{i+t} N_{i+1,i+t} r_{i+t+1} n_{i+t+1}}{r_{i+t}} = r_{i+t+1} N_{i+1,i+t+1},$$ which proves the lemma.

Definition 4.1g:

For positive integer k, $I(k)$ denotes the integer set {0, 1, ..., k-1}.

Lemma 4.1h:

In an EGS network, for $1 \le i < j \le s$, switching module w in stage i has paths to a multi-set $F_{i,j}(w)$ of switching modules in stage j given by

$$F_{i,j}(w) = \{ (wM_{i,j-1} + d)_{\text{mod } r_j} : d \in I(M_{i,j-1}) \}.$$

Proof:

By induction on the stage number. By definition of an EGS network, switching module w in stage i is connected (has paths) to a multi-set $F_{i,i+1}(w)$ of switching modules in stage i+1 given by $F_{i,i+1}(w) = \{ (wm_i + o_i)_{\text{mod } r_{i+1}} : o_i \in I(m_i) \}$. Thus, the lemma is true for $j = i$+1. We now show that if it is true for stage $j = i$+t it is true for stage $j = i$+t+1.

We assume that $F_{i,i+t}(w) = \{ (wM_{i,i+t-1} + d)_{\text{mod } r_{i+t}} : d \in I(M_{i,i+t-1}) \}$. By definition of an EGS network, each of the switching modules in this multi-set will have paths to m_{i+t} switching modules in stage $i+t+1$, resulting in a multi-set in that stage given by

$$\{ \left[m_{i+t}(wM_{i,i+t-1} + d)_{\text{mod } r_{i+t}} + o_{i+t}\right]_{\text{mod } r_{i+t+1}} : d \in I(M_{i,i+t-1}), o_{i+t} \in I(m_{i+t}) \}$$

$= \{ [wM_{i,i+t} + m_{i+t}d + o_{i+t}]_{\text{mod } r_{i+t+1}} : \text{d} \in I(M_{i,i+t-1}), o_{i+t} \in I(m_{i+t}) \}$, by Definition 4.1e and Lemma 4.1c (which applies because r_{i+t+1} divides $m_{i+t}r_{i+t}$ according to the definition of a MIN in Section 2.3). Now the operation $m_{i+t}d + o_{i+t}$ is the same as that described in Lemma 4.1d. Thus that lemma applies and the smallest integer in the resulting $m_{i+t}M_{i,i+t-1} = M_{i,i+t}$ consecutive integer set is zero, occuring when d and o_{i+t} are both zero. Our multi-set thus becomes $\{ [wM_{i,i+t} + d]_{\text{mod } r_{i+t+1}} : \text{d} \in I(M_{i,i+t}) \} = F_{i,i+t+1}(w)$, which proves the lemma.

Theorem 4.1i:

In an EGS network, for $1 \le i < j \le s$, the number of paths between any stage-i switching module (or any stage i-1 link) and any stage-j switching module (or any stage-j link) is either $\left\lfloor \frac{M_{i,j-1}}{r_j} \right\rfloor = \left\lfloor \frac{N_{i+1,j}}{r_i} \right\rfloor$ or $\left\lceil \frac{M_{i,j-1}}{r_j} \right\rceil = \left\lceil \frac{N_{i+1,j}}{r_i} \right\rceil$.

Proof:

First we note that dividing both sides of the equation in Lemma 4.1f by r_ir_j yields $\frac{M_{i,j-1}}{r_j} = \frac{N_{i+1,j}}{r_i}$. Next, by Lemma 4.1h we have $F_{i,j}(w) = \{ [wM_{i,j-1} + d]_{\text{mod } r_j} : d \in I(M_{i,j-1}) \}$. The expression $[wM_{i,j-1} + d]$ (via substitution of all of the possible values of d) represents a set of $M_{i,j-1}$ consecutive integers. Write $M_{i,j-1} = br_j + c$, where b and c are integers with $0 \le b$ and $0 \le c \le r_j - 1$. Since no two integers in a set of $a \le r_j$ consecutive integers belong to the same congruence class modulo r_j, we must have that in any br_j consecutive integers there are exactly b integers in each congruence class modulo r_j and that in any $br_j + c$ consecutive integers there are b+1 integers in each of c different congruence classes and b integers in each of the other $r_j - c$ congruence classes. These b and b+1 occurrences of different congruence classes correspond directly to b and b+1 paths from switching module w in stage i to different switching modules in stage j. Since $b = \left\lfloor \frac{M_{i,j-1}}{r_j} \right\rfloor$ and $b + 1 = \left\lceil \frac{M_{i,j-1}}{r_j} \right\rceil$, we have proved Theorem 4.1i for the switching module cases. The result for links follows immediately from the consideration that the only paths between a stage i-1 link and a stage j link are those which utilize (on a one to one basis) the paths between the stage i and stage j switching modules to which these links are incident, respectively.

4.2 Lowerbound on the Number of Paths between Any Inlet/Outlet Pair

By substituting 1 for i and s for j in Theorem 4.1i, we have immediately that the number of paths between any stage 1 switching module and any stage s switching module (or, equivalently, the number of paths between any inlet/outlet pair) is either

$\left\lfloor \frac{M_{1,s-1}}{r_s} \right\rfloor = \left\lfloor \frac{N_{2,s}}{r_1} \right\rfloor$ or $\left\lceil \frac{M_{1,s-1}}{r_s} \right\rceil = \left\lceil \frac{N_{2,s}}{r_1} \right\rceil$ which is the same as $\left\lfloor \frac{M_{1,s}}{M} \right\rfloor = \left\lfloor \frac{N_{1,s}}{N} \right\rfloor$ or $\left\lceil \frac{M_{1,s}}{M} \right\rceil = \left\lceil \frac{N_{1,s}}{N} \right\rceil$, because $r_s m_s = M$, the total number of network outlets, and $r_1 n_1 = N$, the total number of network inlets. Since these are the only two possible values, a lowerbound on PATHS is provided by simply choosing the floor function expression. And we have just proved:

Theorem 4.2a:

PATHS = $\left\lfloor \frac{M_{1,s}}{M} \right\rfloor = \left\lfloor \frac{N_{1,s}}{N} \right\rfloor$. (The minimum number of paths between any inlet/outlet pair in an EGS network = $\left\lfloor \frac{M_{1,s}}{M} \right\rfloor = \left\lfloor \frac{N_{1,s}}{N} \right\rfloor$.)

An obvious observation is that if M divides $M_{1,s}$ or, equivalently, if N divides $N_{1,s}$ then $\left\lfloor \frac{M_{1,s}}{M} \right\rfloor = \left\lfloor \frac{N_{1,s}}{N} \right\rfloor = \left\lceil \frac{M_{1,s}}{M} \right\rceil = \left\lceil \frac{N_{1,s}}{N} \right\rceil = \frac{M_{1,s}}{M} = \frac{N_{1,s}}{N}$, and this is the single value for number of paths between any inlet/outlet pair. In most practical networks (for uniformity's sake) this will likely be the case. However, our subsequent theoretical results will not require this condition. We will be able to establish nonblocking conditions for EGS networks which do not have the same number of paths between all inlet/outlet pairs.

5.0 The Effect of a Single Intersecting Connection

We have just determined an expression for PATHS and must now consider BLOCKED PATHS in our search for the Strictly Nonblocking Sufficient Condition: PATHS > BLOCKED PATHS. To start this process we will initially limit our considerations to the number of paths between a given inlet/outlet pair which can be blocked by a single additional connection in an EGS network. This corresponds to Module 5.2 in Figure 4.

5.1 Channel Graphs and Intersecting Connections

Recall from Section 2.5 that the channel graph $L(x,y)$ of inlet x and outlet y is defined as the union of all paths between x and y. $L(x,y)$ thus comprises sets of stage-k links, $1 \le k \le s$-1. Any of these links may also be in the channel graph(s) of other inlet-outlet pairs. Thus we may find that one or more links of $L(x,y)$ are busy due to existing connections in the network. Any such existing connection that utilizes one or more links of $L(x,y)$ is said to *intersect* or be an *intersecting connection of* $L(x,y)$. We will frequently shorten this terminology to *intersecting connection* when the reference to $L(x,y)$ is clear.

A link which is common to an intersecting connection and $L(x,y)$ is called an *intersecting link.* The set of intersecting links associated with a given intersecting connection are consecutive in stage number. To see why this is so, assume to the contrary that for $i < k < j$ the stages i and j links of an intersecting connection are intersecting links but the stage k link is not. We then have an immediate contradiction because there is a path from x to the stage i link (from the definition of $L(x,y)$), a path from the stage i link to the stage j link that includes the stage k link (this being part of the overall path of the intersecting connection), and finally a path from the stage j link to y (again from the definition of $L(x,y)$. Thus, the stage k link is part of a path from x to y and must be included in $L(x,y)$.

For an intersecting connection, if i is the smallest stage number intersecting link and j-1 is the largest, we say that the intersecting connection *enters* the channel graph $L(x,y)$ at stage i and *departs* at stage j, corresponding to the switching module stage numbers defining the boundaries of the intersecting links. We denote such an intersecting connection as $C(i,j)$.

5.2 Paths Blocked by a Single Intersecting Connection

In an EGS network, let us identify the stage-k link $(i \le k \le j-1)$ of $C(i,j)$ as λ_k. Let B_k represent the number of paths from inlet x to λ_k. From Theorem 4.1i we know that B_k is either $\left\lfloor \frac{N_{1,k}}{N} \right\rfloor$ or $\left\lceil \frac{N_{1,k}}{N} \right\rceil$. Similarly, let F_{k+1} represent the number of paths from λ_k to outlet y. Again, by Theorem 4.1i, we know that F_{k+1} is either $\left\lfloor \frac{M_{k+1,s}}{M} \right\rfloor$ or $\left\lceil \frac{M_{k+1,s}}{M} \right\rceil$.

Recall from Section 2.6 that a *blocked path* between x and y is one that contains at least one link that is busy from some other connection. An intersecting link is such a link. Consider the stage-i intersecting link λ_i of intersecting connection $C(i,j)$. There are B_i paths from x to λ_i and F_{i+1} paths from λ_i to y. So there are B_iF_{i+1} paths in $L(x,y)$ that include λ_i and all of these paths are blocked by λ_i.

Next we consider how many additional paths are blocked by λ_{i+1}, the stage-i+1 intersecting link of $C(i,j)$. Using the same logic as above, we have that λ_{i+1}, considered singly, blocks $B_{i+1}F_{i+2}$ paths. However, this expression includes some paths already blocked by λ_i; namely, all those paths containing both λ_i and λ_{i+1}. Since there are B_i paths from inlet x to λ_i and F_{i+2} paths from λ_{i+1} to outlet y, there are B_iF_{i+2} paths in $L(x,y)$ that contain both λ_i and λ_{i+1}. Thus, the total number of paths in $L(x,y)$ blocked by λ_i and λ_{i+1} is given by $B_iF_{i+1} + B_{i+1}F_{i+2} - B_iF_{i+2}$.

The logic for three or more contiguous intersecting links is similar to that for two links in that we want to determine how many additional paths are blocked by each additional intersecting link. Theoretically, for $k \ge i+2$, when we compute the number of additional paths blocked by λ_k, we can start with B_kF_{k+1} (the number of paths blocked by λ_k, considered singly) and then subtract the number of blocked paths included in this expression that have already been tabulated for λ_i through λ_{k-1}; namely, all the paths containing both λ_k and any of the intersecting links λ_i through λ_{k-1}. However, at this point we introduce a simplifying approximation by subtracting only the $B_{k-1}F_{k+1}$ paths containing both λ_k and λ_{k-1}. As we shall subsequently see, this approximation greatly facilitates our analysis and, fortunately, has very little effect on the sufficient non-blocking conditions for most practical EGS networks.

The reason this latter point is true is that usually there are very few, if any, blocked paths that do not contain λ_{k-1} but do contain λ_k and any of the links λ_i through λ_{k-2}. The only way any such blocked paths could exist would be if there were at least one path from any of the links λ_i through λ_{k-2} to λ_k that did not include λ_{k-1}. By Theorem 4.1i, this cannot be true unless $\frac{M_{i,k-1}}{r_k} = \frac{N_{i+1,k}}{r_i} > 1$. (Later we will consider this situation in producing a refinement on the sufficient non-blocking condition.)

Since the described approximation counts all blocked paths at least once, it may be used in the determination of an upperbound on the total number of paths blocked by a single intersecting connection. From above we have that $B_iF_{i+1} + B_{i+1}F_{i+2} - B_iF_{i+2}$ paths are blocked by λ_i and λ_{i+1} and for $i+2 \le k \le j-1$ we will add $B_kF_{k+1} - B_{k-1}F_{k+1}$ blocked paths for each λ_k. Thus, an upperbound on the number of paths blocked by $C(i,j)$ is given by $\sum_{k=i}^{j-1} B_kF_{k+1} - \sum_{k=i}^{j-2} B_kF_{k+2}$, which can be rewritten as $\sum_{k=i}^{j-2} B_k(F_{k+1} - F_{k+2}) + B_{j-1}F_j$ or as $\sum_{k=i+1}^{j-1} (B_k - B_{k-1})F_{k+1} + B_iF_{i+1}$. We know there is a path from λ_k to λ_{k+1} because these links are both included in $C(i,j)$. Therefore, $(F_{k+1} - F_{k+2})$ is nonnegative because every path from λ_{k+1} to y is necessarily a part of some path from λ_k to y. Furthermore, since the number of paths between any two entities in a network can never be less than zero, F_j is also nonnegative. Thus, it is evident that $\sum_{k=i}^{j-2} B_k(F_{k+1} - F_{k+2}) + B_{j-1}F_j$ has a maximum value if each B_l is replaced by $\left\lceil \frac{N_{1,l}}{N} \right\rceil$. Applying similar logic, we find that $\sum_{k=i+1}^{j-1} (B_k - B_{k-1})F_{k+1} + B_iF_{i+1}$ has a maximum value if each F_l is replaced by $\left\lceil \frac{M_{l,s}}{M} \right\rceil$.
Thus, we have the result that $B[C(i,j)]$, an upperbound on the number of paths blocked by $C(i,j)$, is given by:

$$B[C(i,j)] = \sum_{k=i}^{j-1} \left\lceil \frac{N_{1,k}}{N} \right\rceil \left\lceil \frac{M_{k+1,s}}{M} \right\rceil - \sum_{k=i}^{j-2} \left\lceil \frac{N_{1,k}}{N} \right\rceil \left\lceil \frac{M_{k+2,s}}{M} \right\rceil \tag{5.2a}$$

5.3 Some Simplifying Observations

Let us consider for a moment how we might use the information in Equation (5.2a). This expression clearly indicates that the number of paths blocked by a single intersecting connection depends on its entry and departure points. Thus, we will eventually have to consider the numbers of connections that can enter and depart a channel graph at the various stages in the network. Our task would seem to be extremely difficult if we needed to consider all of the possible ways that various connections could intersect a channel graph. Fortunately, this will not be necessary as we now explain.

Let us rewrite Equation (5.2a) by adding and subtracting the remaining terms through stage s for both summations. The result is

$B[C(i,j)]$

$$= \sum_{k=i}^{s} \left\lceil \frac{N_{1,k}}{N} \right\rceil \left\lceil \frac{M_{k+1,s}}{M} \right\rceil - \sum_{k=i}^{s} \left\lceil \frac{N_{1,k}}{N} \right\rceil \left\lceil \frac{M_{k+2,s}}{M} \right\rceil - \sum_{k=j}^{s} \left\lceil \frac{N_{1,k}}{N} \right\rceil \left\lceil \frac{M_{k+1,s}}{M} \right\rceil + \sum_{k=j-1}^{s} \left\lceil \frac{N_{1,k}}{N} \right\rceil \left\lceil \frac{M_{k+2,s}}{M} \right\rceil$$

$$= \sum_{k=i}^{s} \left\lceil \frac{N_{1,k}}{N} \right\rceil \left(\left\lceil \frac{M_{k+1,s}}{M} \right\rceil - \left\lceil \frac{M_{k+2,s}}{M} \right\rceil \right) + \left\lceil \frac{N_{1,j-1}}{N} \right\rceil \left\lceil \frac{M_{j+1,s}}{M} \right\rceil - \sum_{k=j}^{s} \left\lceil \frac{N_{1,k}}{N} \right\rceil \left(\left\lceil \frac{M_{k+1,s}}{M} \right\rceil - \left\lceil \frac{M_{k+2,s}}{M} \right\rceil \right) \tag{5.3a}$$

(We should note that here and throughout the paper we adopt the usual convention that, in the event that a summation or product series does not exist because the beginning index exceeds the terminating index, the summation has a value of 0 and the product has a value of 1. Similarly, individual terms in a summation or product series are given values of 0 or 1 respectively if those indivdual terms do not exist. This convention achieves the obvious desired result of not changing summation or product totals if non-existent terms are encountered.)

There are three terms in the third line of Equation (5.3a). The first term is a function of various network parameters and the entry stage i. The last two terms are functions of various network parameters and the departure stage j. Thus, the blocked-path contribution of an intersecting-connection entry point is independent of the departure point, and vice-versa. This means that any two feasible sets of connections that have identical distributions of entry points and departure points will have identical upperbounds on the total number of paths blocked, as determined by Equation (5.3a). This is an extremely important and fortunate result because it removes the need to associate the entry and departure points of a given intersecting connection. We need only be concerned with entry and departure point distributions and not the specifics of the connection sets yielding these distributions.

We can simplify Equation (5.3a) via some relevant definitions and observations. Define t to be the largest stage number satisfying $\left\lceil \frac{N_{1,t}}{N} \right\rceil = 1$ and define u to be the smallest stage number satisfying $\left\lceil \frac{M_{u,s}}{M} \right\rceil = 1$. (Note that since $N_{1,1} = n_1 \le N$ and $M_{s,s} = m_s \le M$, both t and u exist.) According to the definition of t we have that $\left\lceil \frac{N_{1,t+1}}{N} \right\rceil > 1$. This implies that all inlets have at least one path to every stage-t+1 switching module. Now, if the value of i in $C(i,j)$ were greater than t+1, then according to the definition of $C(i,j)$, its stage-t+1link (λ_{t+1}) would <u>not</u> intersect $L(x,y)$. But λ_{t+1} <u>must</u> intersect $L(x,y)$ because there is at least one path from x to all stage-t+1 links and there is a path from λ_{t+1} to the stage-i link of $C(i,j)$ (both links being part of $C(i,j)$) and a path from the stage-i link of $C(i,j)$ to y (because $C(i,j)$ intersects $L(x,y)$). Therefore, $C(i,j)$ must have $i \le t+1$. Using similar logic we can determine that $j \ge u-1$.

The term $\sum_{k=j}^{s} \left\lceil \frac{N_{1,k}}{N} \right\rceil \left(\left\lceil \frac{M_{k+1,s}}{M} \right\rceil - \left\lceil \frac{M_{k+2,s}}{M} \right\rceil \right)$ in Equation (5.3a) has $u \le j+1 = min\{k+1\} < min\{k+2\}$. Thus, in this term all $\left\lceil \frac{M_{k+1,s}}{M} \right\rceil = \left\lceil \frac{M_{k+2,s}}{M} \right\rceil = 1$ and therefore the entire term equals 0. Additionally, the second term of Equation (5.3a) has $\left\lceil \frac{M_{j+1,s}}{M} \right\rceil = 1$ because $u \le j+1$. So we now have

$$B[C(i,j)] = \sum_{k=i}^{s} \left\lceil \frac{N_{1,k}}{N} \right\rceil \left(\left\lceil \frac{M_{k+1,s}}{M} \right\rceil - \left\lceil \frac{M_{k+2,s}}{M} \right\rceil \right) + \left\lceil \frac{N_{1,j-1}}{N} \right\rceil . \tag{5.3b}$$

Next we add and subtract terms for $k = i-1$ in the summation yielding

$$B[C(i,j)] = \sum_{k=i-1}^{s} \left\lceil \frac{N_{1,k}}{N} \right\rceil \left(\left\lceil \frac{M_{k+1,s}}{M} \right\rceil - \left\lceil \frac{M_{k+2,s}}{M} \right\rceil \right) + \left\lceil \frac{N_{1,j-1}}{N} \right\rceil - \left\lceil \frac{N_{1,i-1}}{N} \right\rceil \left(\left\lceil \frac{M_{i,s}}{M} \right\rceil - \left\lceil \frac{M_{i+1,s}}{M} \right\rceil \right) \tag{5.3c}$$

Now since $i-1 \le t$ and $\left\lceil \frac{N_{1,k}}{N} \right\rceil = 1$ for all $k \le t$, we have

$$B[C(i,j)]$$
$$= \sum_{k=i-1}^{t} \left(\left\lceil \frac{M_{k+1,s}}{M} \right\rceil - \left\lceil \frac{M_{k+2,s}}{M} \right\rceil \right) + \sum_{k=t+1}^{s} \left\lceil \frac{N_{1,k}}{N} \right\rceil \left(\left\lceil \frac{M_{k+1,s}}{M} \right\rceil - \left\lceil \frac{M_{k+2,s}}{M} \right\rceil \right) + \left\lceil \frac{N_{1,j-1}}{N} \right\rceil + \left\lceil \frac{M_{i+1,s}}{M} \right\rceil - \left\lceil \frac{M_{i,s}}{M} \right\rceil \tag{5.3d}$$

In the summation $\sum_{k=i-1}^{t} \left(\left\lceil \frac{M_{k+1,s}}{M} \right\rceil - \left\lceil \frac{M_{k+2,s}}{M} \right\rceil \right)$ alternate terms cancel each other, except for the first and last terms, yielding $\left\lceil \frac{M_{i,s}}{M} \right\rceil - \left\lceil \frac{M_{t+2,s}}{M} \right\rceil$. Since the first of these two terms cancels the last term in Equation (5.3d) we have

$$B[C(i,j)] = \left\lceil \frac{M_{i+1,s}}{M} \right\rceil + \left\lceil \frac{N_{1,j-1}}{N} \right\rceil + \sum_{k=t+1}^{s} \left\lceil \frac{N_{1,k}}{N} \right\rceil \left(\left\lceil \frac{M_{k+1,s}}{M} \right\rceil - \left\lceil \frac{M_{k+2,s}}{M} \right\rceil \right) - \left\lceil \frac{M_{t+2,s}}{M} \right\rceil . \tag{5.3e}$$

Now for $k \ge u-1$ we have $\left\lceil \frac{M_{k+1,s}}{M} \right\rceil = \left\lceil \frac{M_{k+2,s}}{M} \right\rceil = 1$ and therefore Equation (5.3e) finally becomes

$$B[C(i,j)] = \left\lceil \frac{M_{i+1,s}}{M} \right\rceil + \left\lceil \frac{N_{1,j-1}}{N} \right\rceil + \sum_{k=t+1}^{u-2} \left\lceil \frac{N_{1,k}}{N} \right\rceil \left(\left\lceil \frac{M_{k+1,s}}{M} \right\rceil - \left\lceil \frac{M_{k+2,s}}{M} \right\rceil \right) - \left\lceil \frac{M_{t+2,s}}{M} \right\rceil \tag{5.3f}$$

5.4 Discussion

Equation (5.3f) is a very important intermediate result in our development of sufficient conditions for non-blocking EGS networks. When we arrived at Equation (5.3a) we were able to demonstrate the separability of the blocked-path contributions of the entry point and departure point of an intersecting connection. This was due to the fact that every term depended on either i or j, but not both. In Equation (5.3f), again no term depends on both i and j, and the terms that depend on them individually are greatly simplified. In addition, in Equation (5.3f) we (surprisingly?) find terms that depend on neither i nor j. This means that for any given EGS network there is a computable constant to be added to the upperbound calculation for blocked paths for every intersecting connection. For many practical EGS networks this constant will be found to have a value of minus 1, thus making Equation (5.3f) much easier to use than might be first evident.

5.5 Examples

At this juncture it would probably be helpful to the reader if we paused to consider some examples of the number of paths between inlets and outlets in an EGS network and the numbers of these paths which can be blocked due to other connections. After these examples we will continue our development of non-blocking conditions.

Figure 5a below illustrates a channel graph $L(x,y)$ of Network η shown in Figure 2. Recall that $L(x,y)$ is defined as the union of all paths between some inlet x and outlet y. In this instance, x is an inlet appearing on switch 4 in stage 1 and y is an outlet appearing on switch 2 in stage 4. (Stage numbers are indicated in parentheses along the top of Figure 5a.) $L(x,y)$ results from deleting all switches and links in Figure 2 that are not part of any path between x and y.

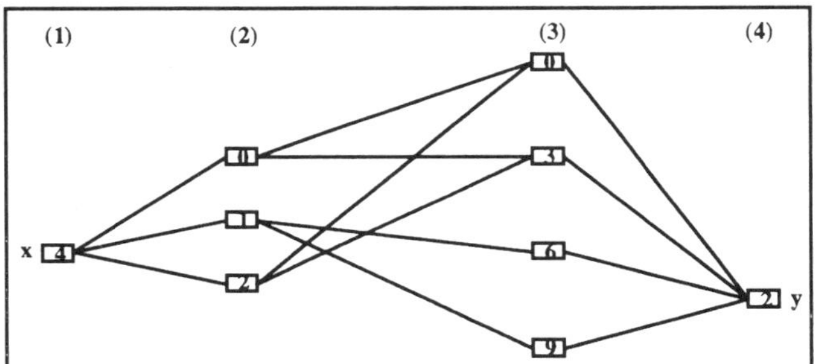

A Channel Graph $L(x,y)$ of Network η

Figure 5a

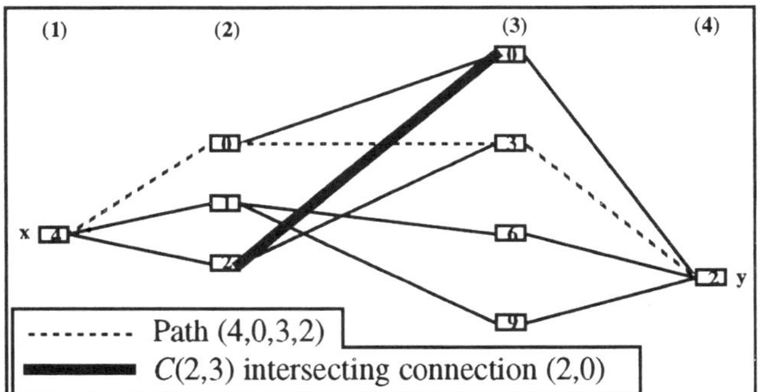

A Channel Graph $L(x,y)$ of Network η

Figure 5b

In Section 4.2 we observed that if M divides $M_{1,s}$ then there is a single value for the number of paths between any inlet/outlet pair as given by $M_{1,s}/M$. In Network η, $M_{1,s}$ =

3x6x3xm_4 = 54 m_4 and $M = 9\ m_4$. Therefore, $M_{1,s}/M = 54\ m_4\ /\ 9\ m_4 = 6$, which is the number of paths we observe between x and y above. We can uniquely identify a particular path with a quadruplet of switch module numbers (one number for each module successively employed in the path from stages 1 through 4). The six paths between x and y are thus given by (4,0,0,2), (4,0,3,2), (4,1,6,2), (4,1,9,2), (4,2,0,2), and (4,2,3,2).

An intersecting connection $C(i,j)$ of $L(x,y)$ can enter $L(x,y)$ at stage i = 1, 2, or 3 and depart at stage j = 2, 3, 4 with the additional proviso that $i < j$. Thus, there are 6 possibilities for $C(i,j)$; namely, $C(1,2)$, $C(1,3)$, $C(1,4)$, $C(2,3)$, $C(2,4)$, and $C(3,4)$. Note that such identifiers do not indicate specific intersecting connections but rather refer to any intersecting connection entering at stage i and departing at stage j. We may uniquely identify a particular intersecting connection of type $C(i,j)$ in a fashion similar to the path identifiers in the previous paragraph; i.e., with the switch module numbers employed successively from stages i through j. Thus, for example, the $C(2,3)$ intersecting connections are (0,0), (0,3), (1,6), (1,9), (2,0), and (2,3). Figure 5b highlights path (4,0,3,2) and $C(2,3)$ intersecting connection (2,0). (We note that (4,0,3,2) could alternatively be interpreted as a $C(1,4)$ intersecting connection between an inlet other than x and an outlet other than y on the stage-1 switch 4 and the stage-4 switch 2, respectively.)

Next, we will consider $B[C(i,j)]$ (an upperbound on the number of paths blocked by $C(i,j)$) as given by Equation (5.3f). In Section 5.3 we defined t to be the largest stage number satisfying $\left\lceil \frac{N_{1,t}}{N} \right\rceil = 1$ and u to be the smallest stage number satisfying $\left\lceil \frac{M_{u,s}}{M} \right\rceil = 1$. From Figure 2 we see that $N = 8n_1$, $N_{1,2} = 4n_1$, $N_{1,3} = 3 \times 4n_1$, and therefore that t = 2. Similarly, we have $M = 9m_4$, $M_{3,4} = 3m_4$, $M_{2,4} = 6 \times 3m_4$, and therefore that u = 3. Thus, the summation term in Equation (5.3f) is equal to 0 because its lowest indexing value $t + 1 = 3$ is greater than its highest indexing value $u - 2 = 1$. Additionally, the last term in Equation (5.3f) has a value of 1 because $M_{t+2,4} = M_{4,4} = m_4 < 4m_4 = M$. Therefore, Equation (5.3f) reduces to

$$B[C(i,j)] = \left\lceil \frac{M_{i+1,4}}{M} \right\rceil + \left\lceil \frac{N_{1,j-1}}{N} \right\rceil - 1 \tag{5.5a}$$

A sample calculation:

$$B[C(1,3)] = \left\lceil \frac{M_{2,4}}{M} \right\rceil + \left\lceil \frac{N_{1,2}}{N} \right\rceil - 1 = \left\lceil \frac{6 \times 3m_4}{9m_4} \right\rceil + \left\lceil \frac{4n_1}{8n_1} \right\rceil - 1 = 2 + 1 - 1 = 2 .$$

Referring to Figure 5a, it is easily verified that example $C(1,3)$ intersecting connection (4,1,6) blocks the two paths (4,1,6,2) and (4,1,9,2). It is also straightforward to verify in Figure 5a that every $C(1,3)$ intersecting connection blocks exactly two paths, thus substantiating our above calculation that $B[C(1,3)] = 2$.

Since $B[C(i,j)]$ is an upperbound on the number of paths blocked by $C(i,j)$, we have in general the possibility that one or more $C(i,j)$ intersecting connections blocks fewer than $B[C(i,j)]$ paths (in $L(x,y)$). As an example of this we first calculate that $B[C(1,4)] = \left\lceil \frac{M_{2,4}}{M} \right\rceil + \left\lceil \frac{N_{1,3}}{N} \right\rceil - 1 = \left\lceil \frac{6 \times 3m_4}{9m_4} \right\rceil + \left\lceil \frac{3 \times 4n_1}{8n_1} \right\rceil - 1 = 2 + 2 - 1 = 3$. In Figure 5b, we can observe that $C(1,4)$ intersecting connection (4,0,3,2) does indeed block the maximum of three paths; namely, (4,0,0,2), (4,0,3,2), and (4,2,3,2). However, $C(1,4)$ intersecting connection (4,1,6,2) blocks only two paths; (4,1,6,2) and (4,1,9,2).

Having completed these examples of paths and blocked paths, we now return to the general development of non-blocking conditions. We next consider a very important attribute of EGS networks that is crucial in the ability to construct efficient non-blocking networks.

6.0 The Forward-Backward Invariance Property

We will use the following preliminary results.

6.1 Two Lemmas

Lemma 6.1a:

If y is a positive integer, then $\left\lfloor \frac{\lfloor x \rfloor}{y} \right\rfloor = \left\lfloor \frac{x}{y} \right\rfloor$.

Proof:

Represent $x = ay + b$, where a is an integer and $0 \le b < y$. Since $0 \le \lfloor b \rfloor \le b < y$, we have $\left\lfloor \frac{\lfloor b \rfloor}{y} \right\rfloor = \left\lfloor \frac{b}{y} \right\rfloor = 0$. Thus,

$$\left\lfloor \frac{\lfloor x \rfloor}{y} \right\rfloor = \left\lfloor \frac{\lfloor ay+b \rfloor}{y} \right\rfloor = \left\lfloor \frac{ay+\lfloor b \rfloor}{y} \right\rfloor = a + \left\lfloor \frac{\lfloor b \rfloor}{y} \right\rfloor = a + \left\lfloor \frac{b}{y} \right\rfloor = \left\lfloor \frac{ay+b}{y} \right\rfloor = \left\lfloor \frac{x}{y} \right\rfloor .$$

Lemma 6.1b:

In a MIN, for $1 \le i < h < j \le s$, (1) if $M_{i,j-1}$ divides r_j then $M_{i,h-1}$ divides r_h and (2) if $N_{i+1,j}$ divides r_i then $N_{h+1,j}$ divides r_h.

Proof:

For Case (1), applying Lemma 4.1f twice we have $\frac{r_h}{M_{i,h-1}} = \frac{r_i}{N_{i+1,h}} = \frac{r_j N_{i+1,j}}{M_{i,j-1} N_{i+1,h}}$ which equals $\left(\frac{r_j}{M_{i,j-1}}\right) N_{h+1,j}$, the product of two integers. Similarly, for Case (2), applying Lemma 4.1f twice we have $\frac{r_h}{N_{h+1,j}} = \frac{r_j}{M_{h,j-1}} = \frac{r_i M_{i,j-1}}{N_{i+1,j} M_{h,j-1}}$ which equals $\left(\frac{r_i}{N_{i+1,j}}\right) M_{i,h-1}$, the product of two integers.

6.2 The Forward-Backward Invariance Theorem

Theorem 6.2:

In an EGS network, for $1 \le i < j \le s$, if $M_{i,j-1}$ divides r_j (or equivalently $N_{i+1,j}$ divides r_i), then for $w \ne v$ either $F_{i,j}(w) = F_{i,j}(v)$ or $F_{i,j}(w) \cap F_{i,j}(v) = \varnothing$.

Proof:

$$F_{i,j}(w) = \{ (wM_{i,j-1} + d)_{\text{mod } r_j} : d \in I(M_{i,j-1}) \} \quad \text{(from Lemma 4.1h)}$$

$$= wM_{i,j-1} + d - r_j \left\lfloor \frac{wM_{i,j-1} + d}{r_j} \right\rfloor \quad \text{(from Definition 4.1b)}$$

$$= wM_{i,j-1} + d - r_j \left\lfloor \frac{w + d/M_{i,j-1}}{r_j/M_{i,j-1}} \right\rfloor = wM_{i,j-1} + d - r_j \left\lfloor \frac{\lfloor w + d/M_{i,j-1} \rfloor}{r_j/M_{i,j-1}} \right\rfloor \text{ (by Lemma 6.1a)}$$

$$= wM_{i,j-1} + d - r_j \left\lfloor \frac{w}{r_j/M_{i,j-1}} \right\rfloor = wM_{i,j-1} + d - r_j \left\lfloor \frac{wM_{i,j-1}}{r_j} \right\rfloor = \{(wM_{i,j-1})_{\bmod r_j} + d : d \in I(M_{i,j-1})\}.$$

Now consider a partition of S_i composed of $r_i/N_{i+1,j}$ subsets $(R_0, R_1, \ldots, R_{r_i/N_{i+1,j}-1})$ each of cardinality $N_{i+1,j}$, where subset R_a is given by

$R_a = \{\frac{br_i}{N_{i+1,j}} + a : b \in I(N_{i+1,j})\}$ for $a \in I\left(\frac{r_i}{N_{i+1,j}}\right)$. Then, $F_{i,j}(R_a)$

$$= \{(R_a M_{i,j-1})_{\bmod r_j} + d : d \in I(M_{i,j-1})\} = \left(\frac{br_i M_{i,j-1}}{N_{i+1,j}} + aM_{i,j-1}\right)_{\bmod r_j} + d = (br_j + aM_{i,j-1})_{\bmod r_j} + d$$

$= (aM_{i,j-1})_{\bmod r_j} + d = \{aM_{i,j-1} + d : d \in I(M_{i,j-1})\}$ because

$aM_{i,j-1} < \left(\frac{r_i}{N_{i+1,j}}\right)M_{i,j-1} = r_j$. Thus, $F_{i,j}(R_a)$ is invariant to the selection of $b \in I(N_{i+1,j})$; i.e., $F_{i,j}(\alpha) = F_{i,j}(\beta)$ for any two elements α and β of R_a.

Now, for $a', a'' \in I\left(\frac{r_i}{N_{i+1,j}}\right)$ where $a' + 1 \le a''$, the largest value of $F_{i,j}(R_{a'})$ $= a'M_{i,j-1} + (M_{i,j-1} - 1)$ and the smallest value of $F_{i,j}(R_{a''}) = a''M_{i,j-1} + 0 \ge (a'+1)M_{i,j-1}$; which is larger than the largest value of $F_{i,j}(R_{a'})$. Thus, none of the values of $F_{i,j}(R_{a'})$ can be the same as any of the values of $F_{i,j}(R_{a''})$.

The theorem is now proven because, for $w \ne v$, if w and v are in the the same subset R_a, then $F_{i,j}(w) = F_{i,j}(v)$. And if w and v are in different subsets, then $F_{i,j}(w) \cap F_{i,j}(v) = \varnothing$.

6.3 Discussion

For ease of discussion, in any EGS network in which, for $1 \le i < j \le s$, $M_{i,j-1}$ divides r_j (or, equivalently, $N_{i+1,j}$ divides r_i); we will say that the *division condition* holds for stages i through j. From Lemma 6.1b, it follows immediately that, if the division condition holds for stages i through j, it also holds for stages g through h, where $i \le g < h \le j$

From Theorem 4.1we know that the number of paths between any stage-i switching module and any stage-j switching module is either $\left\lfloor \frac{M_{i,j-1}}{r_j} \right\rfloor$ or $\left\lceil \frac{M_{i,j-1}}{r_j} \right\rceil$. If $M_{i,j-1}$ divides r_j, then $M_{i,j-1} \le r_j$ and the maximum number of paths between any stage-i switching module and any stage-j switching module is one. Therefore, in Theorem 6.2 $F_{i,j}(w)$ will have only single occurrences of any element. We thus recast the essence of Theorem 6.2 as follows:

The Forward-Backward Invariance Property:

In any EGS network in which the division condition holds for stages i through j, there exists a partition P_i of S_i composed of $r_i/N_{i+1,j} = r_j/M_{i,j-1}$ subsets, each of cardinality $N_{i+1,j}$ and a partition P_j of S_j composed of $r_i/N_{i+1,j} = r_j/M_{i,j-1}$ subsets, each of cardinality $M_{i,j-1}$ where, for each subset ρ of P_i there is a corresponding subset σ of P_j, such that (1) each switching module in ρ has a single path to every switching module in σ (and no paths to any other switching modules in stage *j*) and (2) each switching module in σ has a single path to every switching module in ρ (and no paths to any other switching modules in stage *i*).

The crucial importance of the Forward-Backward Invariance Property in the construction of non-blocking networks will become evident in the next few sections.

6.4 Example

Figure 6 illustrates an EGS Network and highlights all of the paths of a subset pair (ρ,σ) of 9 switching modules from the second stage and 4 switching modules from the fourth stage.

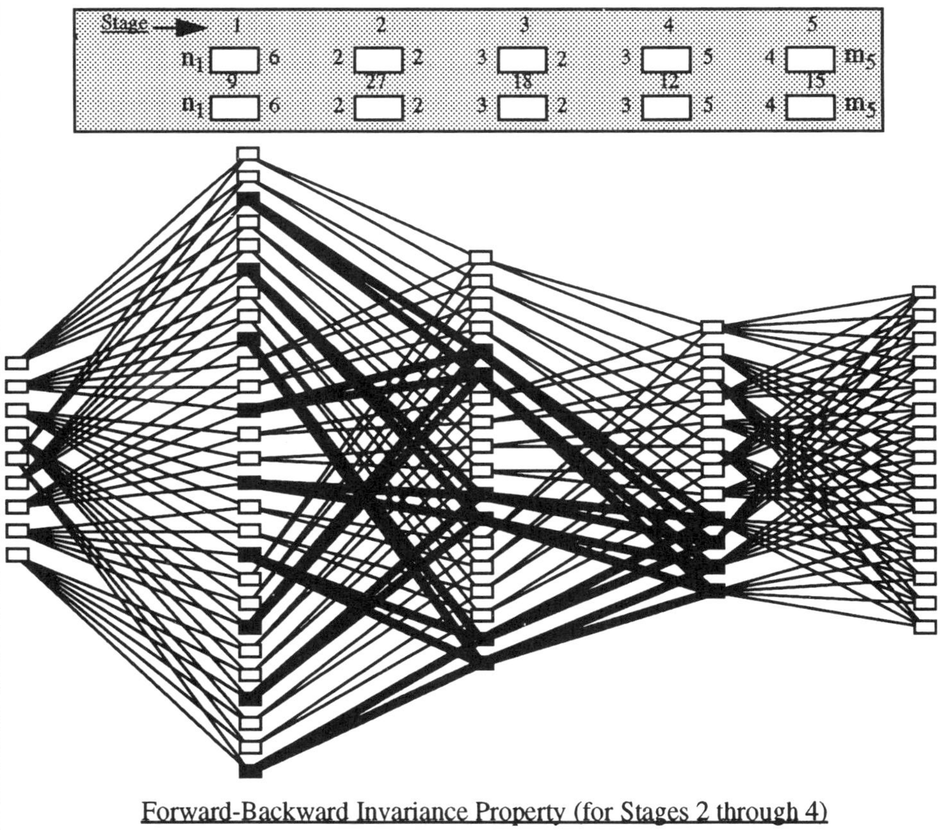

Figure 6

The Forward-Backward Invariance Property applies here for stages 2 through 4 because $m_2m_3 = 2 \times 2 = 4$ which divides $r_4 = 12$ (or, equivalently, $n_3n_4 = 3 \times 3 = 9$ which divides $r_2 = 27$). We note in passing that the Forward-Backward Invariance Property may hold for some successive stages in a network and not hold for others. For example, since $m_1m_2 = 6 \times 2 = 12$ does not divide $r_3 = 18$ and $m_3m_4 = 2 \times 5 = 10$ does not divide $r_5 = 15$, the property does not hold for stages 1 through 3 nor stages 3 through 5, but as we have just seen, it does hold for stages 2 through 4.

7.0 An Upperbound on the Number of Intersecting Connections of a Channel Graph

In Section 5.3 we established an expression for an upperbound on the number of paths blocked by a single intersecting connection in which the blocked-path contributions of the intersecting-connection entry point and departure point are independent of each other. Thus, to determine an upperbound on the number of paths blocked by a group of intersecting connections, we need only be concerned with their entry and departure point distributions. (We don't care which entry point and departure point are associated with any particular intersecting connection.) We will, however, need to know the maximum number of connections that can intersect channel graph $L(x, y)$. We identify this value as ω.

7.1 The Role of the Forward-Backward Invariance Property

Consider an intersecting connection between some inlet $x' \neq x$ and some outlet $y' \neq y$ that enters $L(x, y)$ at stage i, where $1 \leq i \leq k \leq s$. Since every stage-i switching module in $L(x, y)$ has at least one path to at least one stage-k switching module in $L(x, y)$, there is at least one path from x' to some stage-k switching module in $L(x, y)$. Thus, if there are no paths from a given inlet to any stage-k switching modules of $L(x, y)$, then that inlet cannot be part of an intersecting connection that enters $L(x, y)$ from stages 1 through k.

Now suppose that the division condition (and hence the Forward-Backward Invariance Property) holds for stages 1 through k. Then the stage-1 switch, on which idle inlet x appears, is a member of a subset ρ (of cardinality $N_{2,k}$) of S_1, where each switching module in ρ has a single path to every switching module in some subset σ of S_k (and no paths to any other switching modules in stage k). Thus, all stage-k switching modules in $L(x, y)$ must be members of σ. Additionally, each switching module in σ has a single path to every switching module in ρ (and no paths to any other switching modules in stage 1). Therefore, since each of the switching modules of ρ has n_1 inlets, all of the stage-k switching modules in $L(x, y)$ have paths to the same set of $n_1 \times N_{2,k} = N_{1,k}$ inlets (and paths to no other inlets). Removing x from consideration, we have that at most $N_{1,k} - 1$ inlets have paths to any stage-k switches of $L(x, y)$. Using this in conjunction with the result from the previous paragraph gives $N_{1,k} - 1$ as an upperbound on the total number of intersecting connections that can enter $L(x, y)$ from stages 1 through k. (An additional observation, that we will use a little later, is that none of these $N_{1,k} - 1$ inlets can be associated with an intersecting connection that enters <u>after</u> stage k. If such an intersecting connection did exist, its stage-k switching module d would not be a part of $L(x, y)$. But since all of the $N_{1,k}$ inlets (including x) have paths to the same stage-k switching modules, there is a path from x to d. And there is also a path from d to y via the portion of the intersecting connection from d to the entry point and thence via $L(x, y)$. Thus, there is a path from x to y that includes d and so d must be in $L(x, y)$, a contradiction.)

Two important consequences follow from these observations. First, the analysis is simplified because the numbers of switching modules in stages 1 through k in $L(x, y)$ do not have to be considered. Second, and far more importantly, the maximum number of intersecting connections entering from stages 1 through k is potentially much smaller than might otherwise be the case if the Forward-Backward Invariance Property did not limit the number of inlets having paths to any stage-k switches of $L(x, y)$ This is one of the main reasons that we can design efficient non-blocking EGS networks.

For the moment we will proceed on the assumption that the division condition holds, wherever we would like it to hold, throughout a network. After establishing an expression for an upperbound on the number of intersecting connections (ω), we will be able to replace this rather nebulous assumption with a precise sufficient condition.

7.2 Entering and Departing Connections

We have from above that at most $N_{1,k} - 1$ intersecting connections can enter $L(x, y)$ from stages 1 through k and, using similar logic, we can establish that at most $M_{k,s} - 1$ intersecting connections can depart from stages k through s. Let us momentarily focus on stage k, $(1 \le k \le s-2)$, to understand its role in establishing ω.

By definition, an intersecting connection must utilize one or more links of $L(x, y)$. This implies that an intersecting connection which enters at stage i cannot depart any sooner than stage i+1; i.e, it must at least utilize a stage-i link. Thus we have that at most $N_{1,k} - 1$ intersecting connections can utilize stage-1 links through stage-k links of $L(x, y)$ (as limited by the maximum number of connections that can enter from stages 1 through k) and at most $M_{k+2,s} - 1$ intersecting connections can utilize stage-k+1 links through stage-s-1 links of $L(x, y)$ (as limited by the maximum number of connections that can depart from stages k+2 through s). An upperbound on the number of intersecting connections is obtained by assuming that the $N_{1,k} - 1$ entering connections and the $M_{k+2,s} - 1$ departing connections are disjoint; i.e., all of the $N_{1,k} - 1$ entering connections depart at or before stage k+1 and all of the $M_{k+2,s} - 1$ departing connections enter at or after stage k+1.

Thus, considering only the perspective of stage k, there are at most $(N_{1,k} - 1) + (M_{k+2,s} - 1) = N_{1,k} + M_{k+2,s} - 2$ connections that can intersect $L(x, y)$. This expression is valid for $(1 \le k \le s-2)$, since it assumes that intersecting connections can enter as early as stage k and depart as late as stage k+2. Considering all of the possible values for k, we have that at most $\min_{1 \le k \le s-2} \{N_{1,k} + M_{k+2,s} - 2\}$ connections can intersect $L(x, y)$.

7.3 An Expression for ω Plus a Sufficient Condition

Since the number of intersecting connections clearly cannot exceed $\min\{N-1, M-1\}$, we immediately have the following expression for ω:

$$\omega = \min_{1 \le k \le s-2} \{N_{1,k} + M_{k+2,s} - 2, N-1, M-1\} \tag{7.3a}$$

The development of this expression assumed that the division condition held wherever it was necessary to hold. Where is it possibile for the division condition to not hold and yet maintain the validity of the arguments producing expression (7.3a)? We claim it is not required for stages 1 through l, for any l satisfying $N_{1,l} > \omega$; nor for stages r through s, for any r satisfying $M_{r,s} > \omega$.

First of all, for the case where $\omega = \min\{N-1, M-1\}$, the division condition is not required to hold anywhere, because the arguments which employ it yield a larger value for ω than $\min\{N-1, M-1\}$. Next, consider the case where $\omega = \min_{1 \le k \le s-2} \{N_{1,k} + M_{k+2,s} - 2\}$. Since $N_{1,k}$ is monotonically non-decreasing in k, the value of k yielding ω must be less than any l satsifying $N_{1,l} > \omega$, and thus we do not require the division condition to hold for stages 1 through l. Similarly, since $M_{k,s}$ is monotonically non-increasing in k, the value of k+2 yielding ω must be greater than any r satisfying $M_{r,s} > \omega$.

Thus, we have the following condition which is sufficient for expression (7.3a) to be valid:

Condition 7.3b:

$N_{1,k}$ divides N for $N_{1,k} \le \omega$ and $M_{k,s}$ divides M for $M_{k,s} \le \omega$.

7.4 A Review

Let us pause for a moment to highlight some of our results so far and to see what remains in our quest to establish the strictly non-blocking sufficient condition: PATHS > BLOCKED PATHS. (Figure 4 may be helpful during this discussion.)

Theorem 4.2a has given us an expression for PATHS. Expression (5.3f) provides an upperbound on blocked paths per intersecting connection as a separable function of entry point, departure point, plus a constant. Condition 7.3b gives a sufficient condition for which expression (7.3a) yields ω, an upperbound on intersecting connections.

What remains is to allocate these ω intersecting connections according to entry points and departure points; and then to combine these allocations with expression (5.3f) to calculate an upperbound on the number of blocked paths. The aforementioned allocations must be such that they produce the largest possible number of blocked paths, subject to the constraints imposed by Condition (7.3b) and ω.

8.0 Allocation of the Entry and Departure Points of the Intersecting Connections

We will first determine the allocations assuming that Condition 7.3b holds and that there are ω intersecting connections. Then we will justify the assumption of ω intersecting connections by formally establishing the intuitive notion that the maximum number of blocked paths is montonically increasing in the number of intersecting connections; i.e., any number $d < \omega$ of intersecting connections cannot maximally block as many paths as can ω intersecting connections.

8.1 Aggregate Numbers of Entering and Departing Connections

In Section 7.1 we established that at most $N_{1,k} - 1$ inlets can be associated with intersecting connections that enter $L(x, y)$ from stages 1 through k, if the division condition holds for these stages. Assuming Condition 7.3b holds, the division condition will hold for any k such that $N_{1,k} \le \omega$. For any other value of k (i.e., when $N_{1,k} > \omega$), there are at most ω intersecting connections that can enter from stages 1 through k. Therefore we have that at most $\min\{N_{1,k} - 1, \omega\}$ intersecting connections can enter from stages 1 through k. Using similar logic, we find that at most $\min\{M_{k,s}-1, \omega\}$ intersecting connections can depart from stages k through s, assuming Condition 7.3b holds.

8.2 Maximizing the Number of Blocked Paths

We now have expressions for the maximum aggregate numbers of entering and departing intersecting connections from stages 1 through k and stages k through s, respectively. What we need next is the allocation of these intersecting connections on a per stage basis such that we have an upperbound on the total number of blocked paths. This result will follow immediately upon inspection of expression (5.3f).

In this expression, the separable component of the number of blocked paths associated with each intersecting connection <u>entering</u> at stage i is given by $\left\lceil \frac{M_{i+1,s}}{M} \right\rceil$, which is a monotonically non-increasing function in i. Therefore, an upperbound on the component of the number of blocked paths associated with entry points will be achieved if each stage i has as many as possible entering intersecting connections, given that each stage $k < i$ also meets this objective. Thus, we will assume that $\min\{N_{1,1}-1, \omega\}$ intersecting connections enter at stage 1, $\min\{N_{1,2}-1, \omega\}$ - $\min\{N_{1,1}-1, \omega\}$ intersecting connections enter at stage 2, etc.; so that we have in general that $\min\{N_{1,i}-1, \omega\}$ - $\min\{N_{1,i-1}-1, \omega\}$ intersecting connections enter at stage i.

In expression (5.3f), the separable component of the number of blocked paths associated with each intersecting connection <u>departing</u> at stage j is given by $\left\lceil \frac{N_{1,j-1}}{N} \right\rceil$, which is a monotonically non-decreasing function in j. Therefore, an upperbound on the component of the number of blocked paths associated with departure points will be achieved if each stage j has as many as possible departing intersecting connections, given that each stage $k > j$ also meets this objective. This leads to an upperbound of $\min\{M_{j,s}-1, \omega\}$ - $\min\{M_{j+1,s}-1, \omega\}$ intersecting connections departing at stage j. (We should note that the third component term in expression (5.3f) is a function of neither i nor j, and so does not affect this discussion.)

8.3 Justifying the Assumption of ω Intersecting Connections

In Sections 8.1 and 8.2 we developed expressions for the allocation of entry and departure points that gives an upperbound on the number of blocked paths assuming that there are a total of ω intersecting connections. We now show that that the assumption of ω intersecting connections is correct; i.e., there is no number $d < \omega$ of intersecting connections that can maximally block as many paths as can ω intersecting connections. We do this by proving that for any given intersecting connection the upperbound on the number of blocked paths (as given by expression 5.3f) is always positive.

Our first observation is that the sum of the first two terms of expression (5.3f); i.e. $\left\lceil \frac{M_{i+1,s}}{M} \right\rceil + \left\lceil \frac{N_{1,j-1}}{N} \right\rceil$, always has a value of at least 2, since both of these terms always have a value of at least 1. Thus, we need to show that the remaining portion of expression (5.3f); i.e. $\sum_{k=t+1}^{u-2} \left\lceil \frac{N_{1,k}}{N} \right\rceil \left(\left\lceil \frac{M_{k+1,s}}{M} \right\rceil - \left\lceil \frac{M_{k+2,s}}{M} \right\rceil \right) - \left\lceil \frac{M_{t+2,s}}{M} \right\rceil$, never has a value less than minus one.

Case 1: $t+1 \geq u-1$

For this case, the first term (the summation term) has a value of zero because the beginning index exceeds the terminating index. Additionally, the second term equals one, since $t+2 \geq u$, and u (as defined in Section 5.3) is the smallest stage number satisfying $\left\lceil \frac{M_{u,s}}{M} \right\rceil = 1$. Thus, the first term minus the second term equals minus one.

Case 2: $t+1 \leq u-2$

First we note that $\left\lceil \frac{M_{k+1,s}}{M} \right\rceil - \left\lceil \frac{M_{k+2,s}}{M} \right\rceil$ is never negative because $\left\lceil \frac{M_{k+1,s}}{M} \right\rceil$ is monotonically non-increasing in k. Next, for all k within the limits of the summation, we have that $\left\lceil \frac{N_{1,k}}{N} \right\rceil \geq \left\lceil \frac{N_{1,t+1}}{N} \right\rceil \geq 2$, because (as defined in Section 5.3) t is the largest stage number satisfying $\left\lceil \frac{N_{1,t}}{N} \right\rceil = 1$. Therefore,

$$\sum_{k=t+1}^{u-2} \left\lceil \frac{N_{1,k}}{N} \right\rceil \left(\left\lceil \frac{M_{k+1,s}}{M} \right\rceil - \left\lceil \frac{M_{k+2,s}}{M} \right\rceil \right) - \left\lceil \frac{M_{t+2,s}}{M} \right\rceil \geq \sum_{k=t+1}^{u-2} 2 \left(\left\lceil \frac{M_{k+1,s}}{M} \right\rceil - \left\lceil \frac{M_{k+2,s}}{M} \right\rceil \right) - \left\lceil \frac{M_{t+2,s}}{M} \right\rceil$$

$$= 2\left\lceil \frac{M_{t+2,s}}{M} \right\rceil - 2\left\lceil \frac{M_{u,s}}{M} \right\rceil - \left\lceil \frac{M_{t+2,s}}{M} \right\rceil \quad \text{(from alternate term cancellation)}$$

$$= \left\lceil \frac{M_{t+2,s}}{M} \right\rceil - 2\left\lceil \frac{M_{u,s}}{M} \right\rceil = \left\lceil \frac{M_{t+2,s}}{M} \right\rceil - 2 \geq 2 - 2 = 0 \quad \text{(by definitions of } u \text{ and } t\text{)}.$$

This proves that expression (5.3f) always has a positive value, thus justifying our assumption of ω intersecting connections.

9.0 An Upperbound on BLOCKED PATHS

We are now in a position to combine the results of Section 8 with expression (5.3f) to produce the three component terms of BLOCKED PATHS in Figure 4.

9.1 Expressions for the Three Components of BLOCKED PATHS

From Section 8.2 we have $(\min\{N_{1,i}-1, \omega\} - \min\{N_{1,i-1}-1, \omega\})$ intersecting connections <u>entering</u> at stage i and from expression (5.3f) we have $\left\lceil \frac{M_{i+1,s}}{M} \right\rceil$ as the separable blocked path contribution per intersecting connection <u>entering</u> at stage i. Multiplying these two terms together and summing for all i we have

$$\sum_{i=1}^{s} (\min\{N_{1,i}-1, \omega\} - \min\{N_{1,i-1}-1, \omega\}) \left\lceil \frac{M_{i+1,s}}{M} \right\rceil \tag{9.1a}$$

as an upperbound on the intersecting connection <u>entry point</u> contribution of number of blocked paths.

From Section 8.2 we have $(\min\{M_{i,s}-1, \omega\} - \min\{M_{i+1,s}-1, \omega\})$ intersecting connections <u>departing</u> at stage i and from expression (5.3f) we have $\left\lceil \frac{N_{1,i-1}}{N} \right\rceil$ as the separable blocked path contribution per intersecting connection departing at stage i. Multiplying these two terms together and summing for all i we have

$$\sum_{i=1}^{s} (\min\{M_{i,s}-1, \omega\} - \min\{M_{i+1,s}-1, \omega\}) \left\lceil \frac{N_{1,i-1}}{N} \right\rceil \tag{9.1b}$$

as an upperbound on the intersecting connection departure point contribution of number of blocked paths.

The third component term of BLOCKED PATHS is simply the product of ω and the remaining portion of expression (5.3f) (a constant value for every intersecting connection). Thus we have

$$\omega\left(\sum_{k=t+1}^{u-2} \left\lceil \frac{N_{1,k}}{N} \right\rceil \left(\left\lceil \frac{M_{k+1,s}}{M} \right\rceil - \left\lceil \frac{M_{k+2,s}}{M} \right\rceil \right) - \left\lceil \frac{M_{t+2,s}}{M} \right\rceil \right) \tag{9.1c}$$

for this component.

We have finally accounted for all of the elements of Figure 4. And since the sum of expressions (9.1a), (9.1b), and (9.1c) is an upperbound on the total number of paths blocked in $L(x, y)$, we have now established expressions for both PATHS and BLOCKED PATHS. Thus, we are ready to state the main result of this paper---a theorem that gives sufficient conditions for EGS networks to be strictly non-blocking for point-to-point connections. Later we will consider a refinement to our logic that has the potential to reduce the upperbound on the total number of blocked paths for many practical EGS networks. (This refinement is not indicated on Figure 4).

10.0 The Main Result---
A Strictly Non-Blocking Theorem for EGS Networks

We have already proved the theorem we are about to state. That is what the logic map in Figure 4 is all about. We will review this logic after presenting the theorem. We will also discuss some of the attributes and ramifications of the theorem.

Additionally, we will show that the theorem includes (as a small but important subset) strictly non-blocking three stage Clos networks. We will derive some important special case versions of the theorem and provide several examples.

10.1 A Strictly Non-Blocking Theorem

Theorem 10.1:

Any EGS network in which,

$N_{1,k}$ divides N for $N_{1,k} \le \omega$ and $M_{k,s}$ divides M for $M_{k,s} \le \omega$,

where

$$\omega = \min_{1 \le k \le s-2} \{N_{1,k} + M_{k+2,s} - 2, N-1, M-1\} ;$$

and in which,

$$\left\lfloor \frac{M_{1,s}}{M} \right\rfloor > \sum_{i=1}^{s} (\min\{N_{1,i}-1, \omega\} - \min\{N_{1,i-1}-1, \omega\}) \left\lceil \frac{M_{i+1,s}}{M} \right\rceil$$

$$+ \sum_{i=1}^{s} \left(\min \{M_{i,s} - 1, \omega\} - \min \{M_{i+1,s} - 1, \omega\}\right) \left\lceil \frac{N_{1,i-1}}{N} \right\rceil$$

$$+ \omega \left(\sum_{k=t+1}^{u-2} \left\lceil \frac{N_{1,k}}{N} \right\rceil \left(\left\lceil \frac{M_{k+1,s}}{M} \right\rceil - \left\lceil \frac{M_{k+2,s}}{M} \right\rceil \right) - \left\lceil \frac{M_{t+2,s}}{M} \right\rceil \right),$$

where

t is the largest stage number satisfying $\left\lceil \frac{N_{1,t}}{N} \right\rceil = 1$ and u is the smallest stage number satisfying $\left\lceil \frac{M_{u,s}}{M} \right\rceil = 1$;

is strictly non-blocking for point-to-point connections.

Review of Proof:

According to the Strictly Non-Blocking Sufficient Condition in Section 3.0, we needed to show that the minimum number of paths between any inlet/outlet pair exceeds the maximum number of paths which could be blocked between any inlet/outlet pair. In Section 4 (specifically Theorem 4.2a) we established that there are at least $\left\lfloor \frac{M_{1,s}}{M} \right\rfloor$ paths between any inlet/outlet pair. Most of the rest of the paper up to this point has been involved in the process of determining an upperbound on the number of paths which can be blocked between any inlet/outlet pair.

Two results were especially crucial in this endeavor. The first was the formulation of expression (5.3f) as an upperbound on the number of paths blocked by a single intersecting connection. In developing this expression, a simplifying assumption was introduced for those intersecting connections composed of three or more intersecting links. With this simplification, all blocked paths were counted at least once (instead of exactly once), and thus a valid upperbound remained. For most practical networks, the overcount is either zero or very small, preserving the utility of the approximation. However, the most important aspect resulting from this simplifying assumption was that the upperbound became a separable function of the intersecting connection entry point and departure point, plus a constant. This meant that we did not have to associate the entry point and departure point of any particular intersecting connection. We needed only to determine the numbers of entering and departing intersecting connections at each stage, such that we maintained an upperbound on the total number of blocked paths.

This led to our second crucial result; namely, the Forward-Backward Invariance Theorem (Theorem 6.2), in which we established that the Forward-Backward Invariance Property holds for stages i through j if the division condition holds for stages i through j. The importance of this result was that it limited the number of connections that could intersect channel graph $L(x, y)$, assuming that the division condition held where necessary.

Utilizing this information, we next established ω (expression (7.3a)), an upperbound on the number of intersecting connections. We then allocated ω entry points and departure points, guaranteeing that we would maintain an upperbound on the number of blocked paths which result from multiplying the numbers of entering and departing intersecting connections at each stage with their corresponding blocked-path components from expression (5.3f).

A Note on Sufficiency:

We have made two assumptions in this proof that may result in overcounting the maximum possible number of blocked paths. Thus we can assert the sufficiency, but not the necessity, of the given conditions. However, in many (most?) practical networks, the stated upperbounds will either be exact or in error by only a small amount. Let us consider these two assumptions to better understand when overcounting is likely to occur.

The first assumption is the simplification mentioned above regarding the formulation of expression (5.3). It was noted in Section 5.2 that this approximation can only overcount blocked paths if there are at least two paths between a pair of intersecting links of an intersecting connection. This will generally not happen unless there are a relatively large number of stages in the network.

The second assumption is that all of the blocked paths of all of the intersecting connections are different. This assumption is implicit in our process of simply adding the various blocked path components (for ω intersecting connections) to arrive at a total number of blocked paths. We must be careful here to distinguish that we are not talking about the likelihood of all counted blocked paths being different, but whether or not this is even possible. As with the first assumption, we are more likely to have overcounted in a network with a large number of stages. Indeed, a little later, will describe a refinement that shows that we have necessarily overcounted in some networks.

10.2 Observations

The first thing we note about Theorem 10.1 is its generality. The conditions are in terms of the number of inlets and number of outlets (N and M) of the network , the number of inlets and number of outlets (n_i and m_i) on the switching modules in each stage i of the network, and the number of stages (s) of the network. There are no individual constraints on any of these items. There is no required relationship between N and M. There is no symmetry required in the network. There is no relationship required between the size of the switching modules in one stage and any other stage.

Perhaps the most important apsect of the theorem is that nowhere does there appear the ratio of any n_i to any m_i. The implication is that for the most part non-blocking networks can be designed with arbitrarily sized switching modules. Another way of viewing this situation is that the conditions for non-blocking operation are global as opposed to being specific to particular switching modules. The intuitive reason for this is that the non-blocking conditions were established for the network as a whole, and not for subsets thereof; i.e., nothing was defined nor constructed recursively. (There are no imbedded strictly non-blocking subsets in these networks.)

Due in part to its generality, Theorem 10.1 is rather complex. However, much of the complexity disappears for networks exhibiting various constraints, uniformity, and/or symmetry. We will present these simplifications later.

10.3 An Example Non-Blocking Network

Figure 7 depicts a 5-stage EGS network ρ that has a different number of inlets than outlets, and different size switching modules in each stage. We will evaluate the various expressions in Theorem 10.1 to see if this network is strictly non-blocking. We choose this non-uniform and non-symmetric network as our first example so that we may better understand the general capability of this theorem. In the spirit of non-uniformity, note also the numbers of links between stages in this network.

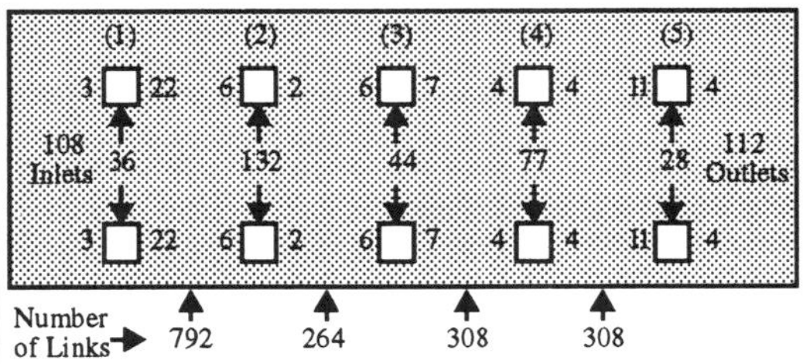

Non-Blocking EGS Network ρ

Figure 7

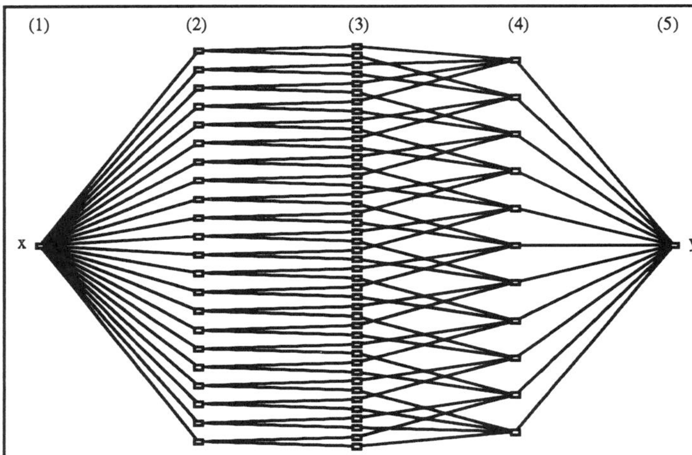

A Channel Graph $L(x,y)$ of Network ρ

Figure 8

The relevant parameters of the network are:
$N = 108, M = 112, n_1$ through $n_5 = 3, 6, 6, 4, 11$ respectively,
m_1 through $m_5 = 22, 2, 7, 4, 4$ respectively, and $s = 5$.

We first calculate $\left\lfloor \frac{M_{1,s}}{M} \right\rfloor$, the minimum number of paths between any inlet/outlet pair. We have

$\left\lfloor \frac{M_{1,s}}{M} \right\rfloor = \left\lfloor \frac{22 \times 2 \times 7 \times 4 \times 4}{112} \right\rfloor = \lfloor 44 \rfloor = 44$. Therefore, there are exactly 44 paths between every inlet/outlet pair. Figure 8 depicts the channel graph $L(x, y)$ of some inlet x and some outlet y of network ρ. (The reader may wish to verify that there are indeed 44 paths between x and y in this channel graph.)

Next, we need to determine ω, the maximum number of connections that can intersect $L(x, y)$. We have

$$\omega = \min_{1 \leq k \leq s-2} \{N_{1,k} + M_{k+2,s} - 2, N-1, M-1\}$$

$$= \min \{N_{1,1} + M_{3,5} - 2, N_{1,2} + M_{4,5} - 2, N_{1,3} + M_{5,5} - 2, 107, 111\}$$

$$= \min\{3 + 112 - 2, 18 + 16 - 2, 108 + 4 - 2, 107, 111\} = \min\{113, 32, 110, 107, 111\} = 32.$$

This limitation on the number of intersecting connections results when our indexing variable $k = 2$. Let us review the logic employed here. When $k = 2$, we are adding the maximum number of intersecting connections that can enter $L(x, y)$ in stages 1 and 2 to the maximum number of intersecting connections that can depart $L(x, y)$ in stages 4 and 5. This addition produces an upperbound by implicitly assuming that the entering and departing intersecting connections are all different; i.e., all of the intersecting connections entering at stages 1 and 2 depart at or before stage 3 and all of the intersecting connections departing at stages 4 and 5 enter at or after stage 3. If this is not so, then one or more of the intersecting connections entering at stages 1 or 2 must depart at stages 4 or 5, thus resulting in fewer than 32 total intersecting connections.

Our result that $\omega = 32$ is provisional at this point. We must additionally verify that $N_{1,k}$ divides N for $N_{1,k} \leq \omega$ and $M_{k,s}$ divides M for $M_{k,s} \leq \omega$. Since $N_{1,3} = 108 = N$ and $M_{3,5} = 112 = M$, these division conditions are satisfied and our result for ω is valid. We should note here that if both division conditions had not been satisfied, we would be unable to establish the non-blocking status of network ρ because of a failure to satisfy all of the conditions of Theorem 10.1.

Next we determine stage numbers t and u. Since $N_{1,3} = 108 = N < 432 = N_{1,4}$, we have $t = 3$. And since $M_{3,5} = 112 = M < 224 = M_{2,5}$, we have $u = 3$. Therefore, $t + 1 > u - 2$ and so

$$\omega\left(\sum_{k=t+1}^{u-2} \left\lceil \frac{N_{1,k}}{N} \right\rceil \left(\left\lceil \frac{M_{k+1,s}}{M} \right\rceil - \left\lceil \frac{M_{k+2,s}}{M} \right\rceil \right) - \left\lceil \frac{M_{t+2,s}}{M} \right\rceil \right) = 32\left(0 - \left\lceil \frac{M_{5,5}}{M} \right\rceil\right) = 32\left(-\left\lceil \frac{4}{112} \right\rceil\right) = -32.$$

Our last step is to determine if

$$\left\lfloor \frac{M_{1,s}}{M} \right\rfloor = 44 > \sum_{i=1}^{s} (\min\{N_{1,i} - 1, \omega\} - \min\{N_{1,i-1} - 1, \omega\}) \left\lceil \frac{M_{i+1,s}}{M} \right\rceil$$

$$+ \sum_{i=1}^{s} \left((\min\{M_{i,s} - 1, \omega\} - \min\{M_{i+1,s} - 1, \omega\}) \left\lceil \frac{N_{1,i-1}}{N} \right\rceil \right) - 32 .$$

Evaluating the right side of this expression we have

$$[(2-0) \times 2 + (17-2) \times 1 + (32-17) \times 1 + (32-32) \times 1 + (32-32) \times 1]$$

$$+ [(32-32) \times 1 + (32-32) \times 1 + (32-15) \times 1 + (15-3) \times 1 + (3-0) \times 4] - 32$$

$$= (2 \times 2 + 15 \times 1 + 15 \times 1) + (17 \times 1 + 12 \times 1 + 3 \times 4) - 32 = 34 + 41 - 32 = 43 < 44$$

Thus, we have at most 43 of the 44 paths between x and y that can be blocked and so network ρ is strictly non-blocking.

This calculation assumed that 2 intersecting connections entered at stage 1, 15 at stage 2, and 15 at stage 3. It also assumed that 17 departed at stage 3, 12 at stage 2, and 3 at stage 1. Part of the power of Theorem 10.1 is that the determination of an upperbound on the number of blocked paths does not require these entrance and departure points be pairwise associated with specific intersecting connections.

Earlier we noted the non-uniformity of the numbers of links between stages in network ρ. We mention it again to underscore the generality it suggests. Observe, for example, that the minimum number of links occurs between stages 2 and 3. Thus, we don't require the number of links to be uniform or increasing as we move toward the center of the network.

This example was chosen to illustrate various points. It is not likely that anyone would ever choose to construct such a network However, it is interesting to note that, even with such "strange" switching module sizes, we have produced a reasonably efficient network. Using the number of crosspoints as the conventional complexity metric, we calculate $3 \times 22 \times 36 + 6 \times 2 \times 132 + 6 \times 7 \times 44 + 4 \times 4 \times 77 + 11 \times 4 \times 28 = 8272$ as compared to $108 \times 112 = 12096$ for a single stage crossbar switch.

10.4 Corollary to Theorem 10.1

For many practical EGS networks it will be the case that $\min_{1 \le k \le s-2} \{N_{1,k} + M_{k+2,s} - 2\} \le \min\{N-1, M-1\}$. For such networks we have the following corollary of Theorem 10.1.

Corollary 10.4:

If $\omega = \min_{1 \le k \le s-2} \{N_{1,k} + M_{k+2,s} - 2\}$, then for the value of k giving ω, Theorem 10.1 becomes

$$\left\lfloor \frac{M_{1,s}}{M} \right\rfloor > \sum_{i=1}^{k} (N_{1,i} - N_{1,i-1}) \left\lceil \frac{M_{i+1,s}}{M} \right\rceil + \sum_{i=k+2}^{s} (M_{i,s} - M_{i+1,s}) \left\lceil \frac{N_{1,i-1}}{N} \right\rceil.$$

Proof:

For the value of k giving ω we must have $k \le t$ and $k+2 \ge u$ for otherwise

$\omega = N_{1,k} + M_{k+2,s} - 2 \ge N_{1,k} - 1 > N - 1$ and/or

$\omega = N_{1,k} + M_{k+2,s} - 2 \ge M_{k+2,s} - 1 > M - 1$, contradicting our assumption on ω. It follows that $u - t \le 2$. Hence,

$$\sum_{k=t+1}^{u-2} \left\lceil \frac{N_{1,k}}{N} \right\rceil \left(\left\lceil \frac{M_{k+1,s}}{M} \right\rceil - \left\lceil \frac{M_{k+2,s}}{M} \right\rceil \right) - \left\lceil \frac{M_{t+2,s}}{M} \right\rceil = -\left\lceil \frac{M_{t+2,s}}{M} \right\rceil = -1.$$

The right side of (10.1) can be written as

$$\sum_{i=1}^{k} [(N_{1,i} - 1) - (N_{1,i-1} - 1)] \left\lceil \frac{M_{i+1,s}}{M} \right\rceil + \sum_{i=k+1}^{t+1} (\min\{N_{1,i} - 1, \omega\} - \min\{N_{1,i-1} - 1, \omega\})$$

$$+ \sum_{i=k+2}^{s} [(M_{i,s} - 1) - (M_{i+1,s} - 1)] \left\lceil \frac{N_{1,i-1}}{M} \right\rceil + \sum_{i=u-1}^{k+1} (\min\{M_{i,s} - 1, \omega\} - \min\{M_{i+1,s} - 1, \omega\}) - \omega$$

$$= \sum_{i=1}^{k} (N_{1,i} - N_{1,i-1}) \left\lceil \frac{M_{i+1,s}}{M} \right\rceil + \min\{N_{1,t+1} - 1, \omega\} - \min\{N_{1,k} - 1, \omega\}$$

$$+\sum_{i=k+2}^{s}(M_{i,s}-M_{i+1,s})\left\lceil\frac{N_{1,i-1}}{N}\right\rceil+\min\{M_{k-1,s}-1,\omega\}-\min\{M_{k+2,s}-1,\omega\}-\omega$$

$$=\sum_{i=1}^{k}(N_{1,i}-N_{1,i-1})\left\lceil\frac{M_{i+1,s}}{M}\right\rceil+\omega-(N_{1,k}-1)+\sum_{i=k+2}^{s}(M_{i,s}-M_{i+1,s})\left\lceil\frac{N_{1,i-1}}{N}\right\rceil+\omega-(M_{k+2,s}-1)-\omega$$

$$=\sum_{i=1}^{k}(N_{1,i}-N_{1,i-1})\left\lceil\frac{M_{i+1,s}}{M}\right\rceil+\sum_{i=k+2}^{s}(M_{i,s}-M_{i+1,s})\left\lceil\frac{N_{1,i-1}}{N}\right\rceil,\text{ because}$$

$(N_{1,k}-1)+(M_{k+2,s}-1)=\omega$.

11.0 Some Special Cases of Theorem 10.1

Using Theorem 10.1 and/or Corollary 10.4, we can develop expressions for three important special-case classes of networks. Due to length considerations, we will show only the results of these developments.

11.1 Three-Stage Clos Networks

Three-stage Clos networks are a subset of the class of networks covered by Theorem 10.1. Figure 9 depicts the generalized Clos network, in which each switching module in the first stage is connected by a link to each switching module in the second stage, and similarly, each switching module in the second stage is connected by a link to every switching module in the third stage. The Clos network thus satisfies our definition of a MIN. It also satisfies the definition of an EGS network in that links from switching modules in stage i may be thought of as being connected consecutively to switching modules in stage $i+1$.

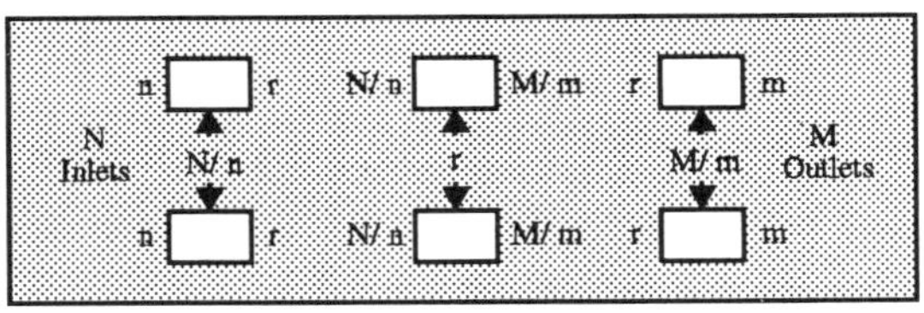

General Three-Stage Clos Network

Figure 9

The usual stated condition for three-stage Clos networks to be strictly non-blocking is $r \geq n+m-1$. Simplification of Theorem 10.1 via the parameters in Figure 9 leads to the same result.

Let us now consider another important special case; i.e., when nearly all of the switching modules are identical and the number of inlets and outlets are equal.

11.2 Uniform Networks

A generalized *uniform network* is depicted in Figure 10 and defined as follows. For $(n \geq 2)$, a uniform network $U(n)$ is an EGS network with $N = n^k$ inlets and outlets and $s = s' + 2$ stages of switching modules. The first stage of the network is composed of N $1 \times F$ switching modules. The next s' stages, $(1 \leq s' \leq 2k-1)$, are each composed of NF/n $n \times n$ switching modules. The last stage is composed of N $F \times 1$ switching modules.

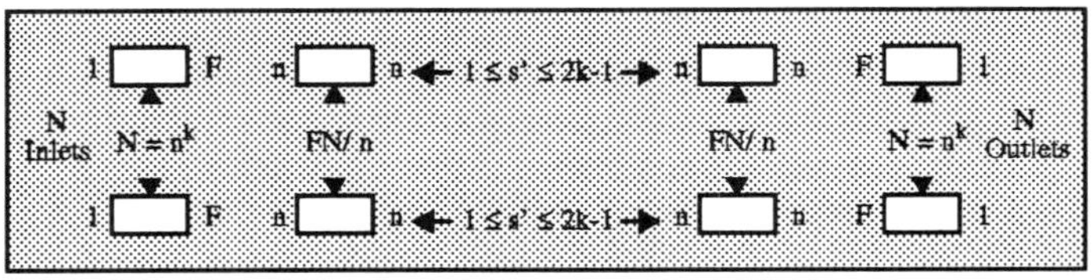

Generalized Uniform Network U(n)

Figure 10

With a bit of effort we can show that

$$F \geq \left\lceil n^{-s+k+2}\left(2\left\lfloor [(s-k-3)(n-1)-1](n^{s-k-3})\right\rfloor +1\right)\right\rceil + n^{k-s/2}\,[2n^{\ddot{s}/2}+(n-1)(1-\ddot{s})] \qquad (11.2a)$$

provides an explicit sufficient condition for F to produce a strictly nonblocking $U(n)$.

11.3 $U(2)$ Networks

$U(2)$ networks are important special cases of $U(n)$ networks. This is because 2x2 switching modules are fundamental building blocks for many types of networks reported in the literature. By substituting the value of 2 for n in expression (11.2a) we immediately have the following result.

A $U(2)$ network is strictly nonblocking if:

$$F \geq \left\lceil 2^{-s+k+2}\left(2\left\lfloor (s-k-4)\,(2^{s-k-3})\right\rfloor + 1\right)\right\rceil + 2^{k-s/2}\,(2\times 2^{\ddot{s}/2} + 1 - \ddot{s}) \qquad (11.3a)$$

We can decompose expression (11.3a) into four simplified specific versions corresponding to the value of s relative to k and the parity of s. The logic and manipulations required to achieve these results are quite straightforward and are thus omitted.

Table 1: F Values for Strictly Nonblocking $U(2)$ Networks

	$3 \leq s \leq \log_2 N + 2$	$\log_2 N + 3 \leq s \leq 2\log_2 N + 1$
s Even	$F \geq 3N/2^{s/2} - N/2^{s-2}$	$F \geq 3N/2^{s/2} - \log_2 N + s - 3$
s Odd	$F \geq 2^{1.5}N/2^{s/2} - N/2^{s-2}$	$F \geq 2^{1.5}N/2^{s/2} - \log_2 N + s - 3$

The refinement to be described next section can improve on some of the values in Table 1.

12.0 A Refinement

In Section 10.1 we mentioned the possibility of overcounting the number of blocked paths. In this section we will develop specific conditions that guarantee that the sum of expressions (9.1a), (9.1b), and (9.1c) does indeed overcount the maximum number of blocked paths. We will not attempt to quantify this overcount (perhaps the topic of another paper), but we will use to advantage the fact that an overcount has occurred.

By Theorem 4.1i, for $1 \le i \le j \le s$, the number of paths beween any stage-(i-1) link and any stage-j link is either $\left\lfloor \frac{M_{i,j-1}}{r_j} \right\rfloor$ or $\left\lceil \frac{M_{i,j-1}}{r_j} \right\rceil$. Consider a $U(2)$ network and let $i = 3$ and $j = s-2$. Then $r_j = F2^{k-1}$ and $M_{i,j-1} = 2^{s-5}$. So if $2^{s-k-4} \ge F$ then $\left\lfloor \frac{M_{i,j-1}}{r_j} \right\rfloor \ge 1$ and there is at least one path between every stage-2 link and every stage-(s-2) link.

Our upperbound on the maximum number of blocked paths assumes that all the paths blocked by entering and departing intersecting connections are different paths. In this case however, if there is at least one path between every stage-2 link and every stage-(s-2) link, then that path will be counted twice in the totalling of blocked paths contributed by stage-2 link intersecting connections and stage-(s-2) intersecting connections. (Both the stage-2 link and stage-(s-2) link will count this path as a blocked path).

Let $s = 2k-1$. Then from Table 1 we have $F \ge 2^{1.5}N/2^{s/2} - k + s - 3 = k$. (Remember that $k = \log_2 N$). It is straightforward to show that $(s-1)/2$ is the value of k giving ω. It can be verified that $(k-1)2^{k-3}$ gives the number of blocked paths in Corollary 10.4 (for $k \ge 3$). If an overcount by one has occurred, we need $F2^{k-3} > (k-1)2^{k-3} - 1$. If $F = k-1$ this inequality will be satisfied. And from above $2^{s-k-4} = 2^{2k-1-k-4} = 2^{k-5} > F = k-1$ will be satisfied for $k \ge 8$.

So, in a $U(2)$ network with $k \ge 8$ and $s = 2k-1$ stages, a value of $F = \log_2 N\text{-}1$ is sufficient for a strictly nonblocking network. This value is one better than Cantor[1] or Shyy and Lea[4]. (Note that our value of s includes the first and last fan-out and fan-in stages and so $s = 2k-1$ means only $2k-3$ stages of 2x2 switch modules.)

With $F = k-1$, each of the $2k-3$ stages has $(k-1)2^{k-1}$ 2x2 switch modules each with 4 crosspoints. So the total crosspoint count is
$(2k-3)4(k-1)2^{k-1} = 2(2k-3)(k-1)N$, thus substantiating our $O(N(\log N)^2)$ construction claim in the Introduction.

13.0 Conclusion

EGS networks are interesting because they are very general, efficient, and can be constructed for strictly nonblocking operation. They include as special cases numerous exisitng classes of networks. The regularity of the interconnection pattern admits to very useful mathematical analyses. This extends to routing algorithms and high speed control mechanisms, both of which are beyond the scope of this already lengthy paper.

14.0 References

1. D. Cantor, "On Non-Blocking Switching Networks", Networks 1 (1972), pp. 367-377.

2. C. Clos, "A Study of Non-Blocking Switching Networks", Bell System Technical J. 32 (1953), pp. 406-424.

3. C.-T. Lea, "Multi-Log_2N Networks and their Applications in High-Speed Electronic and Photonic Switching Systems", IEEE Trans. Commun. C-38 (1990), pp. 1740-1749.

4. D.-J. Shyy and C.-T. Lea, "$Log_2(N,m,p)$ Strictly Nonblocking Networks", IEEE Trans. Commun. C-39 (1991), pp. 1502-1510.

5. C. L. Wu and T. Y. Feng, "On a Class of Multistage Interconnection Networks, IEEE Trans. Commun. C-29 (1980), pp. 694-702.

G. W. Richards is with Bell Laboratories, Lucent Technologies, Naperville, Il 60566.

F.K. Hwang is with Chiaopung University, Taiwan.

DIMACS Series in Discrete Mathematics
and Theoretical Computer Science
Volume 42, 1998

Constructing Nonblocking Multicast Switching Networks with Fanout Reduction

Chun-Fai Chan and Chin-Tau Lea

ABSTRACT. In a multicast switching network, fanout is defined to be the number of outputs that need to be connected to an input. It is well known that fanout sets the complexity of a multicast switch. In the classic three-stage Clos architecture, fanout is reduced to be the number of output switches that need to be reached from an input. Again this number has a strong bearing on the complexity of the entire switch. In real applications, like teleconferencing, three-way call, etc., most multicast connections have small fanouts. But we must design the switch so that it can handle the worst case condition. To improve the efficiency we propose a fanout reduction scheme for the Clos network. The new scheme reduces the complexity of the switch. The price we pay, however, is an increase in path setup time for connections with large fanouts. The new architecture allows us to make a design tradeoff between the two factors in the new architecture.

1. Introduction

Multicast is a required feature for many applications in telecommunication and parallel processing–teleconferencing, video distribution, data retrieval from memory bank, just to name a few [1-3]. Connections in a multicast switching network can be in various forms: one-to-many, many-to-one, and many-to-many. Many-to-one connections are usually realized through different channels, either in time or space domain. Thus, most discussions on multicast focus on one-to-many connections (Figure 1).

Many multicast switches have been proposed [4-8]. The best result by Yang-Masson was described in [9]. A multicast switch has been shown to have a much higher complexity than a unicast switch. One parameter that sets the complexity is fanout, defined to be the maximum number of outputs that can be connected to an input (Figure 1(a)). Although, we like to keep fanout small so as to make the complexity low, a real multicast switching network must have its fanout set as the maximum number of outputs. Yang-Masson's architecture is based on the famous Clos design technique that decomposes a switch into three stages [10-11]. In this approach, fanout can be defined to be the number of the third-stage switches that need to be connected to an input (Figure 1(b)). As explained in section 2, this number again has a strong bearing on the complexity of the network.

1991 *Mathematics Subject Classification.* Primary 94C10.
The research is supported by Hong Kong RGC under the grant HKUST6028/97E, and Hong Kong Telecom under the grant HKTIIT96/97.EG01.

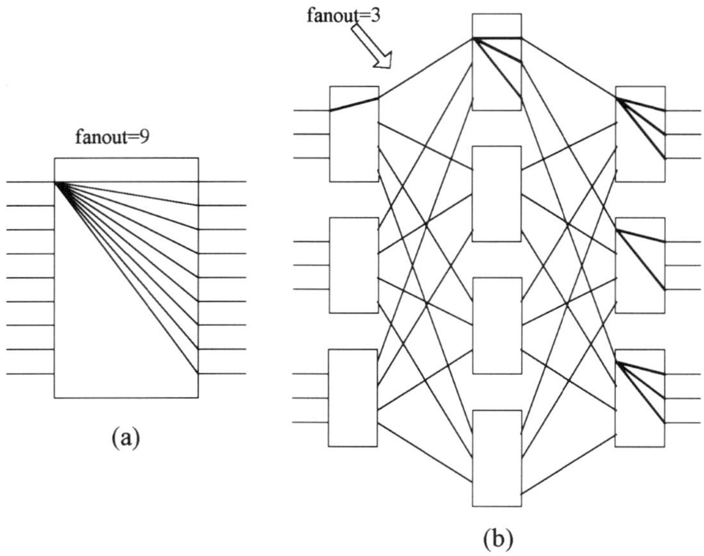

FIGURE 1. Fanout in a (a) single-stage and (b) three-stage Clos multicast switch

Fanouts of most connections in a mutlticast switch are usually small. This is especially true in telecommunications. But we are forced to deal with the largest fanout in designing the multicast switch to maintain the one-to-all multicast capability– however infrequently it is used. To improve the efficiency, we propose in the paper a new scheme that can reduce the fanout in a three-stage Clos network and still maintain the general one-to-all multicast capability inside the switch. We demonstrate that even the best approach available for building a multicast network can be improved by this technique.

The paper is organized as follows. In section 2 we review the three-stage multicast network proposed by Yang and Masson which represents the best result available for designing three-stage multicast interconnection networks. In section 3 we describe the fanout reduction scheme. We also present the theoretical foundation and show how it is used in the three-stage network design. Section 4 presents the numerical results. We conclude our discussion in section 5.

2. Multicast Networks By Yang-Masson

Yang-Masson's approach is based on a three-stage Clos network which is a classic way to reduce the complexity of a large switching network. It was originally proposed for a unicast switch. How to use this approach to reduce the complexity of a multicast switch has been an on-going research topic for many years. For multicast, the major impact of the three-stage approach is in the definition of fanout. In the discussion we assume all switches of the three stages have the multicast capability. Suppose an input needs to be connected to two destination outputs located on the same third-stage switch. Then the input only need to reach one input of the same switch. From there, signals will be sent to the two outputs.

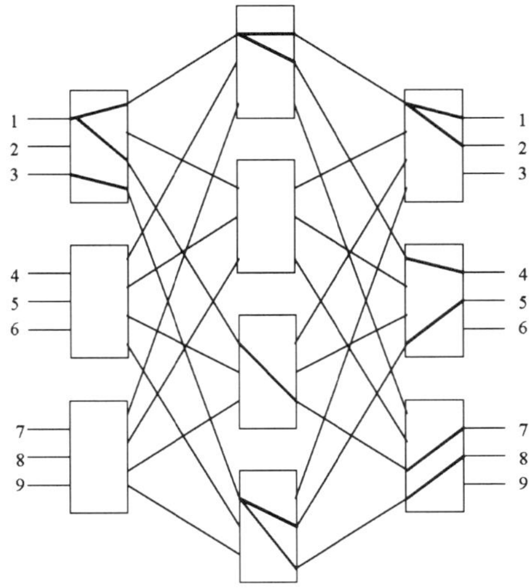

FIGURE 2. A 9×9 three-stage network with $I_1 = \{1,2,3\}$, $I_3 = \{2,3\}$, $M_1 = \{2,3\}$, $M_2 = \phi$, $M_3 = \{3\}$, and $M_4 = \{2,3\}$.

Thus, in the three-stage Clos architecture, fanout is redefined to be the number of third-stage switches that need to be connected to an input.

Consider the example in Figure 2. Given a multicast connection, let I_j be the set of third-stage switches on which the destination outputs are located. Suppose the switch has set up the following connections as shown in Figure 2.

Input 1 $\rightarrow$ connected to Outputs 1,2,4,7

Input 3 $\rightarrow$ connected to Outputs 5,8

Then $I_1 = \{1,2,3\}$ and $|I_1| = 3$ is the fanout for Input 1. Similarly we have $I_3 = \{2,3\}$ and the fanout of Input 3 is 2. Let M_j be the set of third-stage switches which the middle switch j has paths leading to. In the example shown in Figure 2 we have $M_1 = \{1,2\}$, $M_2 = \phi$ (empty set), $M_3 = \{3\}$ and $M_4 = \{2,3\}$. Since each middle switch has a path leading to each third-stage switch, $\overline{M_j}$ is the set of third-stage switches to which middle switch j still have available paths. Suppose we intend to set up a multicast connection from an input to a set, denoted by O, of third-stage switches. If we can find x middle stage switches, denoted by $M_1, M_2, \ldots, M_x$, such that

$$\bigcup_{j=1}^{x} \overline{M_j} = O, \tag{2.1}$$

or equivalently,

$$\bigcap_{j=1}^{x} M_j = \phi \tag{2.2}$$

Then these x switches can be used to set up this multicast connection. Assuming the maximum fanout in the Clos network is f, the authors in [9] showed that whichever the f third-stage switches are, if there are $(n-1)f^{\frac{1}{x}}$ middle switches available upon setting up the multicast connection, we can always find a set of x

middle switches to satisfy Eq. 2. Considering that there are $(n-1)$ other inputs located on the same input switch and each will take x middle switches to set up their connections, we can conclude that the total number of middle switches required, denoted by m, and required for inonblocking is

$$m \geq (n-1)(x + f^{\frac{1}{x}}) \tag{2.3}$$

Under the worst case condition, each input will be connected to all third-stage switches. Thus $f = r$. This result is summarized in the following theorem.

THEOREM 2.1. *A three-stage $C(n, r, m)$ Clos network is wide-sense nonblocking if*

$$m \geq (n-1)(x + f^{\frac{1}{x}}) \quad \text{where } 1 \leq x \leq min\{n-1, r\} \tag{2.4}$$

PROOF. See Ref[9]. □

We need to minimize m with respect to x. Because the maximum fanout $= r$, we have the following formula for m.

$$m = \min_{1 \leq x \leq min\{n-1,r\}} \left\lceil (n-1)(x + f^{\frac{1}{x}}) \right\rceil \tag{2.5}$$

This type of network can be called wide-sense nonblocking since a specific setup procedure must be followed. More discussions on the property and the minimization of m can be found in [9]. It is not difficult to extend the result to a network with unequal numbers of inputs and outputs.

3. Constructing Multicast Switches with Fanout Reduction

Equation (2.4) shows that fanout f strongly affects the complexity of the switching network. Under most circumstances, fanout is much smaller than r. And only occasionally do we need an one-to-all multicast. But if we limit fanout f to reduce the complexity, we will compromise the multicast capability. This causes a dilemma for the designers.

In the following, we propose a scheme that can reduce the fanout in the three-stage approach without compromising the multicast capability of the switching network. It leads to a reduction of the overall network complexity. This approach introduces a new dimension to the design of a multicast switch.

3.1. Fanout Reduction Scheme. Figure 3 describes the idea. Let's add h additional lines to each first and third stage switches. Furthermore, the h lines of a third stage switch are looped back and connected to the h lines of the corresponding first stage switch (Figure 3). In implementation, a first-stage switch and its corresponding third-stage counterpart are co-located and are implemented on the same chip or the same board. Thus the looping scheme shown in figure 3 will not add much wiring complexity. For a digital TST switch the fanout reduction will translate into a lower speed-up ratio requirement for the network. This is explained in section 4.

The looping scheme in Figure 3 allows us to break the setup of a multicast connections into two or even more rounds. We only discuss the two-round procedure in the paper. The immediate result of looping is fanout reduction. Suppose we limit f to 8 and we want to set up a multicast connection in which an input needs to be connected to all third-stage switches (i.e. real fanout = 64). We can break the procedure of connections setup into two steps. In step 1 we arbitrarily select 8

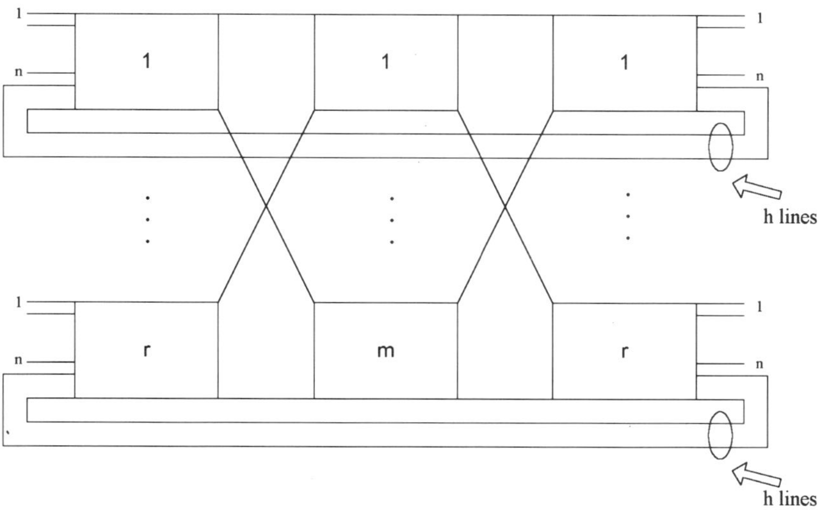

FIGURE 3. The new multicast architecture with fanout reduction

extra output lines added to the third-stage switch and connect them to the input. Recall that all extra output lines are looped back to the input switches. In step 2 we connect each of them to eight third-stage switches that are the real destinations. The fanout of a middle switch is now reduced from 64 to 8. As a result, the amount of hardware is reduced. Of course, this comes at the price of a longer setup time for large fanouts. This scheme represents a tradeoff between the setup time and the amount of hardware. In general if we limit the connection setup time to two-rounds, then $\sqrt{r} \leq f \leq r$ must hold. The relationship between r and h is given in the following theorem.

THEOREM 3.1. *Suppose we limit the connection setup into two rounds. Assuming the maximum fanout is limited to f and $\sqrt{r} \leq f \leq r$, the additional number of outputs added to each third-stage switch, denoted by h, is bounded by the following formula.*

$$h \leq \left\lceil \frac{n}{f-1} \right\rceil \tag{3.1}$$

PROOF. Assume input i is involved in a multicast connection and its real fanout is f_i (i.e. input i needs to be connected to f_i third-stage switches). In the architecture shown in Fig. 3 the connection is implemented in two rounds. Denote the extra number of outputs required for setting up the connection in the two-phase procedure by h_i. Obviously if $f_i \leq f$, $h_i = 0$. If $f_i > f$, we have $h_i = (f - h_i) + f \times h_i$. This leads to

$$h_i = 0 \qquad \text{if } f_i \leq fi \tag{3.2a}$$

$$h_i \geq \left\lceil \frac{f_i - f}{f-1} \right\rceil \qquad \text{if } f_i > f \tag{3.2b}$$

Thus h, the number of extra outputs added to each third-stage switch, becomes

$$\begin{aligned} h &= \left\lceil \frac{\sum_i h_i}{r} \right\rceil \\ &= \left\lceil \frac{1}{r} \sum_i \left\lceil \frac{f_i - f}{f-1} \right\rceil \right\rceil \\ &\leq \left\lceil \frac{1}{r} \sum_i \left(\frac{f_i - f}{f-1} + 1 \right) \right\rceil \\ &= \left\lceil \frac{\sum_i (f_i - 1)}{r(f-1)} \right\rceil \\ &\leq \left\lceil \frac{\sum_i f_i}{r(f-1)} \right\rceil \end{aligned} \tag{3.3}$$

Since $\sum_i f_i$ in Eq.(3.3) cannot exceed the total number of output nr, thus

$$\begin{aligned} h &\leq \left\lceil \frac{nr}{r(f-1)} \right\rceil \\ &= \left\lceil \frac{n}{f-1} \right\rceil \end{aligned} \tag{3.4}$$

We thus prove the bound. □

The total number of crosspoints, denoted by T, can be formulated by including the h lines into input and output switches.

$$\begin{aligned} T &= 2(rm(n+h)) + mr^2 \\ &= 2\left(rm\left(n + \left\lceil \frac{n}{f-1} \right\rceil\right)\right) + mr^2 \end{aligned} \tag{3.5}$$

where m is given in Eq.(2.5). In terms of the total number of crosspoints, there exist optimal values of x, f, and thus h, in Eq.(3.5). This will be explored numerically in the next section.

Yang-Masson's approach also applies to an asymmetric network that has r_1 first-stage switches with size n_1, and r_2 third-stage switches with size n_2. In the following we extend the fanout reduction scheme to cover asymmetric networks of this kind (Figure 4). We further assume that $r_2 \geq r_1$. This is a more likely case in telecommunications than the other way around since the number of customers are usually larger than the number of input channels. But the reverse case can also be easily derived. In this arrangement the number of extra outputs will be distributed among the first r_1 output switches.

THEOREM 3.2. *In an asymmetric Clos network there are r_1 first-stage switches and r_2 third-stage switches. The sizes of a first- and a third-stage switch are n_1 and n_2. Assume $r_2 > r_1$. The extra outputs on each output (i.e. third-stage) switch is bounded by the following formula.*

$$h \leq \left\lceil \frac{n_2 r_2}{r_1(f-1)} \right\rceil \tag{3.6}$$

PROOF. Simlar to Theorem 3.1. □

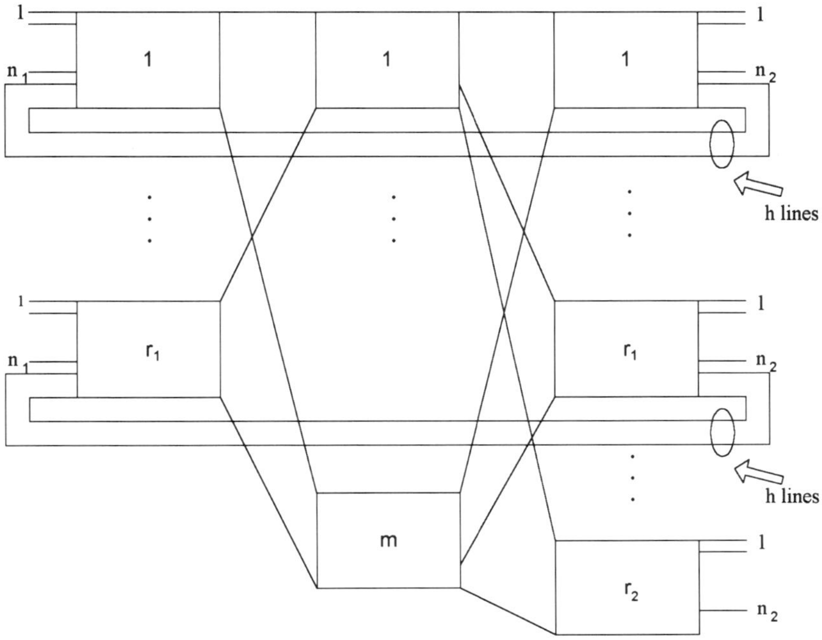

FIGURE 4. An asymmetric three-stage Clos network

4. Numerical Results

In this section we present some numerical results for switches with the same number of inputs and outputs. We will compare the results in terms of number of crosspoints or in terms of the ratio between m and n. In telecommunications, it is well known that a Time-Space-Time (TST) digital switch is equivalent to a 3-stage Clos network in the space domain (Figure 5). The ratio of m over n in the TST, usually called the speed-up ratio, tells how much faster the internal switching speed has to be in order to make the network nonblocking. The larger the value of m, the more costly the system. We choose $n = 24$ which corresponds to a digital TST switch having 24 channels in its incoming links. The commonly used T1 transmission link is such an example.

We first explore the optimal value for f in the range $\sqrt{r} \leq f \leq r$ (see Eq.(3.1)). Recall that for each f, there is a minimum x that minimize m—the number of middle switches. The minimum m is plotted in Figures 6(a). Figure 6(b) plots the number of crosspoints, under the minimum m, against f. They indicate the best performance is achieved when $f = \sqrt{r}$. Under the condition $f = \sqrt{r}$,we plot m for various size of switches in Figure 7(a). Figure 7(b) shows the number of crosspoints for different size of switches as r varies from 32 to 512. Table 1 tabulates the results from which we can see the improvement of the new architecture.

We next explore the time-space tradeoff. As already pointed out, the price we pay in the new architecture is the setup time. We also pointed out that fanouts in most multicast connections are small. The penalty we pay in setup time in the new architecture is analyzed below. We assume the fanout f_i of a connection in the three-stage Yang-Masson network is a random variable. We further assume f_i

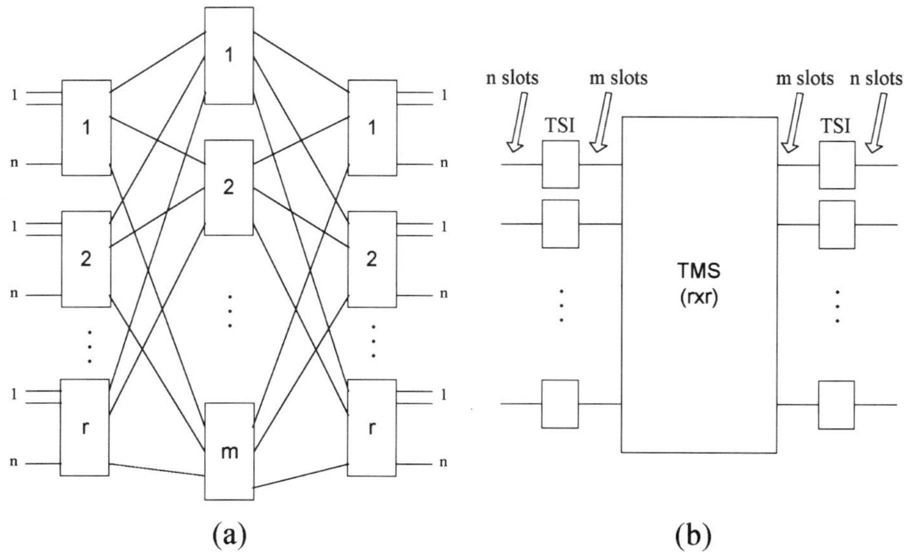

FIGURE 5. A digital TST switch is equivalent to a three-stage Clos network.

TABLE 1. Number of crosspoints in the new and Yang-Masson's architecture.

r	Crosspoints ($Yang-Masson$)	Crosspoints (New)
32	366,080	296,640
64	1,132,554	860,160
128	3,829,760	2,842,112
256	14,319,616	10,013,696
512	56,197,120	38,983,560

has a geometric distribution. Since f_i is within the range [1,r], we lump all the probability mass for $f_i > r$ into $f_i = r$. Thus we have

$$f_i = \begin{cases} p(1-p)^i & \text{if } 1 \le i \le r-1, \\ (1-p)^{r-1} & \text{if } i = r. \end{cases} \tag{4.1}$$

We can easily find p by setting $\overline{f_i}$ to a given average value. The connection setup time was discussed in detail in Ref [9]. Here we only assume it is T_s. Then the ratio of the connection setup time in the new architecture, denoted by T_s', over the connection setup time in the old architecture can be written as

$$\frac{T_s'}{T_s} = 1 + h \tag{4.2}$$

where h, the number of extra outputs involve in the two-phase connection setup, can be found from Eq.(2.5). It can be derived from f_i.

Table 2 tabulates the connection setup time ratio given by Eq.(4.2). It shows that if the average fanout of connections is small, which is usually true, then the

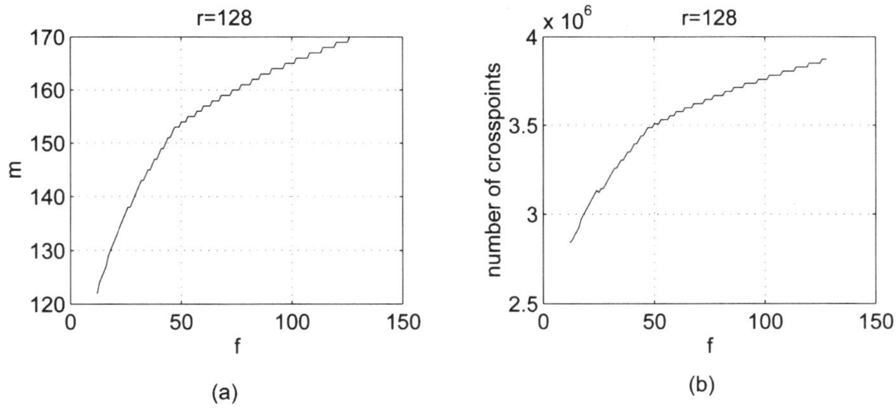

FIGURE 6. (a) The number of middle switches as a function of fanout$f$. (b) The number of crosspoints as a function of fanout f

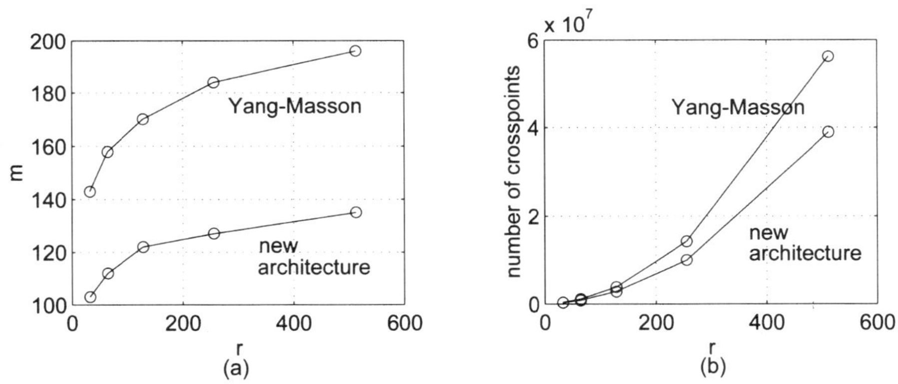

FIGURE 7. (a) The number of middle switches as a function of r, when $f = \lceil \sqrt{r} \rceil$ in the new architecture. (b) The number of crosspoints as a function of r, when $f = \lceil \sqrt{r} \rceil$ in the new architecture.

price we pay in setup time is small. Note again that fanout f_i is defined to be output switches. It can be translate into a much larger number of real outputs that need to be connected to the input.

TABLE 2. The setup time ratio of new architecture over old architecture.

r	ratio of average setup time (f=4)	ratio of average setup time (f=8)
32	1.23	1.92
64	1.11	1.56
126	1.03	1.26
512	1.0001	1.04

5. Concluding Remarks

Multicast is an important feature of interconnection networks for switching or parallel processing. Its most important design issue is the high complexity. While most multicast connections only require a small number of fanouts, we are forced to deal with the largest fanout to maintain the general one-to-all multicast capability. In the paper we proposed a new way to build a multicast switch to improve the efficiency. It is based on a fanout reduction developed in this paper. While the technique can reduce the fanout of a multicast interconnection, it will not compromise the multicast capability. We showed that even the best approach available today can be improved with this technique. The approach offers a new way of designing multicast interconnection networks.

References

[1] John Bellamy, *Digital Telephonyi*, Wiley and Sons, second edition, 1991.
[2] F. T. Leighton, *Introduction to Parallel Algorithms and Architectures*, Morgan Kaufmann, 1992.
[3] K. Hwang and F.A. Briggs, *Computer Architecture and Parallel Processing*, McGraw-Hill, 1984.
[4] F. K. Hwang and A. Jajszczyk, *On nonblocking multiconnection networks*, IEEE Trans. Commun., vol. COM-34, pp. 1038-1041, 1986.
[5] G. M. Masson and B. W. Jordan, *Generalized multi-stage connection networks*, Networks, vol. 2, pp. 191-209, 1972.
[6] F. K. Hwang, *Three stage multiconnection networks which are nonblocking in the wide sense*, Bell Syst. Tech. J., vol. 58, pp. 1283-1287, 1979.
[7] G. W. Richards and F. K. Hwang, *A two-stage rearrangeable broadcast switching network*, IEEE Trans. Commun., vol. com-33, pp. 1025-1034, 1985.
[8] P. Feldman, J. Friedman, and N. Pippenger, *Wide-sense nonblocking networks*, SIAM Journal of Discrete Mathematics, vol. 1 No.2, pp. 158-173, Many 1988.
[9] Y. Yang and G. M. Masson, *Nonblocking broadcast switching networks, IEEE Trans. Computers*,vol. 40, no. 9, 1991, pp.1005-1015.
[10] V.E. Benes, *Mathematical theory of connecting networks and telephone traffic*, Academic press, 1965.
[11] C. Clos, *A study of non-blocking switching networks*, Bell Syst. Tech. J., Vol 32, ppp. 405-424, 1953.

Department of Electrical and Electronics Engineering, The Hong Kong University of Science and Technology, Clear Water Bay, Hong Kong
E-mail address: `eecfchan@ee.ust.hk`

Department of Electrical and Electronics Engineering, The Hong Kong University of Science and Technology, Clear Water Bay, Hong Kong
E-mail address: `eelea@ee.ust.hk`

DIMACS Series in Discrete Mathematics
and Theoretical Computer Science
Volume 42, 1998

Multirate Broadcast Switching Networks Nonblocking in a Wide Sense

Dongsoo S. Kim and Ding-Zhu Du

ABSTRACT. This paper describes wide-sense nonblocking Clos switching networks for multirate broadcast communications, where a compatible connection request can be routed through the networks without disturbing the existing network configuration if all requests are conforming to a given routing algorithm. The routing algorithm partitions middle stage switches into three subsets and routes connection requests by utilizing different subsets of middles according to their normalized weights. The result shows $(2.875n-1)\min(k+r^{1/k})+2$ middle switches for $\mathcal{C}(n,r,m)$ are sufficient for any multirate broadcast requests, whereas strictly nonblocking multirate Clos networks require an unbounded number of middle stage switches when the range of weights become broadly distributed.

1. Introduction

Since a multistage Clos network [**3**] was introduced, it has been heavily used for telephone switching systems in place of crossbar switches because of its asymptotic advantage over the crossbars while providing nonblocking behavior between any pair of idle external ports. The Clos network was initially designed to model a circuit switching, in which a connection request should set up a physical path between its source and destination and the request exclusively occupies links on the path during the entire conversation. Advances in digital technology have made it feasible to allow several telephones and data terminals, which may require different data rates, to share a link at the same time unless the aggregated bandwidths on the link exceed its capacity [**2, 8, 9**].

There has been an increasing interest to integrate a diversity of communication services including teleconferencing, VOD and even entertainment CATV into a network, which need transmit their informations to multiple destinations. The advent of the asynchronous transfer mode(ATM) has further forced to combine dissimilar communication networks so as to provide flexible communication environments and to achieve cost-effectiveness. The current switching networks carrying out broadcast capabilities, however, are based on blocking switching networks, where

1991 *Mathematics Subject Classification.* Primary 94A05, 05C70; Secondary 90B12, 05C50.
Key words and phrases. Clos networks, multirate, broadcast, nonblocking.
Support in part by the National Science Foundation under grant CCR-9530306 and Grant-in-Aid of Research in University of Minnesota.

a compatible broadcast request could not be granted, because the switching device fails to find paths between source and a set of destinations, or the request may cause serious penalties on the existing networks.

In this paper, we will investigate multirate 3-stage Clos networks which are capable of sending informations to a set of destinations without altering the pre-routed paths if all connections conform to the given routing scheme. The routing scheme partitions middle stage switches into three subsets and each connection request is allowed to use middle stage switches out of one or two subsets according to its weight. The number of middle stage switches in the nonblocking multirate networks depends on only the network parameters but not on the pattern of connection requests.

2. Multistage Switching Networks

We represent an interconnection network as a directed weight graph $G = (V, E)$. $V_I \subset V$ is a set of input port nodes, each of which has one outgoing edge and no incoming edge. $V_O \subset V$ is a set of output port nodes, each of which has one incoming edge and no outgoing edge. In multistage networks, nodes between V_I and V_O are divided into stages. For an n-stage network, V_I and V_O are said to be in stage 0 and $n+1$, respectively. Nodes in stage i have directed edges only to nodes in stage $i+1$ for $0 \leq i \leq n$, and there exists only one edge between any pair of nodes. We can construct more uniform 3-stage Clos networks, denoted as $\mathcal{C}(n_1, r_1, n_2, r_2, m)$, by forcing that each node in stage 1 (input stage) has n_1 incoming edges and m outgoing edges to every nodes in stage 2 (middle stage), in which each node has r_1 incoming edges and r_2 outgoing edges to every nodes in stage 3 (output stage). Each node in stage 3 has m incoming edges and n_2 outgoing edges. Each switch corresponding to nodes in stage 1, 2 and 3 is assumed to be a crossbar switch which has nonblocking broadcast capability, even though it can be recursively constructed with smaller Clos subnetworks. A symmetric Clos network, denoted as $\mathcal{C}(n, r, m)$, is induced from the asymmetric network with $n_1 = n_2$, $r_1 = r_2$.

A *connection request* for point-to-point connection is a triple (x, y, ω), where $x \in V_I$, $y \in V_O$ and ω is a normalized bandwidth requirement (weight) of the connection request. A set of connection requests are said to be *compatible* if the sum of all weights passing each external link does not exceed its capacity. A connection request is compatible to the existing network if adding the request does not cause capacity overflows for any external links. A multicast connection request is defined as a triple (x, Y, ω), where $Y \subseteq V_O$ denotes a set of output ports. A point-to-point connection request can easily be represented with the one for multicast by imposing $|Y| = 1$. A *route* is a path connecting an input port to a set of output ports through middle stage switches. A *configuration* of a 3-stage network is a set of all routes. A configuration is said to be *satisfied* if, for all connection requests, routes are found in such a way that the sum of weights on each $e \in E$ is not larger than its capacity. A compatible connection request is *satisfiable* if one can find a set of middle stage switches to route the request without any capacity overflows.

A switching network is said to be *strictly nonblocking*, or simply nonblocking, if a new connection request can be routed without disturbing the existing network no matter how the previous calls were routed. It is well-known that a symmetric

Clos network, $\mathcal{C}(n, r, 2n-1)$, is strictly nonblocking in circuit switching one-to-one communications [**3**]. Melen and Turner [**8**] found a sufficient condition of a nonblocking Clos network for multirate one-to-one communications. They took an advantage of higher speed of internal links over externals to reduce hardware requirements. Chung and Ross [**2**] determined necessary and sufficient conditions for multirate interconnection networks to be nonblocking for both discrete and continuous weights, especially when the external speed is same as the internal.

A network is said to be *nonblocking in a wide-sense* if a new request can be satisfied without interfering the existing network configuration under condition that all requests comply to a given routing algorithm. Beneš [**1**] proved a Clos network $\mathcal{C}(n, 2, m)$ is wide-sense nonblocking in circuit switching if $m = \lfloor 3n/2 \rfloor$. Melen and Turner [**8**] devised a Clos network $\mathcal{C}(n, r, 8n-2)$ to be wide-sense nonblocking in a multirate environment by assembling two nonblocking Clos networks in parallel, each of them has $4n-1$ middles stage switches. All connections with weights more than $1/2$ are routed through only one of subnetworks, and all requests with weights no more than $1/2$ are routed through the other subnetwork. Gao and Hwang [**4**] constructed a wide-sense nonblocking Clos network for multirate traffics with $5.75n$ middle switches by exploiting a quota routing scheme.

Hwang [**5**] had given conditions $m = O(nr)$ of rearrangeable multi-connection 3-stage Clos networks, which is a generalization of interconnection networks such that a set of input ports are able to connect to a set of output ports. His work was based on lack of fan-out capability in middle stage switches so that all duplications must be done in input switches and output switches but not in the middle stage. Masson and Jordan [**7**] proved that an asymmetric Clos network was strictly nonblocking for broadcast if $m \geq r_2 n_1 + n_2 - 1$, which fell into $m = O(nr)$ for a symmetric one. Hwang [**6**] provided two routing strategies for broadcast. One is to use only one middle switch for a broadcast request and all messages are duplicated only in a middle switch. Another is to use k middle switches for a broadcast request with fan-out k, in which all messages are duplicated in an input switch. These two extreme usages of fan-out capabilities — all fan-outs at the input stage or the middle stage but not both — were the main reason of requiring the extraordinarily large number of middle stage switches. The first study that allows any switches to duplicate messages to relieve the severe complexity was carried out by Yang and Masson [**11**]. They showed that a Clos network in circuit switching is nonblocking for broadcast requests when $m > \min(n-1)(k+r^{1/k}) = O(n \ln r / \ln \ln r)$, where $1 \leq k \leq \min(n-1, r)$. Svinnset [**9**] studied a nonblocking 3-stage broadcast switching networks for multirate connection requests with $m = O(nr)$. Yang [**10**] extended her previous result to obtain multirate nonblocking broadcast networks with $m > \min(\lfloor 1/b \rfloor (n-1)(k + r^{1/k}))$. A weakness of the result is that the network requires an unbounded number of middle stage switches when b goes to 0.

3. Preliminaries

It is clear that broadcasting more than two output ports in a same output switch is always satisfiable if it is realizable to one output port in the output switch if all switches in the network have full broadcast capabilities. By this reason, we denote a set of destination as a set of output switches instead of output ports so that all broadcast requests in $\mathcal{C}(n, r, m)$ are assumed to have the maximum fan-out r. For good explanation, we use a vector notation as a set of destinations in

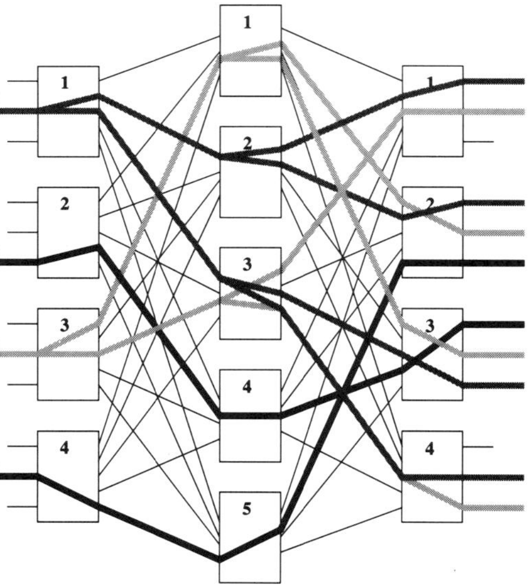

FIGURE 1. $\mathcal{C}(3,4,5)$ Clos Network. Connection requests $C_1 = (2,(1,1,1,1),.2)$, $C_2 = (6,(0,0,1,0),.3)$, $C_3 = (8,(1,1,1,1),.4)$ and $C_4 = (11,(0,1,0,0),.5)$.

such a way that the jth element of a vector is 1 if the destinations contain output switch j, 0 otherwise. For multirate broadcast of 3-stage Clos network $\mathcal{C}(n,r,m)$, a connection request is denoted as $C_i = (x_i, \vec{y}_i, \omega_i)$, where $x_i \in \{1, \dots, nr\}$ is an input port, $\vec{y}_i$ is a vector of size r denoting a set of output stage switches in the form of bit-vector, and ω_i is a required weight. For the connection request, we can define a *connection vector* as $\vec{I}_i = \omega_i \cdot \vec{y}_i$. Figure 1 shows four broadcast connection requests $C_1 = (2,(1,1,1,1),.2)$, $C_2 = (6,(0,0,1,0),.3)$, $C_3 = (8,(1,1,1,1),.4)$ and $C_4 = (11,(0,1,0,0),.5)$. Their connection vectors are $\vec{I}_1 = (.2,.2,.2,.2)$, $\vec{I}_2 = (0,0,.3,0)$, $\vec{I}_3 = (.4,.4,.4,.4)$ and $\vec{I}_4 = (0,.5,0,0)$.

DEFINITION 1 (relational operators of vector). *For given two vectors, $\vec{x}$ and $\vec{y}$ of length n, $\vec{x} \le \vec{y}$ if and only if $x_i \le y_i$ for all $i = 1, 2, \cdots, n$. $\vec{x} < \vec{y}$ if $\vec{x} \le \vec{y}$ and $\vec{x} \ne \vec{y}$. Similarly, we can define $\ge$ and $>$ operators.*

For example, if $\vec{x} = (.3,.2,.5)$, $\vec{y} = (.4,.3,.9)$ and $\vec{z} = (.4,.5,.3)$ then $\vec{x} \le \vec{y}$ but $\vec{x} \not\le \vec{z}$ because $x_3 \not\le z_3$. The k-th element of a connection vector $\vec{I}$ denotes the weight on output switch k so that the sum of connection vectors, $\sum_{all\ i} \vec{I}_i$, represents the sum of weights loaded on each output switch.

To describe the configuration of links between middle stage and output stage, we define *destination vector* $\vec{M}_j$ for each middle switch j such that $\vec{M}_j(k)$ is the sum of weights loaded on link between middle switch j and output switch k. In Figure 1, for example, $\vec{M}_1 = (0,.4,.4,0)$, $\vec{M}_2 = (.2,.2,0,0)$, $\vec{M}_3 = (.4,0,.2,.6)$, $\vec{M}_4 = (0,0,.3,0)$ and $\vec{M}_5 = (0,.5,0,0)$. Like connection vectors, the sum of destination vectors, $\sum_{all\ j} \vec{M}_j$, represents the sum of weights loaded on each output switch. Because each output switch has n output ports, its load is at most n. We can easily obtain the relationship between connection vectors and destination vectors after routing all requests as follows:

$$\sum_{all\ i} \vec{I_i} = \sum_{all\ j} \vec{M_j} \leq \vec{n}$$

Note that we can also describe the configuration of middle stage switches as $m \times r$ matrix, called *destination matrix*, as follows:

$$\mathbf{M} = \begin{bmatrix} \vec{M_1} \\ \vec{M_2} \\ \vdots \\ \vec{M_m} \end{bmatrix}$$

In the destination matrix $\mathbf{M}$, the sum of row i expresses the total loads on middle switch i and the sum of column j describes the total loads on output switch j.

Suppose we want to add a connection with weight ω. If a link between switch u and v has a capacity γ but it is currently carrying weights σ, then $\gamma - \sigma$ is a *residual capacity* of link (u, v) and at most $\gamma - \sigma$ additional weights can be routed through the link. To handle broadcast requests with connection vectors and destination vectors, we introduce other vector operators.

DEFINITION 2 (min and max of vectors). *For two vectors,* $\vec{x} = (x_1, x_2, \ldots, x_n)$ *and* $\vec{y} = (y_1, y_2, \ldots, y_n)$, *operators* min *and* max *are defined as follows:*

$$\mathsf{min}(\vec{x}, \vec{y}) = (\min(x_1, y_1), \min(x_2, y_2), \ldots, \min(x_n, y_n))$$
$$\mathsf{max}(\vec{x}, \vec{y}) = (\max(x_1, y_1), \max(x_2, y_2), \ldots, \max(x_n, y_n))$$

Generally, for k vectors, $\vec{x}_j = (x_{j,1}, x_{j,2}, \ldots, x_{j,n})$ *for* $1 \leq j \leq k$,

$$\mathsf{min}_{1\leq j\leq k}(\vec{x}_j) = (\min_{1\leq j\leq k}(x_{j,1}), \min_{1\leq j\leq k}(x_{j,2}), \ldots, \min_{1\leq j\leq k}(x_{j,n}))$$
$$\mathsf{max}_{1\leq j\leq k}(\vec{x}_j) = (\max_{1\leq j\leq k}(x_{j,1}), \max_{1\leq j\leq k}(x_{j,2}), \ldots, \max_{1\leq j\leq k}(x_{j,n}))$$

For the previous destination vectors,

$$\mathsf{min}_{1\leq j\leq 5}(\vec{M_j}) = (0, 0, 0, 0) \text{ and } \mathsf{max}_{1\leq j\leq 5}(\vec{M_j}) = (.4, .5, .4, .6).$$

4. Nonblocking Multirate Broadcast Clos Networks

Melen and Turner [**8**] introduced *speed advantage*$(1/\beta)$ to refer the ratio of the speeds of internal links to external links, where β is the normalized maximum capacity of external links. We assume that the weights of connection requests are within the range of $[b, B]$. If an internal link is blocking a connection request with weight ω, it means the link is carrying more than $1 - \omega$. Moreover, a blocking link can not carry weight less than b because it is the minimum weight of a connection. Hence, internal blocking links should carry weights at least $s(\omega) = \max\{1 - \omega + \epsilon, b\}$ for $\epsilon > 0$.

LEMMA 1. *Assume that a multirate Clos network $C(n, r, m)$, whose external links can operate on β for $0 < \beta \leq 1$, is in an arbitrary configuration for continuous bandwidth $[b, B]$ to which a broadcast connection request with bandwidth ω is compatible. Then, the* min *of more than* $\lceil \frac{\beta n - \omega}{s(\omega)} \rceil$ *destination vectors* $\vec{M_i}$*'s is less than* $\vec{s}(\omega)$.

PROOF. We will prove this lemma by contradiction. Suppose that

$$(1) \qquad m' > \lceil \frac{\beta n - \omega}{s(\omega)} \rceil \quad \text{and} \quad \vec{M}_{min} = \min_{1 \le i \le m'} (\vec{M}_i) \not< \vec{s}(\omega)$$

Then, there exits at least one column, say j, whose all elements are at least $s(\omega)$ in the $m' \times r$ destination matrix $\mathbf{M}'$. It means that output switch j has m' links connected from the middle stage, all of them are carrying not less than $s(\omega)$ weights. Due to compatibility, the total weights on the output switch should be less than or equal to $\beta n - \omega$. So,

$$(2) \qquad m's(\omega) \le \beta n - \omega$$

By (1) and (2),

$$\lceil \frac{\beta n - \omega}{s(\omega)} \rceil < \frac{\beta n - \omega}{s(\omega)}$$

This contradicts our assumption. □

Let us think a new broadcast request $C_i = (x, \vec{y}, \omega)$ and its connection vector $I_i = \omega \vec{y}$. To satisfy the request, we scan the *available* middle switches whose links from x's input switch ($\lfloor (x-1)/n \rfloor + 1$) are carrying weights less than $1 - \omega$. If a middle switch j whose *capacity vector* $\vec{1} - \vec{M}_j$ is greater than or equal to the connection vector I_i is found, the request can be satisfied through middle switch j. If there is no such middle switch, we could try to find two middle switches j_1 and j_2 whose *maximum capacity vector* $\vec{1} - \min(\vec{M}_{j_1}, \vec{M}_{j_2})$ is not less than the connection vector to satisfy the request through these two middle switches. Generally, If we are able to find k middle switches whose maximum capacity vector $\vec{1} - \min_{1 \le i \le k}(\vec{M}_{j_i})$ is greater than or equal to the connection vector, we can route the request through these k available middle switches. The value k does not need to be greater than r since it is the maximum fan-out of any connection requests. In addition, Lemma 1 instructs us any broadcast requests can be satisfied with no more than $\lceil \frac{\beta n - \omega}{s(\omega)} \rceil$ middle switches, which are maximized when $\omega \to B$ and $\epsilon \to 0$.

LEMMA 2 (number of blockings from an input switch). *For a multirate Clos network $C(n, r, m)$ whose external links can operate on β for $0 < \beta \le 1$, if each connection request with continuous bandwidth of $[b, B]$ duplicates its message to at most k middle switches in any configuration, the number of middle switches blocking a new connection request $C_i = (x, \vec{y}, \omega)$ from the input switch is no more than $\lfloor \frac{\beta n - \omega}{s(\omega)} k \rfloor$.*

PROOF. For the connection request with weight ω, the sum of weights coming to x's input switch is at most $\beta n - \omega$ by compatibility. Because each connection request duplicates its message at most k copies in the input switch, the sum of weights going out of the input switch is no more than $(\beta n - \omega)k$. Let m' be the number of links between the input switch and middle stage switches which are blocking the connection request, i.e. the number of links carrying $s(\omega)$. Then, $m's(\omega) \le (\beta n - \omega)k$. For an integer m', we obtain that

$$m' \le \lfloor \frac{\beta n - \omega}{s(\omega)} k \rfloor$$

□

LEMMA 3 (number of blockings to output stage). *For a multirate 3-stage Clos network $\mathcal{C}(n,r,m)$ with continuous bandwidth of $[b,B]$, whose external links can operate on β for $0<\beta\leq 1$, assume a new connection request $C_i=(x,\vec{y},\omega)$ is compatible to the existing configuration. If $m'>\lfloor\frac{\beta n-\omega}{s(\omega)}r^{1/k}\rfloor$, there exist some k middle switches among m' available middles that satisfy the new connection request.*

PROOF. Let us prove its contrapositive—if any k middle switches among m' are blocking the connection request, then $m'\leq\lfloor\frac{\beta n-\omega}{s(\omega)}r^{1/k}\rfloor$. Consider an $m'\times r$ destination matrix $\mathbf{M}$. Each output switch can carry at most $\beta n-\omega$ due to compatibility. Let $c_1(j)$ be the number of links from middle switch j carrying weights at least $s(\omega)$, i.e. the number of links blocking the new connection request. Then

$$\sum_{j=1}^{m'} c_1(j)s(\omega)\leq(\beta n-\omega)r$$

Let $c_1=\min_{1\leq j\leq m'}\{c_1(j)\}$. Then

$$m'c_1s(\omega)\leq(\beta n-\omega)r$$

Because $c_1\neq 0$ (If so, we can use the middle switch to route the request so that it is not blocking. Contradiction). we can obtain

$$m'\leq\frac{\beta n-\omega}{s(\omega)}\frac{r}{c_1}\tag{3}$$

Without loss of generality, the destination vector of middle switch 1 has the minimum. By using middle switch 1, we can route $r-c_1$ output switches among r. Only c_1 output switches are blocked through the middle switch. The number of columns in the destination matrix can be reduced to c_1 by discarding $r-c_1$ columns which were already satisfied by using the middle switch. The contracted matrix is denoted by $\mathbf{M}^*$. We replace each row vector with the minimum of the row vector itself and the first row vector and denote it as $\mathbf{M}^1$. That is

$$\forall_{1\leq j\leq m'}:\mathbf{M}^1_j=\mathsf{min}(\mathbf{M}^*_1,\mathbf{M}^*_j)$$

In general, for any $m'\times c_{i-1}$ destination matrix $\mathbf{M}^{i-1}$ for $i<k$, there exist only c_{i-1} output switches which are needed to be routed through. Let $c_i(j)$ be the number of links carrying weights not less than $s(\omega)$, i.e. the number of links which are still blocking the new connection request. Then

$$\sum_{j=1}^{m'} c_i(j)s(\omega)\leq(\beta n-\omega)c_{i-1}$$

Let $c_i=\min_{1\leq j\leq m'}\{c_i(j)\}$. Then

$$m'c_is(\omega)\leq\sum_{j=1}^{m'} c_i(j)s(\omega)\leq(\beta n-\omega)c_{i-1}$$

Because of $c_i\neq 0$,

$$m'\leq\frac{\beta n-\omega}{s(\omega)}\frac{c_{i-1}}{c_i}\tag{4}$$

When $i = k$, each row vector in the destination matrix $\mathbf{M}^{k-1}$ has at least one value that is not less than $s(\omega)$. So,

$$(5) \qquad m's(\omega) \le (\beta n - \omega)c_{k-1} \qquad \text{and} \qquad m' \le \frac{\beta n - \omega}{s(\omega)} c_{k-1}$$

A minimum of a sequence is not greater than the geometric mean of the sequence so that, from (3), (4) and (5), we can obtain that

$$\begin{aligned} m' &\le \min\left\{\frac{\beta n - \omega}{s(\omega)}\frac{r}{c_1}, \frac{\beta n - \omega}{s(\omega)}\frac{c_1}{c_2}, \dots, \frac{\beta n - \omega}{s(\omega)}\frac{c_{k-2}}{c_{k-1}}, \frac{\beta n - \omega}{s(\omega)}c_{k-1}\right\} \\ &\le \left[\left(\frac{\beta n - \omega}{s(\omega)}\frac{r}{c_1}\right)\cdot\left(\frac{\beta n - \omega}{s(\omega)}\frac{c_1}{c_2}\right)\cdots\left(\frac{\beta n - \omega}{s(\omega)}\frac{c_{k-2}}{c_{k-1}}\right)\cdot\left(\frac{\beta n - \omega}{s(\omega)}c_{k-1}\right)\right]^{1/k} \\ &= \frac{\beta n - \omega}{s(\omega)} r^{1/k} \end{aligned}$$

Since m' is an integer, we can get $m' \le \lfloor \frac{\beta n - \omega}{s(\omega)} r^{1/k} \rfloor$. □

THEOREM 1. *For a multirate 3-stage Clos network $C(n, r, m)$ with continuous bandwidths of $[b, B]$, whose external links operate on β for $0 < \beta \le 1$, it is non-blocking for broadcasting if*

$$m > \lim_{\epsilon \to 0}\left(\min_{1 \le k \le \min(\lceil \frac{\beta n - B}{s(B)} \rceil, r)}\left(\left\lfloor \frac{\beta n - B}{s(B)} k \right\rfloor + \left\lfloor \frac{\beta n - B}{s(B)} r^{1/k} \right\rfloor\right)\right)$$

PROOF. By Lemma 1, 2 and 3, the number of middle switches blocking from input stage is at most $\lfloor \frac{\beta n - \omega}{s(\omega)} k \rfloor$ and the number of middle switches blocking at output stage is at most $\lfloor \frac{\beta n - \omega}{s(\omega)} r^{1/k} \rfloor$. Moreover, the min of more than $\lceil \frac{\beta n - \omega}{s(\omega)} \rceil$ destination vectors is less than $\vec{s}(\omega)$ and the maximum fan-out is r. We can obtain the maximum number of blocking middle switches when $\omega \to B$ and $\epsilon \to 0$. □

The disadvantage of Theorem 1 is that the Clos network requires infinite number of middle switches when b goes to 0 and B goes to 1. To overcome this, we consider a special case of $B = 1/2$, $\beta = 1$ and $\epsilon \to 0$. Then, we obtain that

$$(6) \qquad m > \min_{1 \le k \le \min(2n-1, r)} (2n-1)(k + r^{1/k})$$

To find out the other case of $\omega > 1/2$, let us consider a special network in which an internal link is blocking a new broadcast requests $C = (x, \vec{y}, \omega)$ if the internal link contains exactly p requests. Then, it is easy to find such a condition as $\frac{1}{p+1} < \omega_i \le \frac{1}{p}$ for all connection requests i, which is called *a harmonic rate network* for some positive p. If external links are operating at the speed of β, the number of calls on the external links can be further limited by the following lemma.

LEMMA 4. *In a harmonic rate network for some positive p, if external links are operating at β for $0 < \beta \le 1$, the external links can carry at most $\lceil \beta(p+1) - 1 \rceil$ calls.*

PROOF. Let h be the maximum number of calls on an external link. The maximum can be achieved when all connections on the external link were carrying the minimum bandwidth, b. Because of $\frac{1}{p+1} < b \le \omega \le \frac{1}{p} \le B$ for all requests and compatibility,

$$h\frac{1}{p+1} < bh \le \beta \qquad \text{hence,} \qquad h \le \lceil \beta(p+1) - 1 \rceil$$

□

Suppose a new broadcast connection request $C = (x, \vec{1}, \omega)$ is compatible to the existing harmonic rate network for some positive p. An internal link blocks the request if and only if it is carrying exactly p calls. In terms of blocking property, we do not need to distinguish the sum of weights and the number of calls in the harmonic rate network. We consider a destination matrix for the network holds the number of calls in the following proofs.

LEMMA 5. *Assume a harmonic rate Clos network $\mathcal{C}(n,r,m)$ for some positive p is in an arbitrary configuration. If a new broadcast request $C = (x, \vec{y}, \omega)$ is compatible to the existing network, then the* min *of more than $\lceil \frac{n\lceil \beta(p+1)-1\rceil - 1}{p} \rceil$ destination vectors is less than $\vec{p}$.*

PROOF. Suppose that

$$m' > \lceil \frac{n\lceil \beta(p+1) - 1\rceil - 1}{p} \rceil \qquad \text{and} \qquad \vec{M}_{min} = \min_{1 \le i \le m'} (\vec{M}_i) \not< \vec{p} \tag{7}$$

By Definition 1 and 2, there exist at least one element j in the destination vector such that $\vec{M}_{min}(j) \ge p$. It implies that all j-th elements of m' destination vectors are at least p. That is,

$$\exists_j \forall_{1 \le i \le m'} \; : \; \vec{M}_i(j) \ge p$$

Hence, the number of calls on output switch j is at least $m'p$. By Lemma 4, each output port of switch j cannot carry more than $\lceil \beta(p+1) - 1\rceil$ calls. Because the new broadcast request is compatible to the existing configuration,

$$m'p \le n\lceil \beta(p+1) - 1\rceil - 1 \tag{8}$$

By (7) and (8),

$$\lceil \frac{n\lceil \beta(p+1) - 1\rceil - 1}{p} \rceil p < m'p \le n\lceil \beta(p+1) - 1\rceil - 1$$

Therefore,

$$\lceil \frac{n\lceil \beta(p+1) - 1\rceil - 1}{p} \rceil < \frac{n\lceil \beta(p+1) - 1\rceil - 1}{p}$$

This contradicts our assumption. □

We have found the compatibility and the maximum fan-out for the harmonic rate Clos network. Next, we will find the maximum number of middle switches blocking from the input stage and from the ouput stage.

THEOREM 2. *For a harmonic rate Clos network $\mathcal{C}(n,r,m)$ for some positive p, if its external links operate on β for $0 < \beta \le 1$, it is nonblocking for broadcast if*

$$m > \min_{1 \le k \le \min(\lceil \frac{n\lceil \beta(p+1)-1\rceil - 1}{p} \rceil, r)} \left(\lfloor \frac{n\lceil \beta(p+1) - 1\rceil - 1}{p} k \rfloor + \lfloor \frac{n\lceil \beta(p+1) - 1\rceil - 1}{p} r^{1/k} \rfloor \right)$$

PROOF. Let $C = (x, \vec{y}, \omega)$ be a new broadcast request. Due to compatibility, the number of calls coming to x's input switch is at most $n\lceil\beta(p+1)-1\rceil - 1$. Each request duplicates its message at most k times in the input switch so that the total number of messages out of the input switch is no more than $(n\lceil\beta(p+1)-1\rceil-1)k$. Let m' be the number of blocking links from the input switch. The blocking links have to carry exactly p calls. Therefore,

$$m'p \le (n\lceil\beta(p+1)-1\rceil - 1)k$$
$$m' \le \lfloor \frac{n\lceil\beta(p+1)-1\rceil - 1}{p} k \rfloor$$

If we are given m'' available middle switches, it is always possible to find at most k middle switches to satisfy the request, where $m'' > \lfloor \frac{n\lceil\beta(p+1)-1\rceil-1}{p} r^{1/k} \rfloor$. To prove this, let us consider its contrapositive — if any k middles among m'' are blocking the request, then $m'' \le \lfloor \frac{n\lceil\beta(p+1)-1\rceil-1}{p} r^{1/k} \rfloor$. Consider an $m'' \times r$ destination matrix $\mathbf{M}$, whose (i, j) element holds the number of calls routed through the link between middle switch i and output switch j. Each output switch can carry at most $n\lceil\beta(p+1)-1\rceil - 1$ calls due to compatibility and is represented as a sum of the column in the matrix. Let $c_1(j)$ be the number of links blocking the new request by carrying exactly p calls in destination vector $\vec{M}_j$. Then,

$$\sum_{j=1}^{m''} c_1(j)p \le (n\lceil\beta(p+1)-1\rceil - 1)r$$

Let $c_1 = \min_{1\le j\le m''}\{c_1(j)\}$. Moreover, $c_1 \ne 0$. If so, we can use the middle switch to route the request so that it is contradicting our assumption.

$$m''c_1p \le (n\lceil\beta(p+1)-1\rceil - 1)r \quad \text{or} \quad m'' \le \frac{n\lceil\beta(p+1)-1\rceil - 1}{p}\frac{r}{c_1} \tag{9}$$

Without loss of generality, the destination vector of middle switch 1 has the minimum. By using the middle switch, we can route $r - c_1$ output switches among r. Only c_1 output switches are left to be blocking the request. The number of columns in the destination matrix can be reduced to c_1 from r by discarding $r - c_1$ columns whose corresponding output switches are already satisfied through middle switch 1. The contracted matrix is denoted by $\mathbf{M}^1$. In general, there are only c_{i-1} output switches which are needed to be routed through for any $m'' \times c_{i-1}$ destination matrix $\mathbf{M}^{i-1}$ for $i < k$. Let $c_i(j)$ be the number of links blocking the new request by carrying exactly p calls in destination vector $\vec{M}_j$. Then, due to compatibility,

$$\sum_{j=1}^{m''} c_i(j)p \le (n\lceil\beta(p+1)-1\rceil - 1)c_{i-1}$$

Let $c_i = \min_{1\le j\le m''}\{c_i(j)\}$. $c_i \ne 0$ because we assume any k middles can not satisfy the new request

$$m''c_ip \le (n\lceil\beta(p+1)-1\rceil - 1)c_{i-1} \quad \text{or} \quad m'' \le \frac{n\lceil\beta(p+1)-1\rceil - 1}{p}\frac{c_{i-1}}{c_i} \tag{10}$$

When $i = k$, each row vector in the destination matrix $\mathbf{M}^{k-1}$ has at least one element whose value is p. So,

$$(11) \qquad m''p \leq (n\lceil\beta(p+1)-1\rceil - 1)c_{k-1} \quad \text{or} \quad m'' \leq \frac{n\lceil\beta(p+1)-1\rceil - 1}{p}c_{k-1}$$

We have to find the minimum of m'' from (9), (10) and (11). Because a geometric mean is not less than a minimum of sequence, we can obtain that

$$m'' \leq \frac{n\lceil\beta(p+1)-1\rceil - 1}{p}r^{1/k}$$

Since m'' is an integer, we can get $m'' \leq \lfloor \frac{n\lceil\beta(p+1)-1\rceil-1}{p} r^{1/k} \rfloor$.

If we are given one more middle switch in addition to $m' + m''$ which are the maximum number of middles blocking from the input stage and the output stage when each connection request uses at most k middle switches. By Lemma 5, we know k does not need to be greater than $\min(\lceil \frac{n\lceil\beta(p+1)-1\rceil-1}{p} \rceil, r)$. □

From Theorem 2, consider when $p = 1$ so that $1/2 < \omega_i \leq 1$ for all requests i, which results in $1/2 < \beta \leq 1$ and $\lceil 2\beta - 1 \rceil = 1$. For this case, we can obtain that

$$(12) \qquad m > \min_{1 \leq k \leq \min(n-1, r)} (n-1)(k + r^{1/k})$$

One can perceive by intuition that there is no difference between circuit switching and multirate environments if all requests have weights greater than 1/2, because a link can not carry any other requests once the link carried one. In that reason, (12) is corresponding to Yang and Masson's result [**11**].

From (6) and (12), we conclude that a Clos network, $\mathcal{C}(n, r, m)$, is nonblocking in a wide-sense for multirate broadcast if

$$m \geq (3n-2) \min_{1 \leq k \leq \min(2n-1, r)} (k + r^{1/k}) + 2$$

The routing algorithm divides the middle switches into two subsets. The first subset has more than $(n-1)\min(k + r^{1/k})$ middle switches and the second has more than $(2n-1)\min(k + r^{1/k})$ middles. It allows broadcast requests of weight greater than 1/2 to use the first subset of middles and other requests to use the second subset. In the following section, we will further reduce the number of middle switches by employing a technique proposed by Gao and Hwang [**4**].

5. Wide-sense Nonblocking Multirate Broadcast Networks

In this section, we inadvertently ignore the integer operations and the bound of k not to stray from the main focus. Let us define $f(r) = \min(k + r^{1/k})$ which was known to have an approximate minimum of $O(\ln r / \ln\ln r)$ [**11**]. The previous wide-sense nonblocking broadcast Clos network can be represented as

$$m \geq (3n-2)f(r) + 2$$

We will consider another special network in which the weight of a new broadcast request is no more than $\frac{1}{p+1}$ for some positive p.

THEOREM 3. *A Clos network $\mathcal{C}(n, r, m)$ is nonblocking for multirate broadcast when $C = (x, \vec{y}, \omega_S)$ where $\omega_S \leq \frac{1}{p+1}$ if*

$$m > \frac{(\beta n - \omega_S)(p+1)}{p} f(r)$$

PROOF. $\omega_S \leq \frac{1}{p+1}$ implies $1-\omega_S \geq \frac{p}{p+1}$. Let m' be the number of middle switches blocking the new broadcast request ω_S from x's input switch. Because each request duplicates its message at most k times at the input switch and the new request is compatible to the existing configuration,

$$m'\frac{p}{p+1} \leq m'(1-\omega_S) \leq \sum(\text{weights on blocking links}) \leq (\beta n - \omega_S)k$$

So, the number of blocking links from the input switch m' is,

$$m' \leq \frac{(\beta n - \omega_S)(p+1)}{p}k \tag{13}$$

We will use the similar approach as Lemma 3 to obtain the number of blocking middle switches to output stage. Suppose that any k middles among m'' available middles can not satisfy the new broadcast with ω_S. Let $c_1(j)$ be the number of links from middle switch j, which are blocking the new request by carrying weight more than $1-\omega_S \geq \frac{p}{p+1}$.

$$m''c_1\frac{p}{p+1} \leq m''c_1(1-\omega_S) \leq \sum_{j=1}^{m''} c_1(j)(1-\omega_S) \leq (\beta n - \omega_S)r,$$

where $c_1 = \min_{1\leq j\leq m''}\{c_1(j)\}$ and $c_1 \neq 0$. If so, the middle switch corresponding the minimum can be used to route the new request. This is contradicting our assumption. Hence,

$$m'' \leq \frac{(\beta n - \omega_S)(p+1)}{p}\frac{r}{c_1} \tag{14}$$

Let middle switch 1 have the minimum c_1. We can route the request to $r-c_1$ output switches by using middle switch 1 so $r-c_1$ columns in the destination matrix can be discarded. The remaining c_1 output switches need to be routed through the rest of middle switches. Each row vector of the contracted matrix is replaced with the minimum of the row vector itself and the first row vector and denote the matrix as $\mathbf{M}^1$. Generally, assume there are only c_{i-1} output switches which are needed to be routed through by using $m'' \times c_{i-1}$ destination matrix $\mathbf{M}^{i-1}$ for $i < k$. Let $c_i(j)$ be the number of links blocking the new broadcast request by carrying more than $1-\omega_S$ and c_i be the minimum of $c_i(j)$ for all j. Then,

$$m''c_i\frac{p}{p+1} \leq m''c_i(1-\omega_S) \leq \sum_{j=1}^{m''} c_i(j)(1-\omega_S) \leq (\beta n - \omega_S)c_{i-1}$$

We know that $c_i \neq 0$ when $i < k$. If so, those i middle switches can be used to route the broadcast request. This contradicts our assumption that any k middle switches can not satisfy the new broadcast request with ω_S.

$$m'' \leq \frac{(\beta n - \omega_S)(p+1)}{p}\frac{c_{i-1}}{c_i} \tag{15}$$

When $i = k$, each row vector has at least one element whose value is more than $1 - \omega_S$. Therefore,

$$m''\frac{p}{p+1} \le m''(1-\omega_S) \le \sum_{j=1}^{m''}(1-\omega_S) \le (\beta n - \omega_S)c_{k-1}$$

$$m'' \le \frac{(\beta n - \omega_S)(p+1)}{p}c_{k-1} \tag{16}$$

Again, a geometric mean is not less than the minimum of sequence so that the minimum from (14), (15) and (16) can be obtained as

$$m'' \le \frac{(\beta n - \omega_S)(p+1)}{p}r^{1/k}$$

If we are given one more middle switch in addition to $m' + m'' = \frac{(\beta n - \omega_S)(p+1)}{p}f(r)$, we can find at most k middle switches to satisfy the new broadcast request with ω_S. □

To provide a general multirate broadcast switching networks, we extend a routing algorithm called a *quota scheme* from Gao and Hwang [**4**]. Connection requests are partitioned into *large* calls (C_L) and *small* calls (C_S) by their weight. In addition, the set of middle switches(M) is also assumed to be partitioned to M_S and M_L whose sizes are m_S and m_L, respectively. The algorithm forces the small calls (C_S) to use only the set of middle switches M_S but allows the large calls (C_L) to use not only the middles M_L but also the middles M_S. Define $p = \lfloor 1/B \rfloor$. To utilize Theorem 3, a broadcast request with $\omega_S \le \frac{1}{p+1}$ for some positive p is classified to a small call and the one with $\omega_L > \frac{1}{p+1}$ be a large call. Of course, ω_L should not be greater than B because it is the upper bound of all connection requests.

THEOREM 4. *The multirate 3-stage Clos networks $\mathcal{C}(n, r, m)$, in which weights are within the range of $[b, B]$ and external links can operate at β, is nonblocking in a wide-sense if*

$$m \ge \begin{cases} \dfrac{\beta n(p+1)(Bp+B+p-1)}{p^2}f(r)+1 & \text{for } B < 23/32 \\ (\dfrac{15\beta n}{8}+n-1)f(r)+2 & \text{for } B \ge 23/32 \end{cases}$$

, where $p = \lfloor 1/B \rfloor$.

PROOF. Assume a large call $C = (x, \vec{y}, \omega_L)$ with $\omega_L > \frac{1}{p+1}$ is compatible to the existing configuration. Let M'_S be the set of middles blocking the large call from x's input switch among M_S by carrying weights more than $1-\omega_L$ and M'_L be the set of middles blocking the request from the input switch among M_L by carrying exactly p calls and their sizes are m'_S and m'_L, respectively. Each request is supposed to multiplicate its message at most k times at the input switch. Because of the compatibility, the maximum weights going to the middle stage out of the input switch is at most $(\beta n - \omega_L)k$. Therefore

$$m'_L\frac{p}{p+1} + m'_S(1-B) \le m'_L\frac{p}{p+1} + m'_S(1-\omega_L) \le (\beta n - \omega_L)k \tag{17}$$

Then $M''_S = M_S \setminus M'_S$ and $M''_L = M_L \setminus M'_L$ will be the set of available middle switches for the large request within the set of M_S and M_L, respectively. Their sizes are denoted as m''_S and m''_L. To find out the maximum number of blocking links to output switches, let us consider $(m''_S + m''_L) \times r$ destination matrix $\mathbf{M}$. Likewise, suppose that any k middles among $m''_S + m''_L$ availables can not satisfy the large broadcast request with ω_L. Let $c_1(j)$ be the number of blocking links between middle switch j and the output stage, which are either carrying exactly p calls if $j \in M''_L$ or carrying weights more than $1 - \omega_L$ if $j \in M''_S$.

$$\sum_{j \in M''_L} c_1(j) \frac{p}{p+1} + \sum_{j \in M''_S} c_1(j)(1-B) \leq (\beta n - \omega_L) r$$

$$m''_L c_1 \frac{p}{p+1} + m''_S c_1 (1-B) \leq (\beta n - \omega_L) r$$

$$m''_L \frac{p}{p+1} + m''_S (1-B) \leq (\beta n - \omega_L) \frac{r}{c_1}, \tag{18}$$

where $c_1 = \min_{j \in M''_L \cup M''_S} \{c_1(j)\}$. We delete $r - c_1$ columns from the destination matrix and obtain a new sub-matrix $\mathbf{M}^1$. In general, consider c_i blocking links from c_{i-1} links in destination matrix $\mathbf{M}^{i-1}$ for $i < k$. Then we obtain that

$$m''_L \frac{p}{p+1} + m''_S (1-B) \leq (\beta n - \omega_L) \frac{c_{i-1}}{c_i} \tag{19}$$

After $k-1$ iterations, each row in the result destination matrix $\mathbf{M}^{k-1}$ has at least one blocking links. So,

$$m''_L \frac{p}{p+1} + m''_S (1-B) \leq (\beta n - \omega_L) c_{k-1} \tag{20}$$

From (18), (19) and (20), we get

$$m''_L \frac{p}{p+1} + m''_S (1-B) \leq (\beta n - \omega_L) r^{1/k} \tag{21}$$

Add (17) and (21). Because $m'_L + m''_L = m_L$ and $m'_S + m''_S = m_S$, we get

$$m_L \frac{p}{p+1} + m_S (1-B) \leq (\beta n - \omega_L) f(r)$$

Hence, we obtain,

$$m_L \leq \frac{[(\beta n - \omega_L) f(r) - (1-B) m_S](p+1)}{p}$$

From Theorem 3, the maximum numbers of blocking middles m^*_S and m^*_L in M_S and M_L are obtained by $\omega_S \to 0$ and $\omega_L \to 0$.

$$m^*_S = \frac{\beta n (p+1)}{p} f(r)$$

$$m^*_L = \frac{\beta n (Bp + B - 1)(p+1)}{p^2} f(r)$$

Hence, the network is wide-sense nonblocking if one more middle switch is provided in addition to $m^*_L + m^*_S$

$$m \geq \frac{\beta n (Bp + B + p - 1)(p+1)}{p^2} f(r) + 1 \tag{22}$$

When $B = 1/2$ so as $p = 2$, a Clos network $\mathcal{C}(n, r, m_1)$ is nonblocking if $m_1 \geq \frac{15\beta n}{8} f(r) + 1$. For the other case of $\omega > 1/2$, a Clos network $\mathcal{C}(n, r, m_2)$ is nonblocking if $m_2 \geq (n-1) f(r) + 1$ as we have seen in Section 4. We can combine these two Clos networks in parallel to construct a general wide-sense nonblocking Clos network $\mathcal{C}(n, r, m)$ for multirate broadcast in which $m \geq (\frac{15\beta n}{8} + n - 1) f(r) + 2$.

In this network, all broadcast requests with $\omega \leq 1/2$ are routed through the first sub-network, and all requests with $\omega > 1/2$ are routed through the other sub-network.

Let us compare $g_1(B) = \beta n(Bp + B + p - 1)(p + 1)f(r)/p^2 + 1$ with $g_2 = (15\beta n/8 + n - 1)f(r) + 2$. It is easy to verify $g_1(B)$ is an increasing function on B and it is always smaller than g_2 for $B \leq 1/2$ because $\beta \leq 1$. For $B > 1/2$, i.e. $p = 1$, $g_1(B)$ is approximately equal to g_2 at $B = 23/32$. □

6. Conclusions

In this paper, we have investigated several nonblocking Clos networks in a wide-sense for multirate broadcast environments. We determined a nonblocking Clos network for multirate broadcast, which needs infinite number of middle switches when $B \to 1$ and/or $b \to 0$. To overcome the obstacle, we combined two Clos networks in parallel, one had more than $(n-1)\min(k + r^{1/k})$ middle switches, the other had more than $(2n-1)\min(k+r^{1/k})$ middle switches. To reduce the hardware complexity further, we extended Gao and Hwang's result on multirate wide-sense nonblocking for one-to-one calls to broadcast communications, which concluded that a Clos network was nonblocking to any multirate broadcast requests when $m \geq (2.875n - 1)\min(k + r^{1/k}) + 2$. In our construction, the required hardware complexity was dependent only on the parameters of the switch device, but not on the patterns of connection requests, which is highly advisable for constructing a real switching network.

References

[1] V. E. Beneš. *Mathematical Theory of Connecting Networks and Telephone Traffic*. Academic Press, New York, 1965.

[2] Shun-Ping Chung and Keith W. Ross. On nonblocking multirate interconnection networks. *SIAM J. on Comp.*, 20(4):726–726, August 1991.

[3] C. Clos. A study of non-blocking switching networks. *Bell Syst. Tech. J.*, 32:406–424, March 1953.

[4] Biao Gao and Frank K. Hwang. Wide-sense nonblocking for multirate 3-stage Clos networks. *to appear in Theoretical Computer Science.*

[5] F. K. Hwang. Rearrangeability of multiconnection three-stage networks. *Networks*, 2:301–306, 1972.

[6] F. K. Hwang. Three-stage multi-connection networks which are nonblocking in the wide sense. *Bell Syst. Tech. J.*, 58(10):1283–1287, December 1979.

[7] G. M. Masson and B. W. Jordan, Jr. Generalized multistage connection networks. *Networks*, 2:192–209, 1972.

[8] Riccardo Melen and Jonathan S. Turner. Nonblocking multirate networks. *SIAM J. on Comp.*, 18(2):301–313, April 1989.

[9] I. Svinnset. Nonblocking ATM switching networks. *IEEE Trans. on Comm.*, 42(2-4):1352–1358, Feb.-Apr. 1994.

[10] Yuanyuan Yang. An analysis model on nonblocking multirate broadcast networks. In *International Conference on Supercomputing*, pages 256–263, 1994.

[11] Yuanyuan Yang and Gerald M. Masson. Nonblocking broadcast switching networks. *IEEE Trans. on Comp.*, 40(9):1005–1015, September 1991.

Department of Computer Science and Engineering, University of Minnesota, Minneapolis, MN 55455
E-mail address: dkim@cs.umn.edu

Department of Computer Science and Engineering, University of Minnesota, Minneapolis, MN 55455
E-mail address: dzd@cs.umn.edu

DIMACS Series in Discrete Mathematics
and Theoretical Computer Science
Volume 42, 1998

Repackable Networks — The Concept and Applications

Andrzej Jajszczyk and Marek Kubale

ABSTRACT. The concept of repackable switching networks is reviewed. In such networks blocking can be avoided by using call repacking control algorithms. Repackable networks are compared with strictly nonblocking, wide-sense nonblocking and rearrangeable networks. The asymptotic number of crosspoints in repackable networks is calculated as well as call repacking algorithm is proposed.

1. Introduction

A *repackable* network is a network in which it is always possible to connect an idle input-output pair if during the operation of the network suitable algorithm is used to assign routes for new calls and suitable repackings of some existing calls are performed after call terminations [1]. This suitable algorithm for assigning new routes uses the most loaded middle-stage switches wherever possible. Such an algorithm avoids all blocking states, similarly as in the case of a wide-sense nonblocking network. However, blocking states can be reached via hang-ups. In repackable networks these blocking states are non-permanent because call repackings are used immediately after a call termination, and the network reaches a state in which no blocking can occur.

In most experiments the call repacking technique was used to improve the traffic performance. However, it has been shown that by using this type of control algorithm in networks containing a sufficient number of middle-stage switches we can avoid any blocking at all [1].

In this paper we review the basic results concerning repackability of switching networks. We also compare repackable networks with networks which are nonblocking in the strict or wide sense as well as with rearrangeable networks. We calculate the asymptotic number of crosspoints in repackable networks and formulate call repacking algorithm under which a three-stage network is repackable.

2. Basic Conditions

To find conditions under which a three-stage Clos network is repackable we introduce the following two notions.

DEFINITION 2.1. A *non-permanent* state of a switching network is a state which when achieved by the network is immediately replaced by another state.

1991 *Mathematics Subject Classification.* Primary 94A40, 94C99.

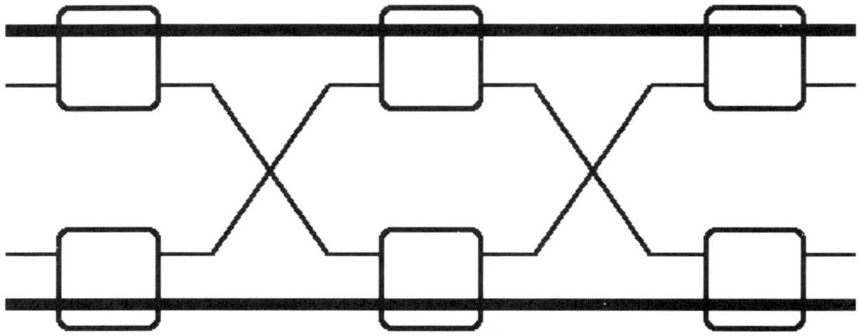

(a)

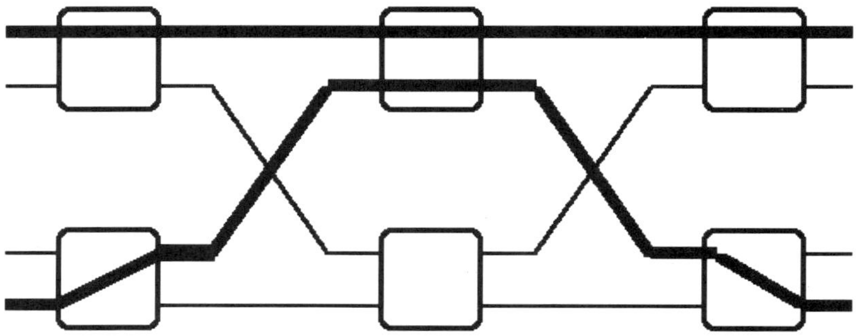

(b)

FIGURE 1. Examples of network's states: (a) an overweight state; (b) non-overweight state.

DEFINITION 2.2. An *overweight* state of a switching network is a state in which there exists a connecting path which can be disconnected and then connected again through another more heavily loaded middle-stage switch than the switch that was used before (loads are compared after the disconnection).

Examples of an overweight and non-overweight states are shown in Fig. 1. Connecting paths 1-1 and 4-4 in Fig. 1a occupy two different middle-stage switches. However, we can route call 4-4 through the same switch as call 1-1, thus reaching a non-overweight state (Fig. 1b).

The following theorem can be formulated [1].

THEOREM 2.3. *No blocking occurs in a three-stage Clos network if and only if*

$$m \geq 2n - \left\lceil \frac{n}{r-1} \right\rceil \tag{2.1}$$

and the network is controlled in such a way that each overweight state of the network is non-permanent.

The proof can be found in reference [1].

It can be noted that for $r > n$ we have $\lceil \frac{n}{r-1} \rceil = 1$, and, therefore, (2.1) is reduced to $m \geq 2n - 1$, i.e., the same result as that given by Clos for strictly nonblocking networks. This means that for $r > n$ call repackings do not save middle-stage switches when nonblocking properties are required (although they can improve the grade of service for blocking networks).

Although call repackings do save middle-stage switches in some networks, it is interesting if they save crosspoints as well. It can be shown that for some network capacities savings can be achieved. For example, for $N = 1369$, the optimum strictly nonblocking Clos network $v(73, 37, 37)$ contains 299811 crosspoints and the optimum repackable network $v(72, 37, 37)$ has 295704 crosspoints. However, 1369 is not a very practical size of a switching network. For $N = 1024$ the costs of strictly nonblocking and repackable networks are identical. More structures of repackable networks containing fewer crosspoints than strictly nonblocking networks of the same size are listed in [2].

Let us consider now a three-stage T-S-T time-division switching network. Such a network contains time-slot interchangers in the first and in the third stages and a time-multiplexed space-division switch in the middle stage. Let us assume that the total capacity of the network is 1024 channels and the maximum capacity of a single time-slot interchanger is 512 time slots at each side. A strictly nonblocking network is shown in Fig. 2a. As can be easily shown a space equivalent of this time-division network is a three-stage Clos network $v(256, 512, 4)$. However, a repackable network of the same total capacity requires only four time-slot interchangers (512×512), as shown in Fig. 2b.

3. Comparison of Various Classes

Table 1 summarizes conditions under which it is always possible to set up a new call between an idle input and an idle output for two-sided Clos networks $v(m, n, r)$. Fig. 3 compares the numbers of required middle-stage switches in repackable, rearrangeable, wide-sense and strictly nonblocking networks for $n = 16$ and different values of r. We can note that repackable networks form a "bridge" between rearrangeable networks ($r = 2$) and strictly nonblocking networks ($r > n$).

TABLE 1. Basic conditions for two-sided Clos network $v(m, n, r)$

Class	Condition
Nonblocking in the strict sense	$m \geq 2n - 1$
Nonblocking in the wide sense	$m \geq \lfloor \frac{3n}{2} \rfloor, r = 2$
Rearrangeable	$m \geq n$
Repackable	$m \geq 2n - \lceil \frac{n}{r-1} \rceil$

4. Asymptotic count of the number of crosspoints in repackable networks [2]

Let us consider a three-stage repackable network $v(m, n, r)$. Let $N = N(v) = nr$ denotes the total number of inputs/outputs, and $C = C(v)$ denotes the number

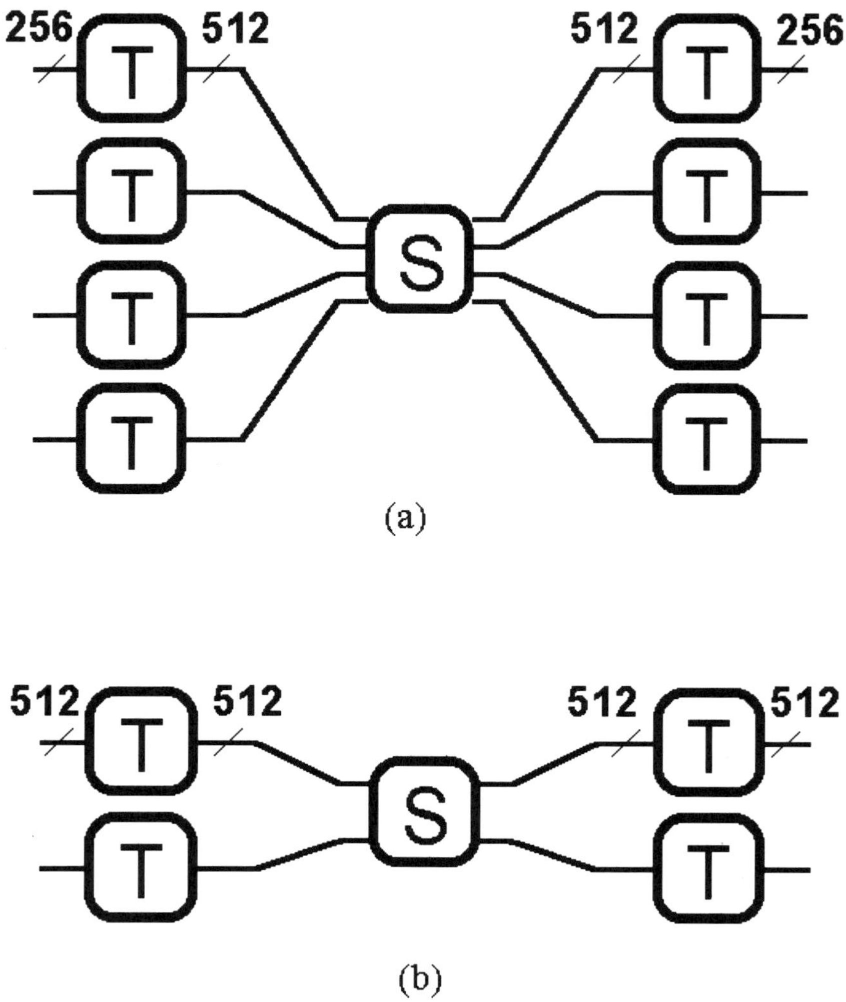

FIGURE 2. T-S-T switching networks of 1024 channels; (a)strictly nonblocking, (b)repackable.

of crosspoints. The number of crosspoints is as follows:

$$C = (mnr) + (mr^2) + (mnr) = mr(2n + r) \tag{4.1}$$

We search for values of m, n, r that minimize C. Let $x = n/(r-1)$. Then, $m \cong 2n - x$, and hence,

$$\begin{aligned} C =& mr(2n + r) \cong (2n - x)r(2n + r) = \\ & (2nr - xr)(2n + r) = (2N - xr)(2n + r) \end{aligned} \tag{4.2}$$

Moreover,

$$N/r = n = x(r - 1) \cong xr, \tag{4.3}$$

since for $r \gg 1$, we have $r \cong r - 1$.
From (4.3) we have $r^2 = N/x$, and hence, $r = \sqrt{N/x}$. Since $n = N/r$ then

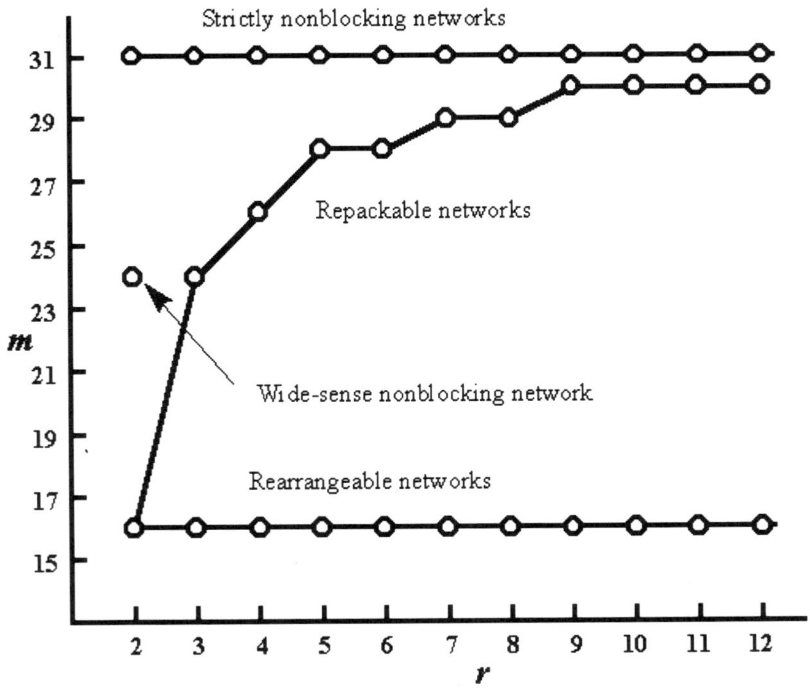

FIGURE 3. Number of middle-stage switches for various network classes; $n = 16$.

$n = N/\sqrt{N/x} = \sqrt{Nx}$. Thus,

$$C = \left(2N - x\sqrt{\frac{N}{x}}\right)\left(2\sqrt{Nx} + \sqrt{\frac{N}{x}}\right) = \\ 4N\sqrt{Nx} + 2N\sqrt{\frac{N}{x}} - 2Nx - N \tag{4.4}$$

To evaluate properties of the above function we will find dC/dx. We obtain

$$\frac{dC}{dx} = \frac{2Nx\sqrt{N} - N\sqrt{N} - 2Nx\sqrt{x}}{x\sqrt{x}} = \frac{2Nx\left(\sqrt{N} - \sqrt{x}\right) - N\sqrt{N}}{x\sqrt{x}} \tag{4.5}$$

If $x < 1$, then $m = 2n - \lceil x \rceil = 2n - 1$, what means that the optimum networks are strictly nonblocking. Let us assume now that $x > 1$. If $2x(\sqrt{N} - \sqrt{x} \geq \sqrt{N}$, then (4.5) is positive. Since $x \leq N/2$ then the above condition is fulfilled for $N \geq 4$. This means that in the range $[1, N/2]$, C grows along with the growth of x. Because for practical reasons we have $x \leq n \leq N/2$, the minimum number of crosspoints for repackable networks is for $x \leq 1$. This result does not allow us to determine the optimum ratio n/r. This ratio can be calculated by investigating the minimum of

C in relation to r. From (4.1) we have

$$\begin{aligned} C =&mr(2n+r) \cong (2n-n/r)r(2n+r) = (2N-n)(2n+r) = \\ &4Nn + 2Nr - 2n^2 - nr = 4N^2/r - 2N^2/r^2 + 2Nr - N \end{aligned} \tag{4.6}$$

Then, we obtain

$$\frac{dC}{dr} = \frac{4N^2 - 4N^2 r + 2Nr^3}{r^3} \tag{4.7}$$

If we assume that $r \gg 1$ we obtain

$$\frac{dC}{dr} \cong \frac{2Nr^3 - 4N^2 r}{r^3} \tag{4.8}$$

When (4.8) is set to zero, we have $2Nr^3 - 4N^2r = 2rN(r^2 - 2N) = 0$, and, therefore, $r^2 = 2N = 2nr$. Hence, $r = 2n$.

5. Call Repacking Algorithm

One of the most important issues concerning repackable networks is to define a control algorithm that is able to remove all overweight states. Unfortunately, it can be shown [3] that an algorithm based on a simple sequential path search from a fixed starting position supplemented by the multiple call repacking [4] is not sufficient. Below we present an algorithm [3] that removes all overweight states and, therefore, makes the network fulfilling condition (2.1) repackably nonblocking.

The key data structure used by the proposed algorithm is the list of all middle-stage switches L ordered by their loads, starting from the most heavily loaded switch. The list is updated at each change of the network state. If, as the result of a new call or call repacking, some switch b_j is "caught" by one of the switches following b_j in the list, say by switch b_i ($j < i$), then b_i is put before b_j (if there were several equally loaded switches, b_i is put in front of all of them).

Let us consider two different situations: a termination of a call, and a new call. All indices are related to the positions of switches in L.

When we have a termination at switch b_i, we scan all connections at switches from b_{i+1} to b_m (independently of the new position of switch b_i) in order to find connections that can be repacked to b_i. All found connections are then repacked. If a scanned switch is equally loaded as b_i and can accommodate some connections from b_i (we should note that connections can be repacked also between equally loaded switches and, therefore, it may be necessary to repack from a switch of a higher position in L to a switch of a lower position) these connections are repacked as well. If, as the result of this operation, a switch at location i or higher was changed the scanning is stopped and then resumed starting from the highest found position. If not, the same operation is performed for switch b_{i+1}.

In case of a new call, we search the middle-stage switches along the list and select the first possible switch that can accommodate this new call. If after the setting up of the new connection the considered switch changes its position in L the same operation as after call termination is performed. The discussed algorithm is presented formally below.

algorithm *Repacking;*
procedure *repack*(*nr*:integer);

```
    {*procedure resumes the network state without overweights; nr is the number
    of the most heavily loaded middle-stage switch for which we suspect a possibility
    to repack a call from a less loaded switch*}
begin for i := nr to m do
    for j := i + 1 to m do
    begin
        if b_i and b_j are loaded identically and it is possible to move a connection
        from b_i to b_j
        then exchange locations of b_i and b_j;
        Repack all possible connections from b_j to b_i;
        Order middle-stage switches in such a way that their loads are not decreasing;
        if b_i has location different than previously then
        begin
            i := new location of b_i; j := i + 1
        end
    end
end;
begin {*main program*}
    if new call then
    begin
        Find the most heavily loaded middle-stage switch b_i that can support this
        new call;
        Connect;
        Sort middle-stage switches by their load;
        if b_i is identically loaded as any other switch then
        begin
            Put b_i before all switches of the same load;
            repack(new location of b_i)
        end
    end
    else
        begin {*call termination; let assume that at b_i*}
            Release the call;
            Sort middle-stage switches in non-decreasing order
            repack(i {*that is the old location of b_i*})
        end
end.
```

Let k denotes the total number of calls served by the network. Let us define the following function: $F = mk_1 + (m-1)k_2 + (m-2)k_3 + \dots k_m$, where k_i is the number of calls served by b_i (i is the position on list L, thus $k_1 \geq k_2 \geq \dots \geq k_m$).

Let us note that each repacking in the above algorithm increases the value of F by at least of 1. This follows from the fact that calls are repacked to only more or equally loaded switches (even if calls are repacked between switches of equal load, the sequence of the switches is changed and the value of F is increased). The maximum of F is reached when all switches from b_1 to $b_{k/n}$ serve exactly n calls, and the remaining switches serve no calls. This maximum value of F is

$$F_{max} = nm + n(m-1) + \cdots + n(m - k/(r+1)) < nmk/n = km$$

Therefore, the number of repackings is always smaller than $km = O(N^{3/2})$. In the worst case, to find one repacking it is necessary to check the complete $r \times m$ state matrix of the network. But if we find a call for repacking on i-th searched switch we repack it up to i locations and, therefore, the value of F will be increased by i. Thus, the total number of operations in the *Repacking* algorithm is never greater than $mkr = O(Nmr) = O(N^2)$.

The estimates given above are asymptotic and probably will not be reached in reality. This is especially true in case of the pessimistic estimate of the number of repackings that, theoretically, can be greater than the number of served connections since the same connection can be repacked several times during a single run of the procedure *repack* (even between the same switches). As an example let us consider network $v(m, n, r)$ with $n > 2$ and $r > 4$ after the following sequence of new calls (the numbers represent the first- and the third-stage switches, respectively): (1,1), (3,3), (4,4), (5,5), (1,4), (4,1), (5,5), (1,2), (2,1) and call termination (1,1). On the other hand, the average number of call repackings is considerably lower than the pessimistic estimate. Computer simulations have shown that the number of call repackings leads to some constant. In this respect the call repacking algorithm is similar to call rearrangements.

6. Conclusion

Although both call repackings and call rearrangements involve reswitching of some existing connections, these two notions are significantly different. The repacking process is intended to "improve" the state of the network to prepare it for a new call, while the rearrangement process changes the network state to unblock a blocked call. Call repackings change the network states globally and call rearrangements act more locally changing only the states closely associated with a blocked call.

Call repackings can be used in real applications to improve the performance of multistage switching networks, can be implemented in smart routing procedures for telecommunication networks or can be applied in the dynamic channel assignment in cellular mobile networks.

References

[1] A. Jajszczyk and G. Jekel, *A New Concept – Repackable Networks*. *IEEE Trans. Commun.*, vol. 41, August 1993, pp. 1232–1236.

[2] M. Kubale, *Analysis of Structures of 3-Stage Repackably Nonblocking Switching Networks*, *Proc. of the National Telecommunications Symposium KST '95*, Bydgoszcz 1995, vol. D, pp. 309–314 (in Polish).

[3] M. Kubale, *On the Control of 3-Stage Repackable Connecting Networks*, *Proc. of the National Telecommunications Symposium KST '97*, Bydgoszcz 1997, vol. D, pp. 199–204 (in Polish).

[4] M. H. Ackroyd, *Call Repacking in Connecting Networks*, *IEEE Trans. Commun.*, vol. 27, March 1979, pp. 589–591.

Institute of Communication and Information Technologies Ltd. (ITTI Ltd.), and University of Technology and Agriculture, Bydgoszcz, Poland
Current address: ITTI Ltd.
ul. Rakoniewicka 20, 60-111 Poznań, Poland
E-mail address: Andrzej.Jajszczyk@itti.com.pl

Technical University of Gdańsk, Gdańsk, Poland
Current address: Technical University of Gdańsk, Gdańsk, Poland
Faculty of Electronics, Telecommunications and Informatics
ul. Narutowicza 11/12, 80-952 Gdańsk, Poland
E-mail address: kubale@eti.pg.gda.pl

DIMACS Series in Discrete Mathematics
and Theoretical Computer Science
Volume **42**, 1998

Modeling the Blocking Behavior of Clos Networks

Yuanyuan Yang and Neil H. Kessler

ABSTRACT. The Clos network has been widely used for data communications and parallel computing systems. It is important to have a thorough understanding of its performance. Several analytical models have been proposed to estimate the blocking probability of Clos networks but, due to the effect of approximations, a common problem with those widely-used models has been that they may not be very accurate in some cases. In particular, the blocking probability in these models contradicts the well-known deterministic nonblocking condition for the Clos network. In this paper, we present a new analytical model assuming a random routing strategy for the blocking probability of the three-stage Clos network. The new model can more accurately describe the blocking behavior of the network and is consistent with the deterministic nonblocking condition. We also extend our model to incorporate a packing strategy. Finally, through simulations we demonstrate that the analytical model presented in this paper is consistent with the simulation results under both random routing and packing strategies.

1. Introduction

The well-known Clos networks [**1, 2**] have been widely used for data communications and parallel computing systems. Some recent applications include the NEC ATOM switch designed for BISDN [**3**], the IBM GF11 multiprocessor [**4**] and ANSI Fibre Channel Standard for interconnection of processors to the I/O system. More recently, it was shown [**5**] that the network in the IBM SP2 [**6**] is functionally equivalent to the Clos network.

In general, a three-stage Clos network has r switch modules of size $n \times m$ in the *input stage*, m switch modules of size $r \times r$ in the *middle stage*, and r switch modules of size $m \times n$ in the *output stage*. The network has exactly one link between every two switch modules in its consecutive stages. We will refer to the interstage links between the input stage and the middle stage as *input-middle interstage links*,

1991 *Mathematics Subject Classification.* Primary: 68M07, 68M10, 68M20, 68R05, 90B12; Secondary: 94A05, 60C05, 94C10.

Key words and phrases. performance analysis, blocking probability, permutation networks, the Clos networks, network simulations, routing algorithm, packing.

This research was supported in part by the U.S. Army Research Office under Grant No. DAAH04-96-1-0234 and by the National Science Foundation under Grant No. OSR-9350540 and MIP-9522532.

and the interstage links between the middle stage and the output stage as *middle-output interstage links*. Figure 1 illustrates a general schematic of the three-stage Clos network.

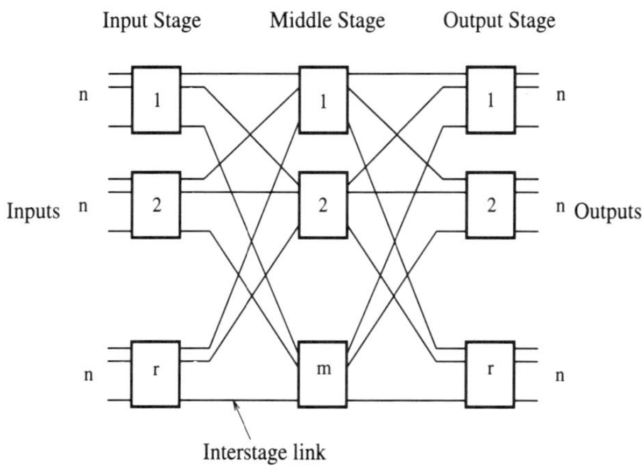

FIGURE 1. A general schematic of a three-stage Clos network.

Clos networks have been extensively studied in the literature [**1, 2, 7, 10, 11, 9, 12, 13**]. Previous work can be generally categorized into two classes: deterministic and probabilistic. In the deterministic study, the main focus has been on finding the minimum number of middle stage switches, m, for a given n and r, such that the network is nonblocking for arbitrary connection requests. Clos showed [**1**] that the network is *strictly nonblocking* if the number of middle stage switches $m \geq 2n - 1$. If the network satisfies this condition, any middle stage switch may be chosen arbitrarily to make a connection without any *rearrangement* of existing connections. In the probabilistic study, the main focus has been on analyzing the blocking behavior of the Clos network that does not satisfy the nonblocking condition (that is, with fewer middle stage switches than the nonblocking condition). Several analytical models have been proposed [**7, 8, 9, 10, 12**] on the blocking probability of the Clos network. However, as pointed out in [**12**], due to the effect of approximations, a common problem with those widely-used models is that in some cases they may not be very accurate. In particular, the blocking probability does not equal zero when $m \geq 2n - 1$, which contradicts the deterministic result that when $m \geq 2n-1$ the network is nonblocking [**1**]. In this paper, we will present a new analytical model for the blocking probability of the Clos network that can more accurately describe the blocking behavior of the network and is consistent with the deterministic nonblocking condition. Besides the random routing strategy, we will also extend our model to incorporate a packing strategy. Through simulations, we will demonstrate that the analytical model proposed in this paper is consistent with the simulation results under both random routing and packing strategies.

The rest of this paper is organized as follows. Section 2 briefly reviews the existing models for the blocking probability of the Clos networks. Section 3 presents the new model for a random routing strategy. Section 4 extends the model to a packing strategy. Section 5 compares the new model for both random routing and packing strategies with previous models. Section 6 describes the simulation results and compares with the analytical results. Finally, Section 7 concludes the paper.

2. Previous Analytical Models

In general, determination of blocking probability in a multistage network is inherently complex and difficult. This is due to the fact that there are many possible paths to consider in a typical large network, and the dependencies among links in the network lead to combinatorial explosion problems. Among the analytical models proposed for the Clos network in the literature that estimate blocking probability, two well-known and widely-used models were proposed by C.Y. Lee [**7**] and C. Jacobaeus [**8**], respectively. C.Y. Lee gave the simplest method for analyzing the blocking probability of the Clos network, in which the events that individual links are busy are assumed to be independent.

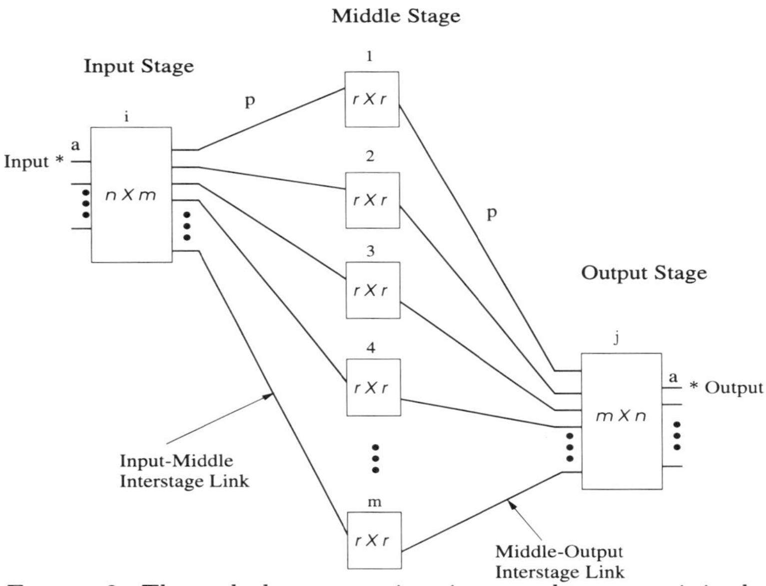

FIGURE 2. The paths between a given input and output pair in the Clos network, where a is the probability that an input link or an output link is busy and and p is the probability that an interstage link is busy.

To see how Lee's model works, let's consider a connection request from an input link of input stage switch i to an output link of output stage switch j in the Clos network shown in Figure 2. Clearly, at most $n-1$ input-middle interstage links from input stage switch i can be already busy. Similarly, at most $n-1$ middle-output interstage links to output stage switch j can be already busy. A path from input stage switch i to output stage switch j is concatenated by an input-middle interstage link and a middle-output interstage link. There are a total of m disjoint paths between input stage switch i and output stage switch j. Let the probability that a typical input link is busy be a and the probability that a typical output link is busy be also a, and assume that the incoming traffic is uniformly distributed over the m interstage links. Then the probability that an interstage link is busy is given by

$$p = \frac{an}{m}, \tag{1}$$

and the probability that an interstage link is idle is given by

$$q = 1 - p. \tag{2}$$

Since one path from the source input to the destination output consists of two interstage links, if any link in the path is busy, the path cannot be used for realizing the new connection from the source input to the destination output. Thus the probability that one path cannot be used is $(1 - q^2)$. It is clear that if all the m paths are busy, the connection request will be blocked. Thus, with the link independence assumption, the probability that no idle path is available for making the connection between the given input and output, or the blocking probability, is given by

$$P_B = (1 - q^2)^m. \tag{3}$$

The problem with the Lee's model is that in some cases it may not be very accurate. Note that with n inputs at each input stage switch, at most n of the m outputs of an input stage switch can be busy. Lee's model does not account for this. A more accurate model was provided by C. Jacobaeus [**8**], in which, this input limitation is taken into account automatically, and the blocking probability of the three-stage Clos network is calculated by the following formula:

$$P_B = \frac{(n!)^2 (2-a)^{2n-m} a^m}{m!(2n-m)!}. \tag{4}$$

Note that Jacobaeus' formula is still an approximation to the blocking probability. If we set $m = 2n - 1$ in (3) and (4), for example, letting

$$n = 32, \ m = 2n - 1 = 63, \text{ and } a = 1$$

Lee' model (3) yields

$$P_B = 2.5 \times 10^{-8},$$

and Jacobaeus' model (4) yields

$$P_B = 3.4 \times 10^{-20}.$$

Apparently, the blocking probability P_B in both models is nonzero, which contradicts the deterministic nonblocking condition for the Clos network. On the other hand, the hypergeometric model proposed by M. Karnaugh [**9**] does not seem to suffer this problem and the expression for the blocking probability turns out to be very close to the one based on Jacobaeus' method, as derived in [**11**]. However, this model is very complex, and it is difficult to extend the model to other routing strategies other than random routing, such as sequential hunting and packing strategies.

3. The New Analytical Model

In this section, we propose a new analytical model for the blocking probability of the Clos network which is consistent with the deterministic nonblocking condition and can be easily extended to packing strategy. We will first present the model assuming random routing.

Let's first consider the network state shown in Figure 3, in which n_1 input-middle interstage links from input stage switch i are busy and n_2 middle-output interstage links to output stage switch j are busy, where $0 \leq n_1, n_2 \leq n-1$. Let $\boldsymbol{n_1}$ denote the event that there are n_1 busy input-middle interstage links from input stage switch i, and $\boldsymbol{n_2}$ denote the event that there are n_2 busy middle-output interstage links to output stage switch j. If a busy input-middle interstage link and a busy middle-output interstage link are in the same path (i.e. share the same middle stage switch), this pair of links is said to be *overlapped*. We have the following lemma concerning the overlapped interstage links.

LEMMA 1. *Given events* $\boldsymbol{n_1}$ *and* $\boldsymbol{n_2}$*, the probability that* k *pairs of interstage links are overlapped in the Clos network is given by*

$$\text{(5)} \qquad \Pr\{k \textit{ pairs of links overlapped} \mid \boldsymbol{n_1}, \boldsymbol{n_2}\} = \frac{\binom{n_1}{k}\binom{m-n_1}{n_2-k}}{\binom{m}{n_2}} = \frac{\binom{n_2}{k}\binom{m-n_2}{n_1-k}}{\binom{m}{n_1}}$$

Proof. As shown in Figure 3, there are a total of $\binom{m}{n_1}\binom{m}{n_2}$ ways to choose n_1 busy input-middle interstage links and n_2 busy middle-output interstage links, among which k pairs of overlapped links can be constructed as follows: $\binom{m}{n_1}$ ways to choose n_1 busy input-middle interstage links; then $\binom{n_1}{k}$ ways to choose k input-middle interstage links which are overlapped with k middle-output interstage links; finally the rest of $n_2 - k$ middle-output interstage links can be chosen from $m - n_1$, and there are $\binom{m-n_1}{n_2-k}$ ways. Thus, the probability that k pairs of links are overlapped is

$$\frac{\binom{m}{n_1}\binom{n_1}{k}\binom{m-n_1}{n_2-k}}{\binom{m}{n_1}\binom{m}{n_2}} = \frac{\binom{n_1}{k}\binom{m-n_1}{n_2-k}}{\binom{m}{n_2}}.$$

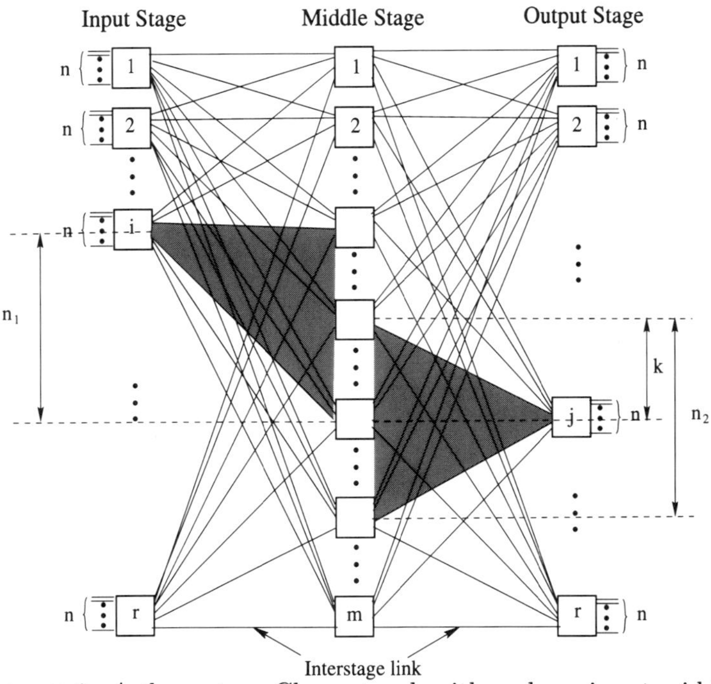

FIGURE 3. A three-stage Clos network with n_1 busy input-middle interstage links from input stage switch i, n_2 busy middle-output interstage links to output stage switch j, and k pairs of interstage links overlapped.

Symmetrically, the probability can also be proved to be

$$\frac{\binom{n_2}{k}\binom{m-n_2}{n_1-k}}{\binom{m}{n_1}}.$$

Q.E.D.

We now derive the relationship between the event that k pairs of links are overlapped and the event that a connection request is blocked. The connection request under consideration is not blocked if and only if there exists one path in which both input-middle interstage link and middle-output interstage link are not busy, that is, if and only if

$$n_1 + n_2 - k < m,$$

which implies

$$k \geq \max\{0, n_1 + n_2 - m + 1\}.$$

On the other hand, we apparently have

$$k \leq \min\{n_1, n_2\}.$$

Therefore, from (5) the probability that the connection request is not blocked, given events $\boldsymbol{n_1}$ and $\boldsymbol{n_2}$, is given by

$$
\text{(6)} \qquad \Pr\{\text{connection not blocked} \mid \boldsymbol{n_1}, \boldsymbol{n_2}\} = \frac{1}{\binom{m}{n_2}} \sum_{k=\max\{0,n_1+n_2-m+1\}}^{\min\{n_1,n_2\}} \binom{n_1}{k}\binom{m-n_1}{n_2-k}
$$

Under the assumption that the events that individual links in the network are busy are independent, we have that the events $\boldsymbol{n_1}$ and $\boldsymbol{n_2}$ are independent, which gives

$$
\text{(7)} \qquad \Pr\{\boldsymbol{n_1}, \boldsymbol{n_2}\} = \Pr\{\boldsymbol{n_1}\} \cdot \Pr\{\boldsymbol{n_2}\}
$$

Now let's calculate $\Pr\{\boldsymbol{n_1}\}$ and $\Pr\{\boldsymbol{n_2}\}$. We still use p to denote the probability that an interstage link is busy, and $q = 1 - p$ to denote the probability that an interstage link is not busy. To start with, we can simply assume that the number of busy input-middle interstage links (or middle-output interstage links) follows the binomial distribution, that is, the probability that n_1 input-middle interstage links are busy is $\binom{m}{n_1} p^{n_1} q^{m-n_1}$. However, given the fact that there are at most $n-1$ busy input-middle interstage links from input stage switch i, we can obtain a more accurate probability distribution. Thus, the probability that n_1 input-middle interstage links are busy now becomes

$$
\text{(8)} \qquad \Pr\{\boldsymbol{n_1}\} = \frac{\binom{m}{n_1} p^{n_1} q^{m-n_1}}{\sum_{j=0}^{n-1} \binom{m}{j} p^j q^{m-j}}
$$

and the probability that n_2 middle-output interstage links are busy is

$$
\text{(9)} \qquad \Pr\{\boldsymbol{n_2}\} = \frac{\binom{m}{n_2} p^{n_2} q^{m-n_2}}{\sum_{j=0}^{n-1} \binom{m}{j} p^j q^{m-j}}
$$

Thus, from equations (6) - (9), the probability that a connection request is not blocked is given by

(10) Pr{connection not blocked} =

$$\frac{\displaystyle\sum_{n_1=0}^{n-1}\sum_{n_2=0}^{n-1}\frac{1}{\binom{m}{n_2}}\sum_{k=\delta_1}^{\delta_2}\binom{n_1}{k}\binom{m-n_1}{n_2-k}\binom{m}{n_1}p^{n_1}q^{m-n_1}\binom{m}{n_2}p^{n_2}q^{m-n_2}}{\left[\displaystyle\sum_{j=0}^{n-1}\binom{m}{j}p^j q^{m-j}\right]^2}$$

$$= \frac{\displaystyle\sum_{n_1=0}^{n-1}\sum_{n_2=0}^{n-1}\sum_{k=\delta_1}^{\delta_2}\binom{m}{n_1}\binom{n_1}{k}\binom{m-n_1}{n_2-k}p^{n_1+n_2}q^{2m-n_1-n_2}}{\left[\displaystyle\sum_{j=0}^{n-1}\binom{m}{j}p^j q^{m-j}\right]^2}$$

where

$$\delta_1 = \max\{0, n_1+n_2-m+1\}$$

and

$$\delta_2 = \min\{n_1, n_2\}.$$

The blocking probability is

$$P_B = 1 - \Pr\{\text{connection not blocked}\} \tag{11}$$

The following theorem demonstrates that the blocking probability we have obtained matches the deterministic nonblocking condition for strictly nonblocking Clos network.

THEOREM 1. *The blocking probability of the Clos network, P_B in (11), becomes zero when the number of middle stage switches $m \geq 2n-1$.*

Proof. We need to use the following combinatorial equality [**18**]

$$\sum_{v=0}^{u}\binom{s}{v}\binom{t}{u-v} = \binom{s+t}{u} \tag{12}$$

This can be easily verified by observing the coefficients of x^u on the both sides of the equation

$$(1+x)^s(1+x)^t = (1+x)^{s+t}$$

Noticing that $\binom{s}{v} = 0$ for $s < v$, the equation (12) can be rewritten as

$$\sum_{v=0}^{\min\{u,s\}}\binom{s}{v}\binom{t}{u-v} = \binom{s+t}{u} \tag{13}$$

Under the condition of Theorem 1, $m \geq 2n-1$, for any n_1 and n_2 where $0 \leq n_1, n_2 \leq n-1$, we have

$$n_1+n_2 \leq 2(n-1) \leq m-1,$$

which implies

$$\max\{0, n_1 + n_2 - m + 1\} = 0.$$

Thus by using the equality (13), the right hand side of (6) becomes

$$\frac{1}{\binom{m}{n_2}} \sum_{k=0}^{\min\{n_1,n_2\}} \binom{n_1}{k}\binom{m-n_1}{n_2-k} = 1.$$

Therefore, from (10)

$$\Pr\{\text{connection not blocked}\} = \frac{\sum_{n_1=0}^{n-1}\sum_{n_2=0}^{n-1} \binom{m}{n_1} p^{n_1} q^{m-n_1} \binom{m}{n_2} p^{n_2} q^{m-n_2}}{\left[\sum_{j=0}^{n-1} \binom{m}{j} p^j q^{m-j}\right]^2} = 1$$

and from (11) we obtain $P_B = 1 - 1 = 0$.

Q.E.D.

4. Extending the Model to Packing Strategy

In the previous section, we assume a random routing strategy. Another commonly used routing strategy is the so-called *packing* strategy [**2**]. Under the packing strategy, a connection is realized on a path found by trying the most used part of the network first and the least used part last. For a Clos network, packing means that when choosing a middle stage switch for satisfying a connection request, an empty middle stage switch is not used unless there is not any partially filled middle stage switch that can satisfy this connection request. It is generally believed that packing can lower the blocking probability of the network. First, let us see how the packing strategy was introduced into Lee's model [**13, 14**]. The assumption now is that a certain number of middle stage switches d ($d \leq n-1$) are fully-packed, that is, all interstage links to or from these middle stage switches are busy.

As we saw earlier in Figure 2, there are a total of m distinct paths from the source input to the destination output. Since any path from the source input stage switch to the destination output stage switch through the d fully-packed middle stage switches is busy, we cannot use the d fully-packed middle stage switches for realizing a new connection request. The incoming traffic is now uniformly distributed over the $m-d$ interstage links corresponding to the middle stage switches that are not fully-packed. This results in the probability that an interstage link is busy being given by

$$p = \begin{cases} 1, & \text{for the } d \text{ links to the fully-packed middle stage switches} \\ \frac{an-d}{m-d}, & \text{for the } m-d \text{ links to the non fully-packed middle stage switches} \end{cases}$$

Thus, the probability that all paths are busy, or the blocking probability, is given by

$$P_B = \left[1 - \left(1 - an - \frac{d}{m-d}\right)^2\right]^{m-d} \tag{14}$$

Note that when $d = 0$ (no packing) (14) reduces to (3).

Now let's consider how the packing strategy can be introduced into the model proposed in the last section. Let's consider the network state shown in Figure 4, in which there are d fully-packed middle stage switches, n_1 input-middle interstage links from input stage switch i are busy and n_2 middle-output interstage links to output stage switch j are busy, where $d \le n_1, n_2 \le n-1$ and k interstage links connected to non fully-packed middle stage switches are overlapped. It is clear from the definition of a fully-packed middle stage switch that there is a busy input-middle interstage link from every input stage switch in the network to a fully-packed middle stage switch and there is a busy middle-output interstage link to every output stage switch from a fully-packed middle stage switch. Therefore, the d fully-packed middle stage switches cannot be used for realizing a new connection request. Let us denote $\boldsymbol{n_1'}$ and $\boldsymbol{n_2'}$ to be the events that there are $n_1' = n_1 - d$ busy input-middle interstage links to the non fully-packed middle stage switches from some input stage switch i and there are $n_2' = n_2 - d$ busy middle-output interstage links from the non fully-packed middle stage switches to some output stage switch j, respectively.

By removing the d fully-packed middle stage switches from consideration in equation (5) the probability of k interstage links connected to non fully-packed middle stage switches being overlapped given $\boldsymbol{n_1'}$ and $\boldsymbol{n_2'}$ becomes

$$\begin{aligned}
(15)\quad & \Pr\{k \text{ pairs of links overlapped} \mid \boldsymbol{n_1'}, \boldsymbol{n_2'}\} \\
&= \frac{\binom{n_1-d}{k}\binom{m-d-(n_1-d)}{n_2-d-k}}{\binom{m-d}{n_2-d}} = \frac{\binom{n_2-d}{k}\binom{m-d-(n_2-d)}{n_1-d-k}}{\binom{m-d}{n_1-d}} \\
&= \frac{\binom{n_1-d}{k}\binom{m-n_1}{n_2-d-k}}{\binom{m-d}{n_2-d}} = \frac{\binom{n_2-d}{k}\binom{m-n_2}{n_1-d-k}}{\binom{m-d}{n_1-d}}
\end{aligned}$$

In a Clos network with d fully-packed middle stage switches, a connection request is not blocked if and only if there exists one path from the remaining $m-d$ possible paths where both the input-middle interstage link and the middle-output interstage link are not busy. We represent that by

$$(n_1 - d) + (n_2 - d) - k < (m - d),$$

which implies

$$k \ge \max\{0, n_1 + n_2 - m - d + 1\}.$$

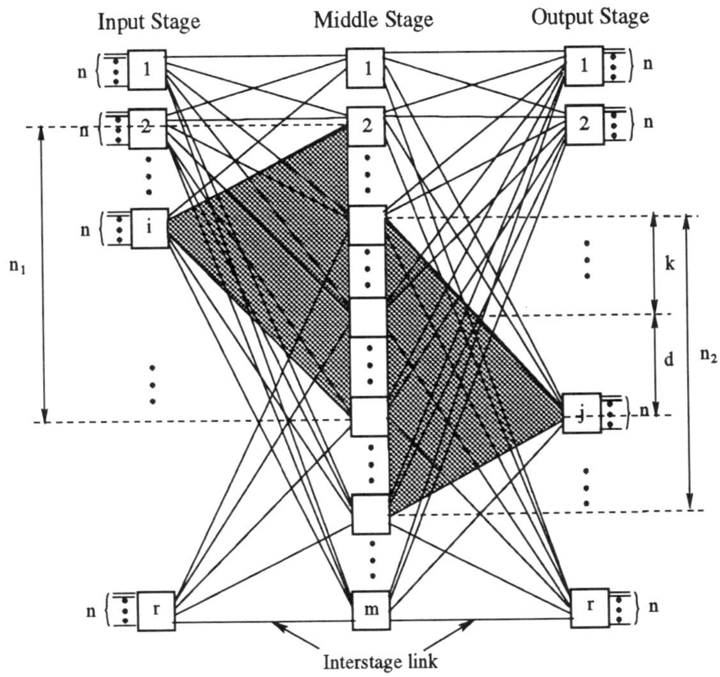

FIGURE 4. A three-stage Clos network with n_1 busy input-middle interstage links from input stage switch i, n_2 busy middle-output interstage links to output stage switch j, d fully-packed middle stage switches, and k pairs of overlapped links connected to non fully-packed middle stage switches.

From (15) we see that k is now being chosen from the $n_1 - d$ busy input-middle interstage links and from the $n_2 - d$ busy middle-output interstage links. Therefore the following condition now holds for k

$$k \leq \min\{(n_1 - d), (n_2 - d)\}.$$

Thus, from (15), the probability that the connection request is not blocked, given events $\boldsymbol{n_1'}$ and $\boldsymbol{n_2'}$ is given by

$$\text{(16)} \qquad \Pr\{\text{connection not blocked} \mid \boldsymbol{n_1'}, \boldsymbol{n_2'}\} = \frac{1}{\binom{m-d}{n_2-d}} \sum_{k=\max\{0,n_1+n_2-m-d+1\}}^{\min\{n_1-d,n_2-d\}} \binom{n_1-d}{k}\binom{m-n_1}{n_2-d-k}$$

Now, let's look at $\Pr\{\boldsymbol{n_1'}, \boldsymbol{n_2'}\}$. We assume that events $\boldsymbol{n_1'}$ and $\boldsymbol{n_2'}$ are independent, which gives

$$\text{(17)} \qquad \Pr\{\boldsymbol{n_1'}, \boldsymbol{n_2'}\} = \Pr\{\boldsymbol{n_1'}\} \cdot \Pr\{\boldsymbol{n_2'}\}$$

and note that the probability that an interstage link is busy is given by

$$p = \begin{cases} 1, & \text{for the } d \text{ links to the fully-packed middle stage switches} \\ \frac{an-d}{m-d}, & \text{for the } m-d \text{ links to the non fully-packed middle stage switches.} \end{cases}$$

We know that there are at most $n-1$ busy input-middle interstage links from input stage switch i, but assuming d fully-packed middle stage switches, we only take into account all possible ways of choosing up to $n-1-d$ busy input-middle interstage links from the $m-d$ middle stage switches. Therefore, the probability distribution for the event $\boldsymbol{n_1'}$ is

$$\Pr\{\boldsymbol{n_1'}\} = \frac{\binom{m-d}{n_1-d} p^{n_1-d} q^{(m-d)-(n_1-d)}}{\sum_{j=d}^{n-1} \binom{m-d}{j-d} p^{j-d} q^{(m-d)-(j-d)}} \tag{18}$$

and the probability distribution for $\boldsymbol{n_2'}$ is

$$\Pr\{\boldsymbol{n_2'}\} = \frac{\binom{m-d}{n_2-d} p^{n_2-d} q^{(m-d)-(n_2-d)}}{\sum_{j=d}^{n-1} \binom{m-d}{j-d} p^{j-d} q^{(m-d)-(j-d)}} \tag{19}$$

Then the probability that a connection is not blocked using the packing strategy is represented as

$$\Pr\{\text{connection not blocked}\} = \frac{\sum_{n_1=d}^{n-1} \sum_{n_2=d}^{n-1} \sum_{k=\delta_1}^{\delta_2} \binom{m-d}{n_1-d} \binom{n_1-d}{k} \binom{m-n_1}{n_2-d-k} p^{n_1+n_2-2d} q^{2m-n_1-n_2}}{\left[\sum_{j=d}^{n-1} \binom{m-d}{j-d} p^{j-d} q^{m-j} \right]^2} \tag{20}$$

where

$$\delta_1 = \max\{0, n_1+n_2-m-d+1\}$$

and

$$\delta_2 = \min\{n_1-d, n_2-d\}.$$

The blocking probability is

$$P_B = 1 - \Pr\{\text{connection not blocked}\} \tag{21}$$

Finally, note that when $d=0$, (20) reduces to the case of random routing (10).

5. Comparisons

In this section, we give some numerical comparisons of the newly proposed model and the existing models, Lee's model (3) and Jacobaeus' (Karnaugh) model (4) under network input link utilization $a = 0.8$.

In Figure 5, we plot the curves of these three models for a Clos network with $n = r = 32$ and $n \le m \le 2n-1$. In Table 1, we list the blocking probabilities for some typical larger network sizes. From both Figure 5 and Table 1, we can see that the newly proposed model matches well with previous models with the added benefit that it can accurately reflect the deterministic nonblocking condition.

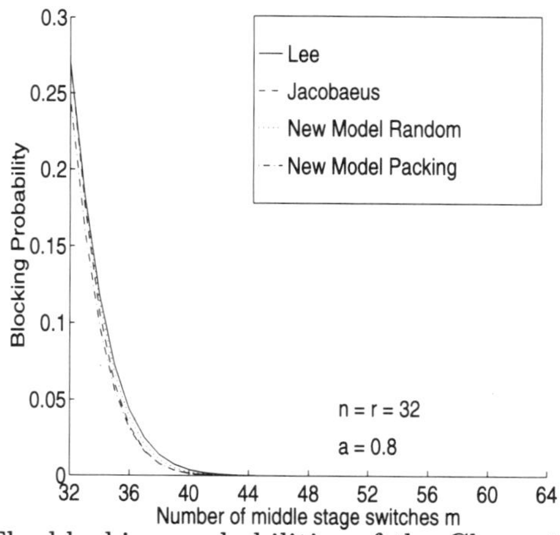

FIGURE 5. The blocking probabilities of the Clos network in Lee, Jacobaeus (Karnaugh) and the new (random routing and packing strategies) models. for a network with $n = r = 32$ and $n \leq m \leq 2n - 1$, under network input link utilization $a = 0.8$.

TABLE 1. Blocking probability comparison, $a = 0.8$.

n	m	P_B(Lee)	P_B(Jacobaeus/ Karnaugh)	P_B(New) Random routing	P_B(New) Packing($d = 2$)
64	64	0.073	0.073	0.073	0.067
64	68	0.014	0.011	0.014	0.011
64	72	0.0019	0.0011	0.0018	0.0015
64	76	1.92×10^{-4}	5.99×10^{-5}	1.75×10^{-4}	1.35×10^{-4}
120	128	2.58×10^{-4}	1.71×10^{-4}	2.58×10^{-4}	2.25×10^{-4}
240	256	6.68×10^{-8}	2.92×10^{-8}	6.68×10^{-8}	5.81×10^{-8}

6. The Experimental Simulations

An experimental study was performed in order to validate the newly proposed analytical model for the three-stage Clos network under random routing and packing strategies.

The network simulator we developed consists of two major modules: the request generator module and the request processor module. The request generator module randomly generates a list of connection requests for the network. The request processor attempts to route the connection request through the network using the specified strategy. If the connection request is not satisfied the connection request is recorded as blocked, otherwise, if the connection request can be satisfied, the connection is established in the network. During the simulation, a certain workload is maintained. The workload is measured by the *network utilization,* which is defined as the ratio of the number of busy output ports to the total number of output ports of the network.

6.1. Simulation Results. In this subsection we will examine the approximations for blocking probability obtained from the analytical model and the data resulting from the simulation of the Clos network under the random routing and packing strategies.

We have examined three network configurations, $n = r = 16$ with $16 \leq m \leq 31$, $n = r = 32$ with $32 \leq m \leq 63$ and $n = r = 64$ with $64 \leq m \leq 127$. For each network configuration, blocking probabilities were generated for the analytical model for $50\% \leq a \leq 90\%$ in increments of 5%. For the simulation, a translates to network utilization, therefore blocking probabilities were obtained for the same percentages.

Using all of the network configurations for the analytical model, blocking probabilities for the packing strategy were obtained for $1 \leq d \leq 4$. Four was chosen as an upper bound for d for the data examined from the analytical model because preliminary simulation results showed an actual average d of no greater than approximately 1.5. In the simulation for the packing strategy, to maintain the maximum number of fully packed middle stage switches, we also employed *repacking* strategy. Repacking means when a connection is released one or more existing connections are moved to the most filled non-empty middle stage switch. Finally, for each instance of the network simulation for both random routing and packing strategies 10,000 requests were processed.

6.1.1. *Random routing strategy: analytical model vs. simulated results.* The blocking probability curves in Figure 6 show the relationship between the analytical model data for the random routing strategy and the simulated results for the random routing strategy. First, for $n = r = 16$, $n = r = 32$ and $n = r = 64$ we see that the simulated results yield a lower blocking probability than the analytical model. We should also see the analytical data follow closely with the deterministic nonblocking condition in [**1**]. For $n = r = 16$ the blocking probability for the analytical model goes to zero at $m = 31$. The blocking probability for the simulated network goes to zero at $m = 24$. The comparisons are similar for the other two network states $n = r = 32$ and $n = r = 64$.

We also see in Figure 6 that both blocking probability curves exhibit approximately the same shape, that is, they fall most sharply between $m = 16$ and $m = 18$ for $n = r = 16$, between $m = 32$ and $m = 34$ for $n = r = 32$ and between $m = 64$ and $m = 68$ for $n = r = 64$. This shows that the analytical model is a good approximation of the actual blocking probability of a Clos network using a random routing strategy.

6.1.2. *Packing strategy: analytical model vs. simulated results.* In Figure 7 we compare the blocking probability for the simulated packing strategy with the blocking probability obtained from the analytical model for the packing strategy for $d = 1$ and $d = 2$. We chose the analytical data from $d = 1$ and $d = 2$ for the comparison because for the simulated network using the packing strategy the range for average d for $n = r = 16$ was $0.7024 \leq d \leq 0.8129$, for $n = r = 32$, $0.7154 \leq d \leq 0.9458$ and for $n = r = 64$, $0.7717 \leq d \leq 1.1882$. It seemed that $d = 1$ and $d = 2$ would give the best basis for comparison. The blocking probability

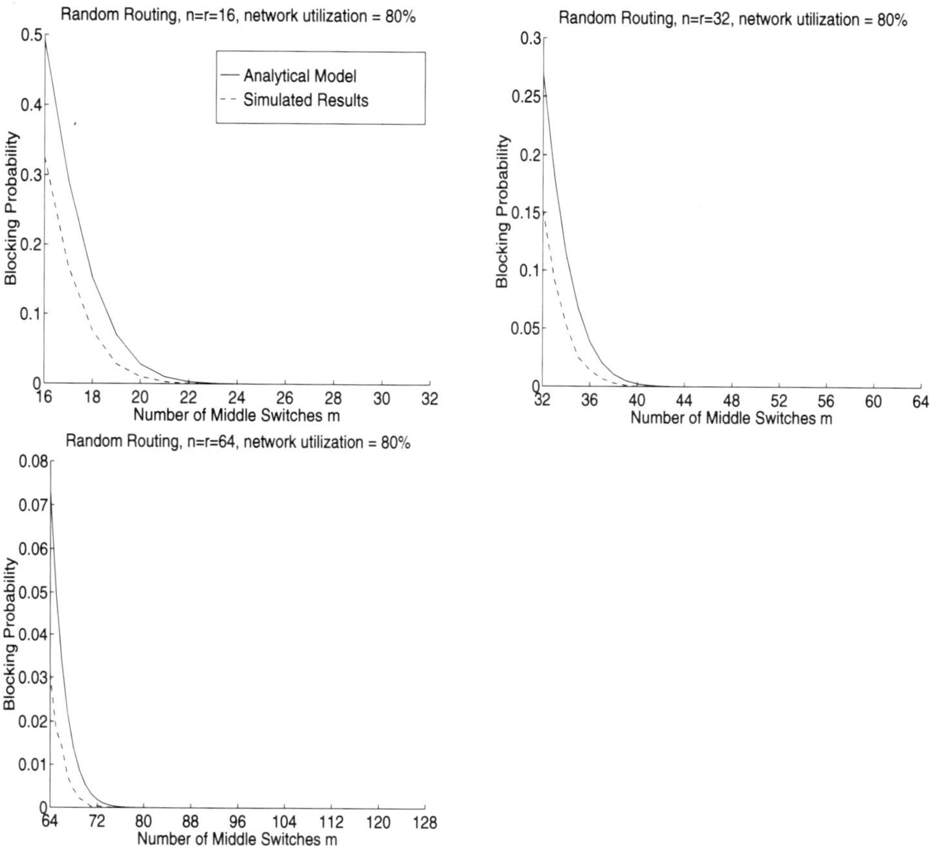

FIGURE 6. Blocking probability of the Clos network for the random routing strategy comparing the analytical model results with the simulated results for $n = r = 16$ with $16 \leq m \leq 31$, $n = r = 32$ with $32 \leq m \leq 63$ and $n = r = 64$ with $64 \leq m \leq 127$ at 80% network utilization.

obtained from simulating the packing strategy over a given range of m follows the same pattern of decrease as the analytical data as m increases. But, the blocking probabilities from the simulated results for a given n and m are significantly lower than the blocking probabilities from the model data for both $d = 1$ and $d = 2$ for the same n and m. This is due to the effect of repacking in the simulation. Also, we can see that the blocking probability is slightly higher for $d = 1$ compared to $d = 2$ for any given m value. For $n = r = 16$, the blocking probability from the simulation goes to zero at $m = 19$, from the analytical model at $d = 1$ at approximately $m = 30$, and from the analytical model at $d = 2$ at approximately $m = 29$. The same trends can be seen for $n = r = 32$ and $n = r = 64$.

6.1.3. *Analytical model: random routing strategy vs. packing strategy.* Figure 8 shows the blocking probability for the analytical model for the packing strategy for $n = r = 16$ with $16 \leq m \leq 31$, $n = r = 32$ with $32 \leq m \leq 63$ and $n = r = 64$ with

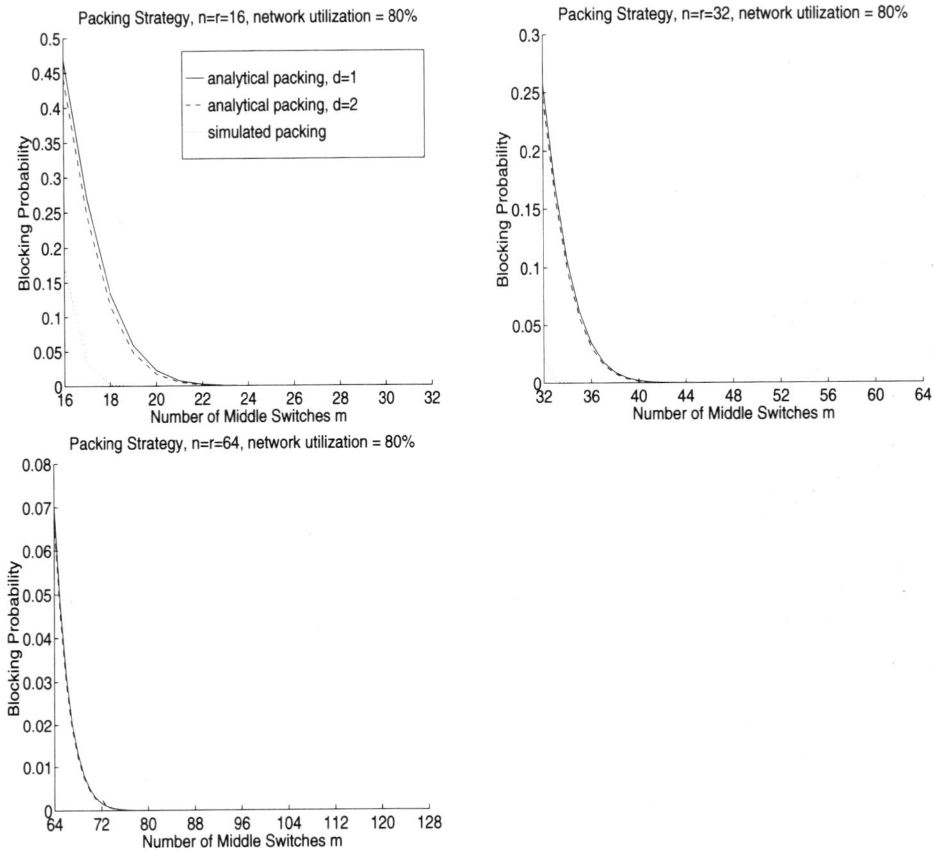

FIGURE 7. Blocking probability of the Clos network for the packing strategy comparing the analytical model results with the simulated results for $n = r = 16$ with $16 \leq m \leq 31$, $n = r = 32$ with $32 \leq m \leq 63$ and $n = r = 64$ with $64 \leq m \leq 127$ at 80% network utilization.

$64 \leq m \leq 127$ at 80% network utilization for $d = 0$ and $d = 3$. In order to evaluate how reliable the analytical model for the packing strategy is as compared to the model for random routing we examine the blocking probabilities for a given set of network parameters for different d values. Since we saw in Section 4 that when $d = 0$ the formula reverts back to the model for random routing, we will compare the blocking probability curve shown for $d = 0$ to the blocking probability curve shown for $d = 3$. We see that the blocking probability is slightly lower for $d = 3$ compared to $d = 0$. This is expected since packing should produce a lower blocking probability than random routing. The analytical data over the range of d values showed that as d increased, the blocking probability decreased slightly.

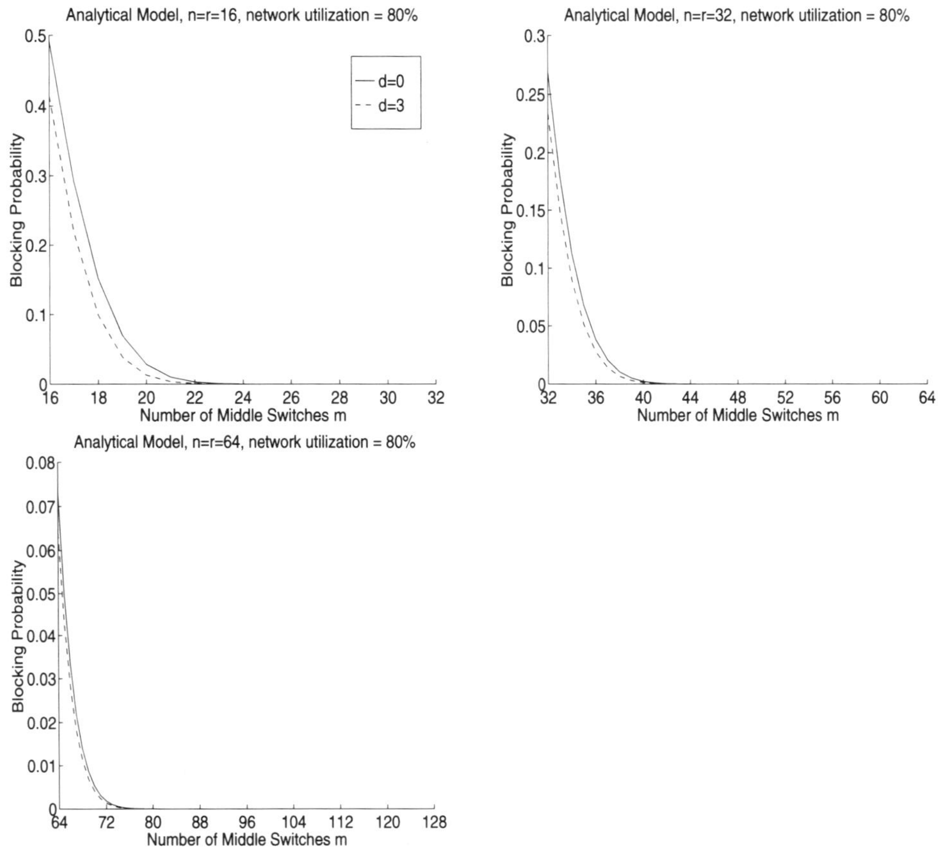

FIGURE 8. Blocking probability of the Clos network for $n = r = 16$ with $16 \leq m \leq 31$, $n = r = 32$ with $32 \leq m \leq 63$ and $n = r = 64$ with $64 \leq m \leq 127$ at 80% network utilization for $d = 0$ and $d = 3$.

7. Conclusions

In this paper, we have proposed a new analytical model to estimate the blocking probability for the three-stage Clos networks using a random routing and a packing strategy. We have shown that the new model can more accurately describe the blocking behavior of the network and is consistent with the deterministic non-blocking condition. We have also simulated the Clos networks for both a routing strategy and a packing strategy in order to derive blocking probabilities for comparison to the analytical model. We have seen that the results produced from the analytical model under both random routing and packing strategy is consistent with the blocking probabilities acquired through simulation. This indicates that the model is a reliable predictor of blocking probability for the networks it is intended to model. We believe that based on the model presented in this paper, more accurate analytical models for other routing strategies can be obtained as well.

References

[1] C. Clos, "A study of non-blocking switching networks," *The Bell System Technical Journal,* vol. 32, pp. 406-424, 1953.

[2] V.E. Benes, "Heuristic remarks and mathematical problems regarding the theory of switching systems," *The Bell System Technical Journal,* vol. 41, pp.1201-1247, 1962.

[3] A. Itoh *et al.,* "Practical implementation and packaging technologies for a large-scale ATM switching system," *Journal of Selected Areas in Communications*, vol. 9, No. 8, pp. 1280-1288, 1991.

[4] J. Beetem, M. Denneau and D. Weingarten, "The GF11 supercomputer," *Proceedings of the 12th Annual International Symposium on Computer Architecture*, pp. 108-115, 1985.

[5] M.T. Bruggencate and S. Chalasani, "Equivalence between SP2 high-performance switches and three-stage Clos networks," *Proceedings of the 25th International Conference on Parallel Processing*, pp. I-1–I-8, 1996.

[6] G.B. Stunkel, D.G. Shea, B. Abali, *et al.,* "The SP2 high-performance switch," *IBM System Journal*, 34(2), pp. 185-204, 1995.

[7] C.Y. Lee, "Analysis of switching networks," *The Bell System Technical Journal,* vol. 34, no. 6, pp. 1287-1315, Nov. 1955.

[8] C. Jacobaeus, "A study on congestion in link systems," *Ericsson Technics*, vol. 51, No. 3, 1950.

[9] M. Karnaugh, "Loss of point-to-point traffic in three-stage circuit switches," *IBM Journal of Research and Development,* vol. 18, pp. 204-216, 465, 1974.

[10] P.M. Lin, B.J. Leon, C.R. Stewart, "Analysis of circuit-switched networks employing originating office control with spill forward," *IEEE Transactions on Communications,* vol. 26, No. 6, pp. 754-765, 1978.

[11] A.A. Collins and R.D. Pedersen, *Telecommunications: A Time for Innovation,* Merle Collins Foundation, Dallas, 1973.

[12] M. Schwartz *Telecommunication Networks: Protocols, Modeling, and Analysis*, Addison-Wesley Publishing Company, Reading, 1987.

[13] Y. Mun and H.Y. Youn, "Performance modeling and evaluation of circuit switching using Clos networks," *IEEE Transactions on Computers*, vol. 43, pp. 854-861, 1994.

[14] Y. Mun, Y. Tang and V. Devarajan, "Analysis of call packing and rearrangement in a multi stage switch," *IEEE Transactions on Communications*, vol. 42, pp. 252-254, 1994.

[15] M.H. Ackroyd, "Call repacking in connecting networks," *IEEE Transactions on Communications*, vol. COM-27, pp. 589-591, 1979.

[16] C.-K. Kim and T.T. Lee "Call scheduling algorithms in a multicast switch," *IEEE Transactions on Communications*, vol. 40, pp. 625-635, 1992

[17] C.M. Melas and A. Milewski, "The effect of call routing rules in non-blocking three-stage switching networks," *IEEE Transactions on Communications*, vol. COM-27, pp. 150-153, 1979.

[18] K.P. Bogart, *Introductory Combinatorics, 2nd Edition*, Harcourt Brace Jovanovich, 1990.

(Yuanyuan Yang) DEPARTMENT OF COMPUTER SCIENCE, UNIVERSITY OF VERMONT, BURLINGTON, VT 05405 USA

E-mail address, Yuanyuan Yang: `yang@cs.uvm.edu`

(Neil H. Kessler) DEPARTMENT OF COMPUTER SCIENCE, UNIVERSITY OF VERMONT, BURLINGTON, VT 05405 USA

E-mail address, Neil H. Kessler: `nkessler@emba.uvm.edu`

DIMACS Series in Discrete Mathematics
and Theoretical Computer Science
Volume **42**, 1998

Performance of a New Decomposition Algorithm for Rearrangeable Fault-tolerant Clos Interconnection Networks under Sub-maximal and No-fault Conditions

John D. Carpinelli
and
Catherine B. Wang

Abstract

This paper presents the results of an analysis of a decomposition algorithm for the Fault Tolerant Clos (FTC) network. The algorithm, slightly modified for this research, makes use of the additional fault-tolerant hardware to attempt to improve the performance of the routing algorithm when few or no faults are present. The analysis indicates that the number of successive swaps, the bottleneck of the algorithm, decreases significantly, even with the addition of only one extra switch. However, the total number of swaps actually increases as extra switches are added. The fault-tolerant algorithm offers the promise of increased performance as it lends itself to a more easily parallelized implementation.

I INTRODUCTION

1.1 Clos Interconnection Networks

The Clos interconnection network has received attention due to its simple structures and rearrangeability and nonblocking properties. Improving the runtime has been one of the major issues with Clos networks since the decomposition process is highly serial in nature.

1991 Mathematics Subject Classification. Primary 94C

The three-stage Clos interconnection network is shown in Figure 1. It consists of two symmetrical outer stages of rectangular switches, with an inner stage of square switches. It is completely determined by the integer parameters *n, m,* and *r* that give the switch dimensions. Each switch is a crossbar which can realize any mapping of its inputs onto its outputs on a one-to-one basis. The number of inputs to the network is $N = nr$.

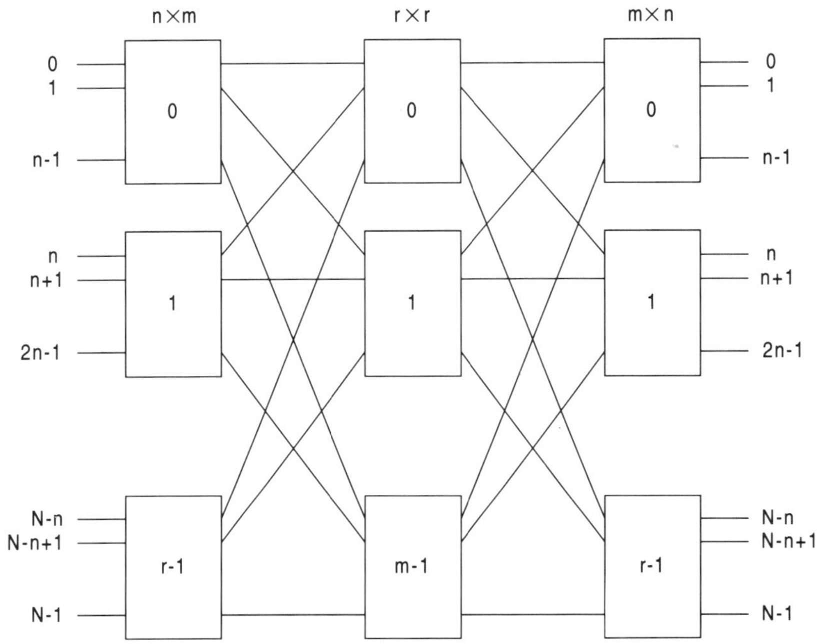

Figure 1 A three-stage Clos network

1.2 Lee, Hwang and Carpinelli's New Decomposition Algorithm

Several matrix decomposition algorithms for Clos networks have been reported. Unfortunately, many of the early matrix decomposition algorithms are incomplete. Gordon and Srikanthan [4] published a simple and fast algorithm referred to as the GS algorithm. This algorithm uses two matrices: the specification matrix *S* and the count matrix *C*. However, the GS algorithm does not work in some cases, which is shown by Chiu and Siu [5], and Lee and Carpinelli [2]. Then, a new simple algorithm for the control of rearrangeable Clos networks was proposed by Lee, Hwang and Carpinelli [1] based on the GS algorithm. The new decomposition algorithm also swaps elements of the specification matrix *S* until *S* is complete, but uses a different approach.

This algorithm consists of three kinds of swaps: simple swap, next simple swap, and successive swap. The successive swap enables the algorithm to find alternate paths by successively swapping two elements in two selected columns until it finds a free path. For a given element *e*, a successive swap iterative step can be completed in at most *e* swappings. This algorithm works for all permutations.

The time complexity of the new algorithm for the ordinary Clos network is $O(nr^2)$. It is faster than all existing decomposition algorithms. The simple swap dominates the other two kinds of swaps, and the next simple swap also dominates successive swaps. The simple swap does not require much time to swap two elements. On the other hand, the successive swap is not frequent, but take relatively long since up to $r-1$ swaps must be made in order to be completed. Therefore, successive swaps have considerable effects on the overall runtime although they are less frequent.

The new algorithm can be extended to the fault tolerant Clos (FTC) network. As the ordinary Clos network becomes complicated, fault tolerant schemes are required since a single fault in the Clos network can cause a severe degradation in performance unless schemes are provided to make the network tolerant to such faults. The Clos network can not realize all possible permutations when a fault occurs in the system. The Clos network can be made fault-tolerant as shown in Figure 2 by adding extra

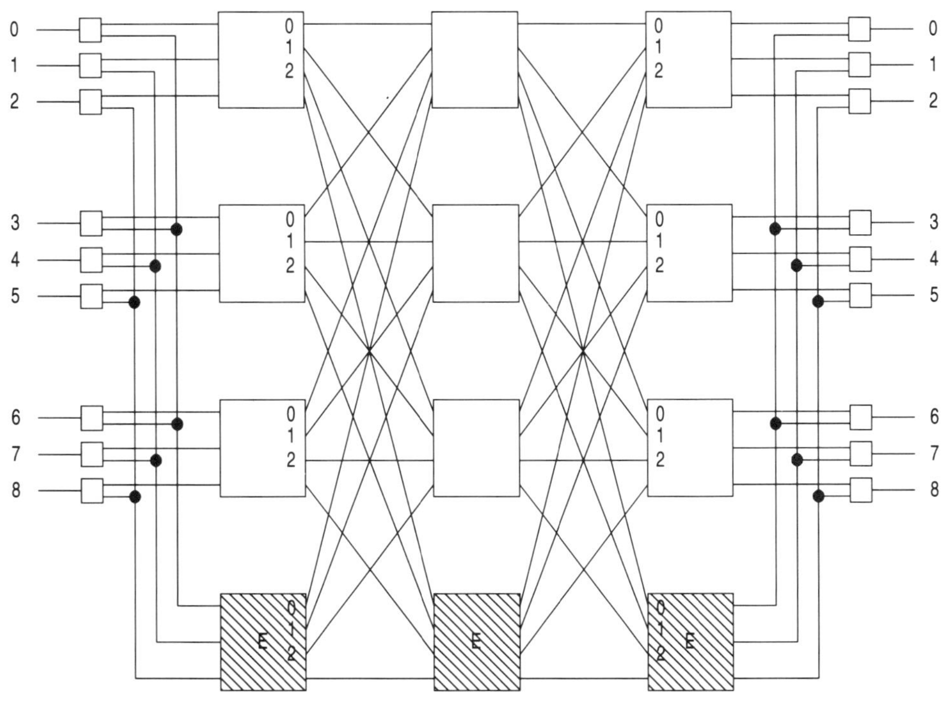

Figure 2 A fault-tolerant 9×9 Clos network with one extra switch in each stage

switches in all stages and multiplexers/demultiplexers before the first stage and after the last stage. The extra switches in the FTC network can provide alternative paths when faults occur in the network. However, when there are no faults in the system, these extra switches can be utilized as additional routing paths, which simplify the requirements of the routing process and reduce the run time. When there is a fault in the outer stage switches, the FTC achieves fault tolerance by redirecting inputs to multiplexers, demultiplexers and extra switches in the outer stages. On the other hand, inputs bound to a faulty middle stage switch are directed to one of the extra switches in the middle stage. FTC networks with y extra switches in each outer stage and x extra switches in the inner stage can tolerate up to $2y + x$ switch failures.

The algorithm for the general Clos network is extended to the fault-tolerant cases for following reasons:

- The structure of the FTC network is basically the same as the ordinary Clos network except for the extra switches, so the representation of the FTC network does not become complicated.
- The spare switches can greatly simplify the routing process, which is an obvious advantage when there are few or no faults in the system.
- The same routing algorithm for the FTC network can be used for the ordinary Clos network which is a special case of the fault tolerant network when there is not any extra switch.

The algorithm for the FTC networks also swaps elements of S until S is complete. This algorithm consists of four kinds of swaps: wild swap, simple swap, next simple swap, and successive swap. Wild swap refers to swappings associated with extra switches. Additionally, some changes have been made in simple swap, next simple swap and successive swap because of the extra switches. A successive swap iterative step may be completed in at most $e + y$ swaps because of the additional y rows.

The number of swappings for the FTC network without wild swaps is at most $O(r(n + x)(r + y))$. The wild swap can suppress at most $rx + ny$ simple, next simple, or successive swaps. Therefore, as the number of extra switches increases, the algorithm may suppress more possible swaps, and thus can significantly improve the run time when there are few or no faults.

1.3 Outline

The rest of this paper is organized as follows. In section 2, the notation used in the description of the algorithms is presented. Then, in section 3, the new algorithm for routing on the fault tolerant Clos (FTC) networks is described in detail with some examples. The performance analyses based on the simulation results are given in section 4. Finally, conclusions are presented in section 5.

II NOTATION

2.1 The *S* Matrix

Assume there is an ordinary Clos network with r $n \times n$ switches in each outer stage, and n $r \times r$ switches in the center stage. Generally, the connections between every input and every output of the network are represented by a permutation

$$P = \begin{bmatrix} 0 & 1 & \Lambda & i & \Lambda & N-1 \\ \pi(0) & \pi(1) & \Lambda & \pi(i) & \Lambda & \pi(N-1) \end{bmatrix}$$

where, input i is connected to output $\pi(i)$, $0 \le i \le N\text{-}1$, and $N = rn$. We can transform permutation P into a mapping P^* between input switches and output switches. For example, for $r = 4$ and $n = 3$,

$$P = \begin{bmatrix} 0 & 1 & 2 & 3 & 4 & 5 & 6 & 7 & 8 & 9 & 10 & 11 \\ 3 & 11 & 4 & 6 & 5 & 10 & 1 & 9 & 2 & 0 & 8 & 7 \end{bmatrix}$$

Then, P can be transformed into

$$P^* = \begin{bmatrix} 0 & 0 & 0 & 1 & 1 & 1 & 2 & 2 & 2 & 3 & 3 & 3 \\ 1 & 3 & 1 & 2 & 1 & 3 & 0 & 3 & 0 & 0 & 2 & 2 \end{bmatrix}$$

The specification matrix S can be obtained from P^*. S is an $r \times n$ matrix = (s_{ij}), where s_{ij} is the jth element in the ith row of S, and $0 \le i \le r - 1$, $0 \le j \le n - 1$. The rows of S are indexed by input switches and the columns by center switches, while entries represent output switches. Then, $s_{ij} = e$ implies that a connection from the ith input switch to the eth output switch is routed through the jth center switch. From the P^* given before, the matrix S is as follows:

$$S = \begin{bmatrix} 1 & 3 & 1 \\ 2 & 1 & 3 \\ 0 & 3 & 0 \\ 0 & 2 & 2 \end{bmatrix}$$

The routing represented by S is feasible if and only if each column is complete, that is, containing each output switch exactly once. S is complete if each column is complete. Obviously, the matrix S in the example is not complete. The algorithm that will be introduced in next section gives a method to yield a complete S through iterative swapping of two entries in the same raw.

For the FTC network, the S matrix is basically the same as in the ordinary Clos network except the spare switches are considered in the matrix. The outer stage spare switches generate additional rows in the S matrix and the inner stage switches creates additional columns. Therefore, the extra y switches in each of the outer stage add y additional rows to the S matrix, thus providing ny extra elements denoted as α. The extra x switches in the middle stage add x extra columns to the S matrix, which generate rx extra elements represented as β.

For the permutation given previously, an FTC network with $x = y = 1$ would have the following matrix:

$$S = \begin{bmatrix} 1 & 3 & 1 \\ 2 & 1 & 3 \\ 0 & 3 & 0 \\ 0 & 2 & 2 \end{bmatrix}$$

In order to speed up the algorithm for the Ordinary Clos Networks, we need to do some preprocessing which consists of constructing the following three types of sets based on the matrix S.

1) (j,e), $j = 0, 1, \ldots, n-1$, $e = 0, 1, \ldots, r-1$, is the set of rows $\{i\}$ such that $s_{ij} = e$.
2) $0(e)$, $e = 0, 1, \ldots, r-1$, is the set of columns $\{j\}$ such that S_j does not contain e.
3) $2(e)$, $e = 0, 1, \ldots, r-1$, is the set of columns $\{j\}$ such that S_j contains e at least twice.

For the algorithm for the Fault-tolerant Clos Network, we also construct the sets $0(\alpha)$, $0(\beta)$, $2(\alpha)$ and $2(\beta)$. When $e = \alpha$ (respectively, β), $0(e)$ is the set of columns such that the column contains less than y α's (respectively, x β's), and $2(e)$ is the set of columns such that the column contains more than y α's (respectively, x β's).

III THE NEW DECOMPOSITION ALGORITHM

The algorithm for the FTC networks consists of four kinds of swaps: wild swap, simple swap, next simple swap, and successive swap. Wild swap refers to swappings between elements and α or β in the additional α rows or β columns. The rules for the wild swap are as follows. The α elements may swap with any element in the same column except β

elements. On the other hand, the β elements can be swapped with any element in the same row except α elements.

Fault checking is performed before the execution of the algorithm. When an input stage extra switch is faulty, one of the α rows in the matrix is not used. When a middle stage extra switch is faulty, one of the β columns is removed. When an output stage switch is faulty, its corresponding e value is not used. If any nonspare switch is faulty, an available extra switch in the corresponding stage must be assigned to replace the faulty switch. Then, the algorithm is applied as if there is no fault in the system.

The basic idea of the algorithm for FTC networks introduced in [1] is correct. However, it lost some details during its implementation. The algorithm for FTC networks is proposed as follows.

Initialize by setting $e = 0$.

Step 1) If $2(e)$ is empty, i.e., $|2(e)| = 0$, set $e = e + 1$. Stop if $e = r$; otherwise repeat Step 1.

Step 2) If $2(e)$ is not empty, i.e., $|2(e)| > 0$, take its first element j. Also, take the first element k of $0(e)$.

Step 3) (Wild Swap) Set i to be the first element of (j, e). If u is an element of (j, α), i.e., $u \in (j, \alpha)$, and $u \geq r$, swap s_{ij} with s_{uj}. Remove i from (j, e), u from (j, α) and add i to (j, α), u to (j, e). For any $i \in (v, \beta)$, and $i < r$, $v \geq n$, swap s_{ij} with s_{iv}. Remove i from (j, e) and (v, β) and add i to (j, β) and (v, e). If $|(j, \beta)| = x$, remove j from $0(\beta)$. If $|(j, \beta)| = x + 1$, add j to $2(\beta)$. If $|(v, \beta)| = x - 1$, add v to $0(\beta)$. If $|(v, \beta)| = x$, remove v from $2(\beta)$. Remove v from $0(e)$. If $|(j, e)|^* = 1$ in any of the two cases, remove j from $2(e)$. Go to Step 1. If no spares are available, go to Step 4.

Step 4) (Simple Swap) Set i to be the first element of (j, e). If $e < s_{ik}$, or $s_{ik} = \beta$, swap s_{ij} with s_{ik}. Suppose $s_{ik} = e'$. Remove i from (j, e) and (k, e'), and add i to (j, e') and (k, e). If $|(j, e)|^* = 1$, remove j from $2(e)$. If $s_{ik} = \beta$, do { If $|(j, \beta)| = x$, remove j from $0(\beta)$. If $|(j, \beta)| = x + 1$, add j to $2(\beta)$. If $|(k, \beta)| = x - 1$, add k to $0(\beta)$. If $|(k, \beta)| = x$, remove k from $2(\beta)$.} else do {If $|(j, e')|^* = 1$, remove j from $0(e')$. If $|(j, e')|^* = 2$, add j to $2(e')$. If $|(k, e')|^* = 0$, add k to $0(e')$. If $|(k, e')|^* = 1$, remove k from $2(e')$.} Remove k from $0(e)$ and go to Step 1.

Step 5) (Next Simple Swap) If $e > s_{ik}$, repeat Step 4) on the second element i' of (j, e). If $e > s_{i'k}$ or $s_{i'k} = \alpha$, go to Step 6.

Step 6) (Successive Swap) Divide into substeps.

A) Set $u = e$. Remove k from $0(u)$. If $|(j, u)|^* = 2$, remove j from $2(u)$.
B) Set $v = s_{ik}$. Swap s_{ij} with s_{ik}. Remove i from (j, u) and (k, v), and add i to (j, v) and (k, u).
C) Suppose $e < v$ or $v = \beta$ or $|(j, v)|^* = 1$. If $v = \beta$, do { If $|(k, \beta)| = x - 1$, add k to $0(\beta)$. If $|(j, \beta)| = x$, remove k from $2(\beta)$. If $|(j, \beta)| = x$, remove j from $0(\beta)$. If $|(j, \beta)| = x + 1$, add j to $2(\beta)$.} else do {If $|(k, v)|^* = 0$, add k to $0(v)$. If $|(k, v)|^* = 1$, remove k from $2(v)$. If $|(j, v)|^* = 1$, remove j from $0(v)$. If $|(j, v)|^* = 2$, add j to $2(v)$.} Go to Step 1.
D) Suppose $e > v$ or $v = \alpha$. Set $u = v$. Take i from (j, v) and go to Step 6B.

Annotation:

$|(j, e)|$ is the cardinality of (j, e).

$|(j, e)|^*$ is the number of entries which are less than r in the set (j, e).

As an example, consider the same 4×3 S matrix in the previous section with $x = y = 1$. That is, there is one extra switch in each outer stage and one extra switch in the inner stage. Assume that no switches are faulty.

Construct three types of sets of rows or columns from the matrix S as follows:

$$S = \begin{bmatrix} 1 & 2 & 3 & \beta \\ 0^* & 3 & 3 & \beta \\ 0^* & 2 & 1 & \beta^* \\ 0 & 2 & 1 & \beta \\ \alpha^* & \alpha & \alpha & \end{bmatrix}$$

$0(0) = \{ 1, 2, 3 \}$ $\quad 2(0) = \{ 0 \}$
$0(1) = \{ 1, 3 \}$ $\quad 2(1) = \{ 2 \}$
$0(2) = \{ 0, 2, 3 \}$ $\quad 2(2) = \{ 1 \}$
$0(3) = \{ 0, 3 \}$ $\quad 2(3) = \{ 2 \}$
$0(\alpha) = \{ 3 \}$ $\quad 2(\alpha) = \{ \}$
$0(\beta) = \{ 0, 1, 2 \}$ $\quad 2(\beta) = \{ 3 \}$

$(0,0) = \{ 1,2,3 \}$	$(1,0) = \{ \}$	$(2,0) = \{ \}$	$(3,0) = \{ \}$
$(0,1) = \{ 0 \}$	$(1,1) = \{ \}$	$(2,1) = \{ 2,3 \}$	$(3,1) = \{ \}$
$(0,2) = \{ \}$	$(1,2) = \{ 0,2,3 \}$	$(2,2) = \{ \}$	$(3,2) = \{ \}$
$(0,3) = \{ \}$	$(1,3) = \{ 1 \}$	$(2,3) = \{ 0,1 \}$	$(3,3) = \{ \}$
$(0, \alpha) = \{ 4 \}$	$(1, \alpha) = \{ 4 \}$	$(2, \alpha) = \{ 4 \}$	$(3, \alpha) = \{ \}$
$(0, \beta) = \{ \}$	$(1, \beta) = \{ \}$	$(2, \beta) = \{ \}$	$(3, \beta) = \{ 0,1,2,3 \}$

1. Wild Swap: e = 0, j = 0, No. of alpha swaps = 1, No. of beta swaps = 1

$$S = \begin{bmatrix} 1 & 2 & 3 & \beta \\ \alpha & 3 & 3 & \beta \\ \beta & 2 & 1^* & 0 \\ 0 & 2 & 1 & \beta \\ 0 & \alpha & \alpha^* & \end{bmatrix}$$

0(0) = { 1, 2 }
0(1) = { 1, 3 }
0(2) = { 0, 2, 3 }
0(3) = { 0, 3 }
0(α) = { 3 }
0(β) = { 1, 2 }

2(0) = { }
2(1) = { 2 }
2(2) = { 1 }
2(3) = { 2 }
2(α) = { }
2(β) = { 3 }

(0,0) = { 3, 4 }
(0,1) = { 0 }
(0,2) = { }
(0,3) = { }
(0, α) = { 1 }
(0, β) = { 2 }

(1,0) = { }
(1,1) = { }
(1,2) = { 0, 2, 3 }
(1,3) = { 1 }
(1, α) = { 4 }
(1, β) = { }

(2,0) = { }
(2,1) = { 2, 3 }
(2,2) = { }
(2,3) = { 0, 1 }
(2, α) = { 4 }
(2, β) = { }

(3,0) = { 2 }
(3,1) = { }
(3,2) = { }
(3,3) = { }
(3, α) = { }
(3, β) = { 0, 1, 3 }

2. Wild Swap: e = 1, j = 2, No. of alpha swaps = 1, No. of beta swaps = 0

$$S = \begin{bmatrix} 1 & 2^* & 3 & \beta \\ \alpha & 3 & 3 & \beta \\ \beta & 2 & \alpha & 0 \\ 0 & 2^* & 1 & \beta^* \\ 0 & \alpha^* & 1 & \end{bmatrix}$$

0(0) = { 1, 2 }
0(1) = { 1, 3 }
0(2) = { 0, 2, 3 }
0(3) = { 0, 3 }
0(α) = { 3 }
0(β) = { 1, 2 }

2(0) = { }
2(1) = { }
2(2) = { 1 }
2(3) = { 2 }
2(α) = { }
2(β) = { 3 }

(0,0) = { 3, 4 }	(1,0) = { }	(2,0) = { }	(3,0) = { 2 }
(0,1) = { 0 }	(1,1) = { }	(2,1) = { 3, 4 }	(3,1) = { }
(0,2) = { }	(1,2) = { 0, 2, 3 }	(2,2) = { }	(3,2) = { }
(0,3) = { }	(1,3) = { 1 }	(2,3) = { 0, 1 }	(3,3) = { }
(0, α) = { 1 }	(1, α) = { 4 }	(2, α) = { 2 }	(3, α) = { }
(0, β) = { 2 }	(1, β) = { }	(2, β) = { }	(3, β) = { 0, 1, 3 }

3. Wild Swap: e = 2, j = 1, No. of alpha swaps = 1, No. of beta swaps = 1

$$S = \begin{bmatrix} 1 & \alpha & 3^* & \beta^* \\ \alpha & 3 & 3 & \beta \\ \beta & 2 & \alpha & 0 \\ 0 & \beta & 1 & 2 \\ 0 & 2 & 1 & \end{bmatrix}$$

0(0) = { 1, 2 }	2(0) = { }
0(1) = { 1, 3 }	2(1) = { }
0(2) = { 0, 2 }	2(2) = { }
0(3) = { 0, 3 }	2(3) = { 2 }
0(α) = { 3 }	2(α) = { }
0(β) = { 2 }	2(β) = { 3 }

(0,0) = { 3, 4 }	(1,0) = { }	(2,0) = { }	(3,0) = { 2 }
(0,1) = { 0 }	(1,1) = { }	(2,1) = { 3, 4 }	(3,1) = { }
(0,2) = { }	(1,2) = { 2, 4 }	(2,2) = { }	(3,2) = { 3 }
(0,3) = { }	(1,3) = { 1 }	(2,3) = { 0, 1 }	(3,3) = { }
(0, α) = { 1 }	(1, α) = { 0 }	(2, α) = { 2 }	(3, α) = { }
(0, β) = { 2 }	(1, β) = { 3 }	(2, β) = { }	(3, β) = { 0, 1 }

4. Wild Swap: e = 3, j = 2, No. of alpha swaps = 0, No. of beta swaps = 1

$$S = \begin{bmatrix} 1 & \alpha & \beta & 3 \\ \alpha & 3 & 3 & \beta \\ \beta & 2 & \alpha & 0 \\ 0 & \beta & 1 & 2 \\ 0 & 2 & 1 & \end{bmatrix}$$

0(0) = { 1, 2 }	2(0) = { }
0(1) = { 1, 3 }	2(1) = { }
0(2) = { 0, 2 }	2(2) = { }
0(3) = { 0 }	2(3) = { }
0(α) = { 3 }	2(α) = { }
0(β) = { }	2(β) = { }

(0,0) = { 3, 4 }	(1,0) = { }	(2,0) = { }	(3,0) = { 2 }
(0,1) = { 0 }	(1,1) = { }	(2,1) = { 3, 4 }	(3,1) = { }
(0,2) = { }	(1,2) = { 2, 4 }	(2,2) = { }	(3,2) = { 3 }
(0,3) = { }	(1,3) = { 1 }	(2,3) = { 1 }	(3,3) = { 0 }
(0, α) = { 1 }	(1, α) = { 0 }	(2, α) = { 2 }	(3, α) = { }
(0, β) = { 2 }	(1, β) = { 3 }	(2, β) = { 0 }	(3, β) = { 1 }

The number of alpha swaps: 3
The number of beta swaps: 3
The number of simple swaps: 0
The number of times that successive swap is called: 0

IV PERFORMANCE ANALYSES

For the FTC network, the runtime of the algorithm is dominated by the number of swaps. Wild and simple swaps do not require much time to swap two elements. On the other hand, the successive swaps are not frequent, but take a relatively long time. The wild swaps can greatly reduce the number of simple swaps, next simple swaps, and successive swaps. The new algorithm for the FTC networks was simulated to see how wild swaps improve the runtime of the algorithm with respect to various numbers of extra switches.

The decomposition algorithm for the FTC network is programmed using the C language. The program generates the matrix S automatically and randomly to achieve the simulation. That is, a basic (no extra switches) matrix S with r rows and n columns is constructed first. Then the α and β elements are added into S according to the given parameters x and y. In one simulation run, the parameters r, n, x, and y are the inputs and are kept the same for all the samples.

In the simulation run, we count the number of different kinds of swaps for a given matrix S determined by parameters r, n, x, and y. We keep a count of the number of α and β swaps separately. This allows us to see if there is a large difference between them. We also count the number of simple swaps. Since the runtime of the next simple swap is almost the same as the simple swap, it is appropriate to consider the next simple swap as the simple swap. Therefore, the number of simple swaps includes the number of next simple swaps.

The counting of the successive swaps is a little complicated since each successive swap can make several swaps. First, we keep a count of the number of times the successive swap is called. We also keep a count of the number of swaps each successive swap takes.

In consideration of the reliability of the simulation results, simulations with 100 samples and 500 samples are processed respectively for each set of { r, n, x, y }.

Some information is taken from each simulation output based on all the samples. These include:

1. Average number of α swaps
2. Average number of β swaps
3. Average number of simple swaps
4. Average number of successive swaps. Actually, this value is the average number of swaps that the successive swaps take.
5. Total number of swaps, which is the summation of the proceeding four items.

The simulation results obtained are used for performance analysis.

Through the observation of the figures obtained from the tabular results, we note that the trend of the 100-sample simulation is consistent with the 500-sample simulation for the same input set $\{r, n, x, y\}$. There is only a small quantitative difference between them.

The average number of α wild swaps increases steadily with increases of the y parameter until it reaches some maximum. After that, it won't increase any more. For a matrix S determined by a specific $\{r, n\}$, the total number of swaps has an upper bound. This can be seen from Figure 3. When the number of extra switches changes from $x = y = 2$ to $x = 6$, $y = 5$,

Figure 3 Number of swaps when r=5, n=6 (500 samples)

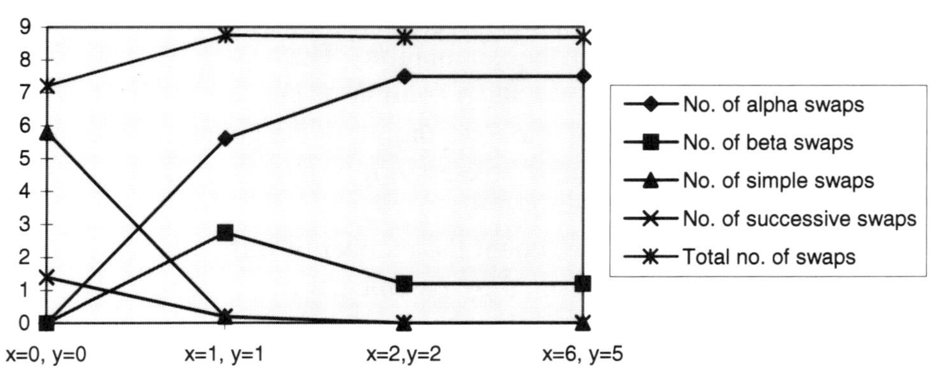

the average number of each kind of swap does not change. For a specific matrix S, this implies that the runtime can not be improved any more by adding excessive extra switches after it reaches a lower bound.

On the other hand, the average number of β swaps is determined not only by the value of the x parameter but also by the value of the y parameter, since we process α swaps before β swaps in the algorithm. This can be observed from Figures 4 and 5. The average number of β swaps increases with increases of the x parameter, and decreases with increases of the y parameter. The increasing of β swaps also has an upper bound.

Figure 4 Number of swaps when r=4, n=3 (100 samples)

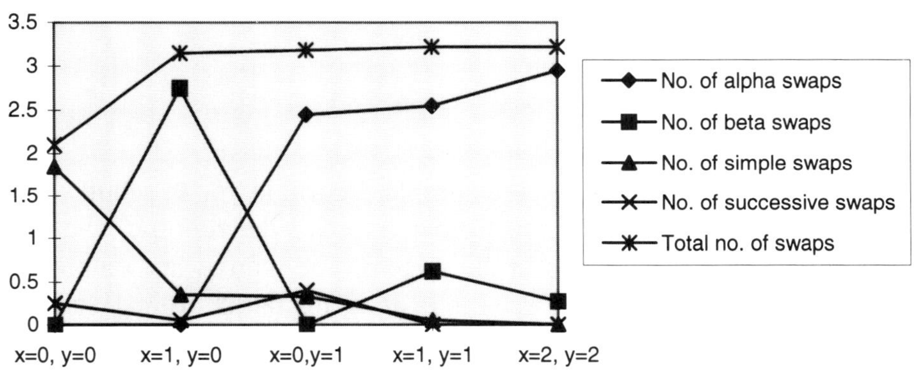

Figure 5 Number of swaps when r=4, n=6 (100 samples)

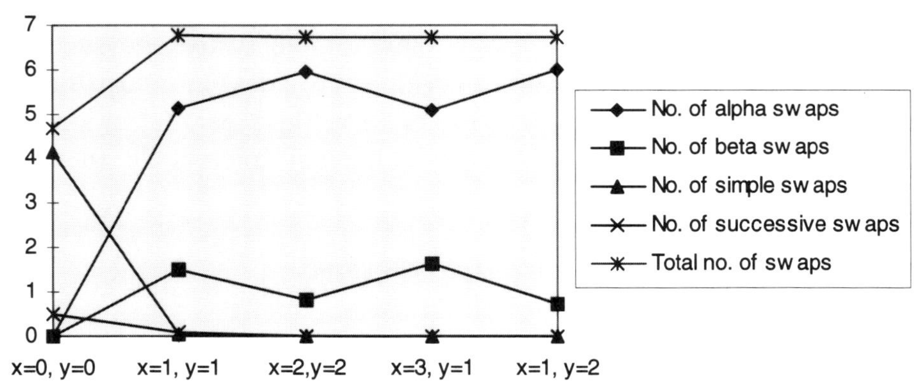

As expected, simple swaps and successive swaps decrease rapidly as wild swaps increase in all the figures, which implies the improvement in runtime.

The total number of swaps increases as x or y increases. It reaches an upper bound very quickly. For the matrix S with small size in the simulation runs, the total number of swaps reaches its upper bound when x or y is not zero. This can be seen from Figures 3, 4 and 5.

Although not presented here, we ran the simulation for various values of n, r, x and y with similar results.

When there are extra switches, the total number of swaps is dominated by the number of wild swaps. Wild swaps do not require much time to swap two elements. On the other hand, the successive swaps take a relatively much longer time. Therefore, the run time will decrease as extra switches increase until it is saturated to a certain value, although the total number of swaps increases. In addition, the fault-tolerant algorithm offers the promise of increased performance as it lends itself to a more easily parallelized implementation.

V CONCLUSIONS

This paper introduced the new simple algorithm for the fault tolerant Clos (FTC) network, which can further improve the runtime when there are relatively few or no faults in the system. The wild swaps employed in the algorithm, in addition to three kinds of swaps in the ordinary case, can reduce the chances of entering into the time-consuming successive swaps.

Programs based on the two new algorithms were written using the C language under UNIX to process simulation runs. In the simulation, the average number of α swaps, β swaps, simple swaps and successive swaps are recorded. By the observation of the trends of different kinds of swaps and the total number of swaps, we can see that wild swaps can effectively reduce the number of simple swaps, next simple swaps and successive swaps. Thus the fault-tolerant algorithm provides the possibility of parallelized implementation. It is easier to achieve parallelization when performing the fault-tolerant algorithm on the fault-tolerant Clos network.

REFERENCES

1. H. Y. Lee, F. K. Hwang and J. D. Carpinelli, "A New Decomposition Algorithm for Rearrangeable Clos Interconnection Networks," *IEEE Transactions on Communications*, vol. 44, no. 11, pp. 1572-1578, November 1996.
2. H. Y. Lee and J. D. Carpinelli, "Routing Algorithm in Fault Tolerant Clos Networks," *Conf. Inform. Sci. Syst.*, Princeton University, NJ, pp. 227-231, March 1994.

3. H. Y. Lee, Algorithms in Fault-tolerant Clos Networks, Ph.D. Thesis, New Jersey Institute of Technology, Newark, NJ, July 1994.
4. J. Gordon and S. Srikanthan, "Novel Algorithm for Clos-type Networks," *Electron. Lett.*, vol. 26, no. 21, pp. 1772-1774, October 1990.
5. Y. K. Chiu and W. C. Siu, "Comment: Novel Algorithm for Clos-type Networks," *Electron. Lett.*, vol. 27, no. 6, pp. 524-526, March 1991.
6. H. J. Siegel, "Interconnection Networks for Large-scale Parallel Processing," 2nd Edition, McGraw-Hill, 1990.
7. C. B. Wang, "The Simulation of A New Decomposition Algorithm for Rearrangeable Clos Interconnection Networks," Master Project, New Jersey Institute of Technology, Newark, NJ, May 1997.

John D. Carpinelli
Department of Electrical and Computer Engineering
New Jersey Institute of Technology
Newark, NJ 07102-1982

Catherine B. Wang
Lucent Technologies
Murray Hill, NJ 07974

DIMACS Series in Discrete Mathematics
and Theoretical Computer Science
Volume 42, 1998

Non-Blocking Multistage Interconnection Networks with Limited Depth

Italo Busi and Achille Pattavina

ABSTRACT. A new class of interconnection networks built by means of banyan networks with reduced size is defined. These networks represent the basic tool for studying the non-blocking conditions of a very wide set of multistage networks, in which the number of switching stages can range up to that of a banyan network. This network class, which includes the banyan networks just as a limiting case, has the same topological properties of the these latter networks: there is a unique path between any inlet and any outlet, whose descriptor is always given by the outlet address. The rearrangeability and strict-sense non-blocking conditions are given for the vertical replication of these networks as well as for the networks built by replicating banyan networks of reduced size. Using the Omega network as the basic banyan structure, an equivalence between the vertical replication of reduced Omega networks and the vertical replication of Shuffle/Exchange networks is found. Non-blocking conditions for these networks are provided. Finally new results about the rearrangeability of EGS networks are found.

1. Introduction

Banyan networks [1] have been defined in the seventies in the framework of distributed processing systems as devices capable of interconnecting processors and memories by achieving a very high degree of parallel processing. Banyan networks are multistage structures built of elementary $b \times b$ switching elements (SEs) that provide a single path between any inlet and outlet of the network. Typically the SE size $b = 2$ is considered. Since different I/O paths share one or more internal links, only some of all the inlet/outlet permutations can be set-up by the network. In spite of such limitations, banyan networks are very attractive as interconnection devices for two reasons: (i) each SE can be implemented in hardware by a few tens of gates, by thus making it easy the integration of a large network on a single chip, and (ii) the SEs of the network can operate concurrently in a distributed fashion to set-up a given network permutation. Therefore it becomes important to understand how banyan networks can be used to build a (multi-path) non-blocking interconnection network capable of setting up arbitrary permutations.

1991 *Mathematics Subject Classification.* Primary 94C10, 05C60, 94A99.

Basically two types of non-blocking conditions can be defined for an interconnection network by looking at the set-up of a new connection between an idle inlet and an idle outlet in a network with an arbitrary set of inlet/outlet connections already set-up, *strict-sense non-blocking* and *rearrangeability*. In the former case the new connection can be always set-up independently from the current network state determined by the on-going connections, whereas in the second case the connections already set-up may have to be re-established in a different way to allow the required new connection to be set-up too.

In the literature two approaches have been described to build a non-blocking interconnection network $N \times N$ using a banyan network with $\log_2 N$ stages as basic building block: the vertical replication and the horizontal expansion of a banyan network. Multiple I/O paths are provided in the former case by stacking together several banyan networks and in the latter case by adding stages to a single banyan network. The number of switching stages crossed by a connection is equal to $\log_2 N$ with vertical replication and larger than $\log_2 N$ with horizontal expansion. The rearrangeable and strict-sense non-blocking conditions of these networks have been described in Refs. [**2**], [**3**], [**4**]. In this paper we give very general results about non-blocking multistage networks whose number of switching stages per I/O connection is less than or equal to $\log_2 N$.

A new class of interconnection networks is defined in Section 2, called *Generalized Banyan Networks* (GBN) that includes splitters $1 \times k$ and combiners $k \times 1$ in addition to banyan networks with reduced size. This very wide network class, characterized by a single-I/O path, includes the banyan networks as a particular case. Interconnection structures of this type that are formed by splitters, a few switching stages and combiners are very promising to build photonic switching networks [**5**]. Rearrangeability and strict-sense non-blocking conditions are given in Section 3 for a network built by vertical replication of GBNs. By means of topological equivalencies between networks, these conditions are found also for two classes of networks built with "reduced" banyan networks: the vertical replication of banyan networks with reduced size (Section 4) and the replication of shuffle-exchange networks [**6**], [**7**], that is omega networks [**8**] with a reduced number of stages (Section 5). Finally, using again topological equivalencies, the non-blocking properties of EGS networks with 2×2 SEs [**9**] are studied by giving new results on network rearrangeability (Section 6).

2. Definition of the Generalized Banyan Networks

The Extended Generalized Shuffle (EGS) pattern [**9**] interconnecting two generic switching stages is first defined. In an EGS connection of size H between two adjacent switching stages, H adjacent outlets of the upstream stage are connected to H inlets of the downstream stage. Let k be the size of the switching blocks

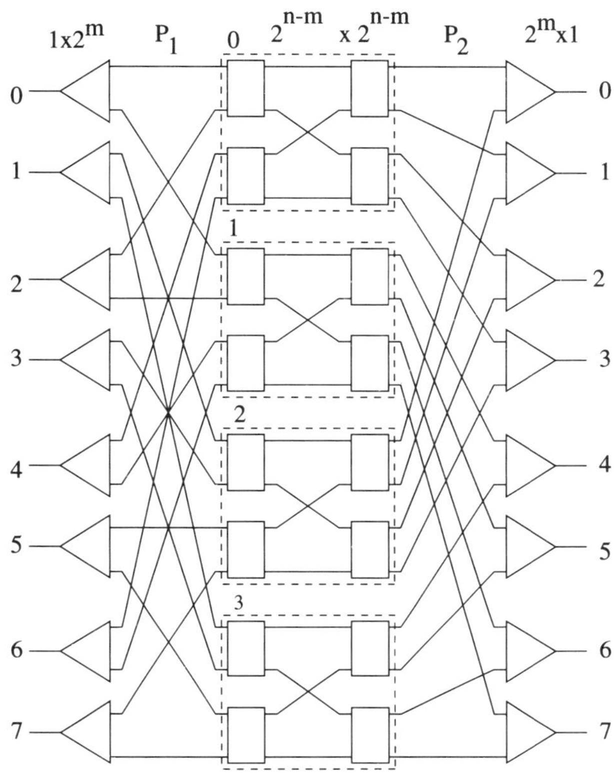

FIGURE 1. An 8×8 GBN with $m = 1$

(combiners, single SEs or subnetworks) in the downstream stage. The interstage pattern is built connecting the first of the H outlets to the first of the H inlets, and then sequentially each of the other outlets to the k-th inlet modulo H after the last that has been connected. If the selected inlet is already connected, the following one is selected.

The building elements of an $N \times N$ generalized banyan network, with $N = 2^n$ (n integer), are banyan networks of size $2^{n-m} \times 2^{n-m}$, with $0 \leq m \leq n$. A GBN $N \times N$ includes a stage of N splitters 1×2^m , a stage of 2^{2m} banyan networks $2^{n-m} \times 2^{n-m}$ and a stage of N combiners $2^m \times 1$. The interconnection patterns P_1 between the splitters and the banyan networks and P_2 between the banyan networks and the combiners are EGS permutations of size N. An 8×8 GBN with $m = 1$ is shown in Figure 1; it is made of four 4×4 banyan networks, 1×2 splitters and 2×1 combiners. Let (i, j) be the outlet j of the i-th splitter, with $0 \leq i \leq N-1$ and $0 \leq j \leq 2^m - 1$ and (i', j') be the inlet j' of the banyan network i' with $0 \leq i' \leq 2^{2m} - 1$ and $0 \leq j' \leq 2^{n-m} - 1$. The interconnection pattern P_1 is such that the splitter output

(i, j) is connected with the banyan input (i', j') given by

$$P_1 : \begin{cases} j' = \left\lfloor \dfrac{i}{2^m} \right\rfloor \\ i' = j + 2^m \,(i \bmod 2^m) \end{cases} \tag{1}$$

Analogously, by labelling (i, j) the outlet j of the i-th banyan network and (i', j') the inlet j' of the i'-th combiner, the interconnection pattern P_2 can be described as

$$P_2 : \begin{cases} j = i' \bmod 2^{n-m} \\ i = 2^m j' + \left\lfloor \dfrac{i'}{2^{n-m}} \right\rfloor \end{cases} \tag{2}$$

GBNs have the main properties of banyan networks: they have a unique path between any inlet and any outlet and the path descriptor associated to the path is always given by a set of n bits identifying the network outlet address. The first m bits identify the splitter outlet and the other $n - m$ bits select the banyan network outlet and hence the selected GBN outlet (the last stage is just a combiner).

It is worth noting how wide is the class of GBNs by looking at the two limiting cases given by $m = 0$ and $m = n$. In the former case the GBN reduces to a classical banyan network, whereas in the latter case it becomes a (non-blocking) crossbar-tree network including a stage of N splitters $1 \times N$ followed by a stage of N combiners $N \times 1$ interconnected by an EGS pattern. We note also that the number of crosspoints required in a GBN is always higher than in a banyan network of the same size, unless the limiting case $m = 0$ is considered. Nevertheless, our objective is not to find less complex networks, rather to develop tools necessary to prove the non-blocking conditions of a very wide class of networks whose number of stages is less than or equal to $\log_2 N$.

3. Vertical Replication of GBNs

GBNs have the main properties of the banyan networks and thus in general are blocking networks. So they can be made strictly or rearrangeably non-blocking by using vertical replication. An 8×8 network made by vertical replication of two GBNs is shown in Figure 2.

The interconnection pattern P_3 between the first and the second stage of splitters is EGS. By labelling (i, j) the outlet j of the i-th splitter in the first stage and (i', j') the inlet j' of the i'-th GBN, the interconnection patter P_3 can be described by

$$\begin{cases} j' = i \\ i' = j \end{cases}$$

Each splitter in the second stage can be labeled from 0 to $NK - 1$ from top to bottom, so that the splitter associated with (i', j') is labeled with $l = Ni' + j'$. The pattern P_3 connects the outlet (i, j) of the first splitter stage with the splitter in the second stage labelled $l = Nj + i$. The permutation P_4 between the two stages

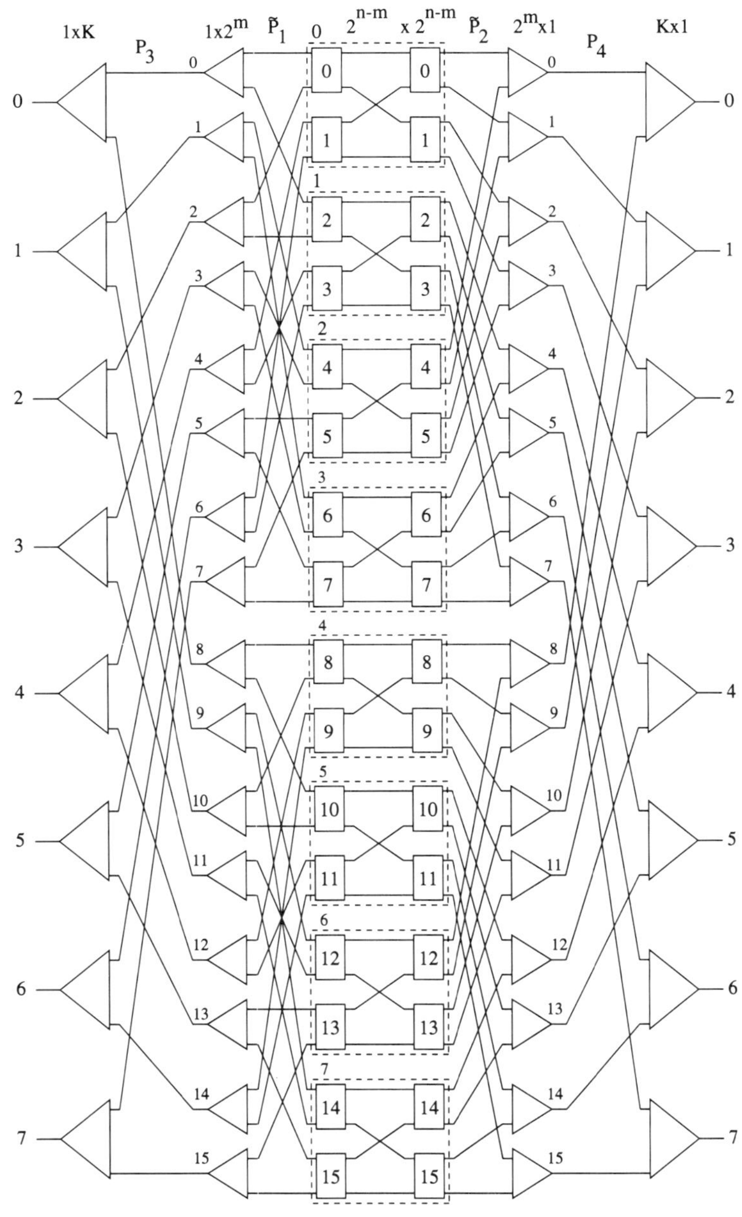

FIGURE 2. An 8×8 network made by vertical replication of two GBNs 8×8 with $m = 1$

of combiners is an EGS pattern. The inlet (i', j') of the last stage of combiners is connected to the output (i, j) of the GBN with

$$\begin{cases} j = i' \\ i = j' \end{cases}$$

So the inlet (i', j') of the last stage of combiners is connected with the combiner $l = Nj' + i'$ in the previous stage.

The interconnection pattern $\tilde{P}_1$ between the second stage of splitters and the banyan networks can be described by

$$\tilde{P}_1 : \begin{cases} j' = \left\lfloor \dfrac{i \bmod N}{2^m} \right\rfloor \\ i' = j + 2^m (i \bmod 2^m) + 2^{2m} \left\lfloor \dfrac{i}{N} \right\rfloor \end{cases} \tag{3}$$

and the interconnection pattern $\tilde{P}_2$ between the banyan networks and the first stage of combiners by

$$\tilde{P}_2 : \begin{cases} j = i' \bmod 2^{n-m} \\ i = 2^m j' + 2^{2m} \left\lfloor \dfrac{i'}{N} \right\rfloor + \left\lfloor \dfrac{i' \bmod N}{2^{n-m}} \right\rfloor \end{cases} \tag{4}$$

Necessary and sufficient conditions that make rearrangeable or strictly non-blocking a network built replicating K times a GBN are now given.

THEOREM 1. *A network $N \times N$ made by vertical replication of generalized banyan networks is rearrangeable if and only if the replication factor K is*

$$K \geq 2^{\left\lfloor \frac{n-m}{2} \right\rfloor} \tag{5}$$

PROOF. The network structure is such that the permutation $N \times N$ is split into 2^{2m} subpermutations $2^{n-m} \times 2^{n-m}$, each of these being set-up by a banyan network. These reduced banyan networks have to be made rearrangeable. Because there are $\left\lfloor \frac{n-m}{2} \right\rfloor$ shells in each banyan network, according to the graph coloring technique used in Ref. [**2**], they have to be replicated by a factor K given in Eq. 5. □

THEOREM 2. *A network $N \times N$ made by vertical replication of K generalized banyan networks is strictly non-blocking if and only if the replication factor K is*

$$K \geq \begin{cases} \dfrac{3}{2} 2^{\frac{n-m}{2}} - 1 & (n-m) \text{ even} \\ 2^{\frac{n-m+1}{2}} - 1 & (n-m) \text{ odd} \end{cases} \tag{6}$$

PROOF. The structure of the GBN is such that the permutation $N \times N$ is split into 2^{2m} subpermutations $2^{n-m} \times 2^{n-m}$, realized by the central banyan networks, and blocking can occur inside each of these networks. The worst case is given when all the 2^{n-m} inlets of one banyan network have to be connected, so the count of the maximum number of blocked path is the same as that shown in Ref. [**2**]. There is a double-tree structure with only $(n-m)$ stages, so that the maximum number of blocked paths for any inlet-outlet pair is

$$n_b = \begin{cases} \dfrac{3}{2} 2^{\frac{n-m}{2}} - 2 & (n-m) \text{ even} \\ 2^{\frac{n-m+1}{2}} - 2 & (n-m) \text{ odd} \end{cases}$$

The replication factor K has to be $n_b + 1$ and therefore Eq. 6 holds. □

	N=8	16	32	64	128	256	512	1024
m=0	2	4	4	8	8	16	16	32
1	2	2	4	4	8	8	16	16
2	1	2	2	4	4	8	8	16
3	1	1	2	2	4	4	8	8
4		1	1	2	2	4	4	8
5			1	1	2	2	4	4
6				1	1	2	2	4
7					1	1	2	2
8						1	1	2
9							1	1
10								1

TABLE 1. Replication factor K of a rearrangeable vertically replicated GBN.

	N=8	16	32	64	128	256	512	1024
m=0	3	5	7	11	15	23	31	47
1	2	3	5	7	11	15	23	31
2	1	2	3	5	7	11	15	23
3	1	1	2	3	5	7	11	15
4		1	1	2	3	5	7	11
5			1	1	2	3	5	7
6				1	1	2	3	5
7					1	1	2	3
8						1	1	2
9							1	1
10								1

TABLE 2. Replication factors K of a strictly non-blocking vertically replicated GBN.

The values of K making rearrangeable and strictly non-blocking a network built by vertical replication of GBNs are shown in Tables 1 and 2, respectively. The first row ($m = 0$) gives the same values as in Ref. [**2**] that make non-blocking (rearrangeable or strict-sense) the vertical replication of banyan networks. On the diagonals $m = n$ and $m = n - 1$ the value $K = 1$ is found. In fact GBNs with $m = n$ are crossbar-tree networks with $1 \times N$ splitters and the GBNs with $m = n-1$ are crossbar-tree networks with $1 \times \frac{N}{2}$ splitters and one switching stage. They are strictly non-blocking networks, so they do not have to be replicated to become rearrangeable or strictly non-blocking.

4. Generalized Vertical Replication of Reduced Banyan Networks

The network made by vertical replication of K GBNs includes two stages of splitters that can be merged into one stage. It is rather easy to see that a new network is so found including splitters $1 \times F$, $2^m F$ banyan networks $2^{n-m} \times 2^{n-m}$, combiners $F \times 1$ with $F = 2^m K$. By properly exchanging the position of the internal banyan networks, the pattern between the splitters and the banyan networks, as well as that between the banyan networks and the combiners can be made of the EGS type. The network in Figure 3 is thus obtained from that shown in Figure 2.
This network has the structure of the vertical replication of reduced banyan networks: in fact both the interconnection patterns between splitters and banyan networks as well as between these and the combiners are EGS. Because banyan networks are used whose size is smaller than that of the GBNs originally used these networks are said to be the generalized vertical replication of reduced banyan networks.

THEOREM 3. *A network N_1 with size $N \times N$ made by vertical replication of K generalized banyan networks with $n - m$ stages is functionally equivalent to a network N_2 of the same size made by generalized vertical replication of reduced banyan networks with a splitter fanout $F = 2^m K$ if the internal networks are moved according to*

$$i_l = i + 2^m(2^m - 1)\left\lfloor \frac{i}{2^m} \right\rfloor + 2^m(1 - F)\left\lfloor \frac{i}{F} \right\rfloor \tag{7}$$

PROOF. The complete proof of this theorem can be found in Ref. [**10**]. A trace of the proof is given also in Appendix A. □

The two networks N_1 and N_2 are functionally equivalent (they perform the same permutations); so, if one is non-blocking (either rearrangeably or strict-sense), the other one is non-blocking too. Therefore the non-blocking conditions for networks built by generalized vertical replication of reduced banyan networks are easily obtained by means of Theorems 1 and 2 developed in the previous section for the vertical replication of GBNs.

THEOREM 4. *A network $N \times N$ made by generalized vertical replication of reduced banyan networks $2^{n-m} \times 2^{n-m}$, with $0 \le m \le n$ is rearrangeable if and only if the splitter fanout is*

$$F \ge 2^{\left\lfloor \frac{n+m}{2} \right\rfloor} \tag{8}$$

PROOF. A network made by vertical replication of K GBNs is functionally equivalent to a network made by generalized vertical replication of reduced banyan networks, with a splitter fanout $F = 2^m K$. According to Theorem 1 the vertical replication factor K of the former network has to satisfy Eq. 5; so the splitter fanout

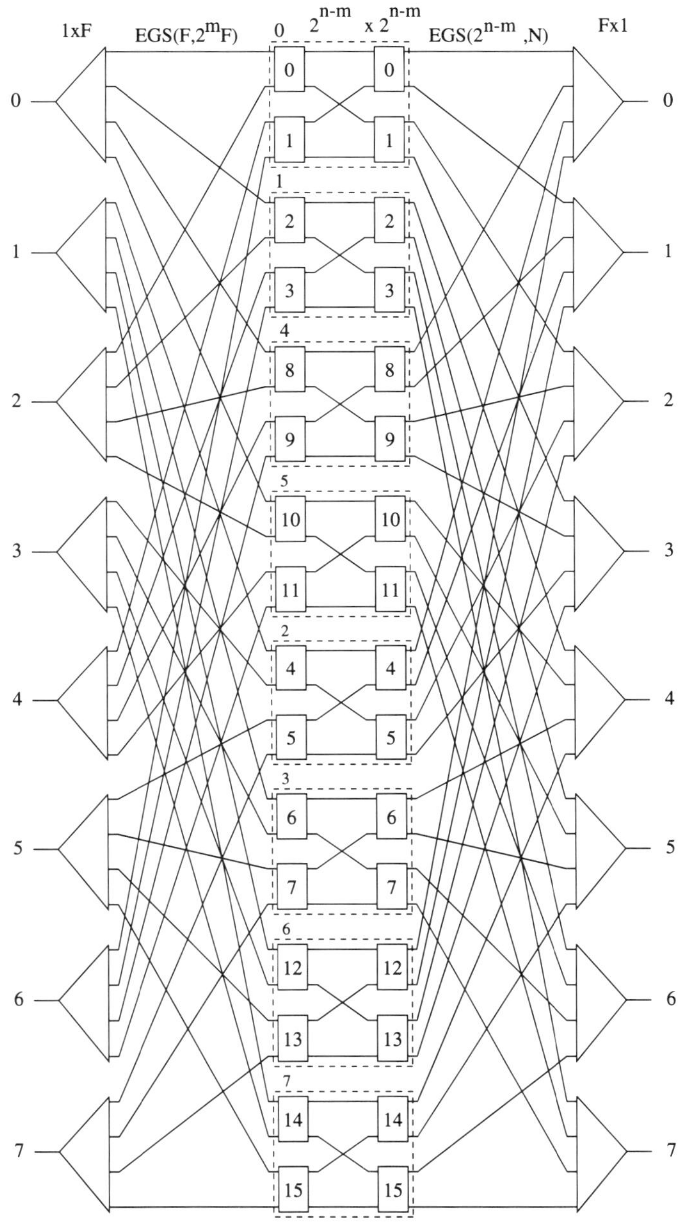

FIGURE 3. A network 8×8 made by vertical replication of 8 reduced banyan networks with size 4×4

F in the latter network is

$$F = 2^m K \geq 2^m 2^{\left\lfloor \frac{n-m}{2} \right\rfloor} = 2^{\left\lfloor m + \frac{n-m}{2} \right\rfloor} = 2^{\left\lfloor \frac{n+m}{2} \right\rfloor}$$

□

THEOREM 5. *A network $N \times N$ made by generalized vertical replication of reduced banyan networks $2^{n-m} \times 2^{n-m}$ is strictly non-blocking if and only if the splitter*

	N=8	16	32	64	128	256	512	1024
m=0	2	4	4	8	8	16	16	32
1	4	4	8	8	16	16	32	32
2	4	8	8	16	16	32	32	64
3	8	8	16	16	32	32	64	64
4		16	16	32	32	64	64	128
5			32	32	64	64	128	128
6				64	64	128	128	256
7					128	128	256	256
8						256	256	512
9							512	512
10								1024

TABLE 3. Splitter fanout F of a rearrangeable network make by generalized vertical replication of reduced banyan networks.

fanout is

$$F \geq \begin{cases} \frac{3}{2} 2^{\frac{n+m}{2}} - 2^m & (n-m)\ \textit{even} \\ \\ 2^{\frac{n+m+1}{2}} - 2^m & (n-m)\ \textit{odd} \end{cases} \tag{9}$$

PROOF. A network made by vertical replication of K GBNs is functionally equivalent to a network made by generalized vertical replication of reduced banyan networks with splitter fanout F. The replication factor K has to satisfy Eq. 6 to make the former network strictly non-blocking; so the splitter fanout F in the latter network has to be

$$F = 2^m K \geq \begin{cases} \frac{3}{2} 2^{\frac{n+m}{2}} - 2^m & (n-m)\ \textit{even} \\ \\ 2^{\frac{n+m+1}{2}} - 2^m & (n-m)\ \textit{odd} \end{cases}$$

□

The fanout value F making rearrangeable and strictly non-blocking a network made by generalized vertical replication of reduced banyan networks is shown in Tables 3 and 4, respectively. The row $m = 0$ gives the same values reported in Ref. [**2**] that make non-blocking (rearrangeable or strict-sense) the vertical replication of banyan networks. On the diagonals $m = n$ and $m = n - 1$ the values $F = N$ and $F = \frac{N}{2}$ are found: as in the previous section, these networks are the crossbar tree network without switching stages and the crossbar tree network with one switching stage.

5. Vertical Replication of Shuffle/Exchange networks with $s \leq n$ stages

Non-blocking networks can be built also by vertical replication of Shuffle-Exchange networks with $s < n$ stages, even if EGS patterns cannot be used to interconnect

	N=8	16	32	64	128	256	512	1024
m=0	3	5	7	11	15	23	31	47
1	4	6	10	14	22	30	46	62
2	4	8	12	20	28	44	60	92
3	8	8	16	24	40	56	88	120
4		16	16	32	48	80	112	176
5			32	32	64	96	160	224
6				64	64	128	192	320
7					128	128	256	384
8						256	256	512
9							512	512
10								1024

TABLE 4. Splitter fanout F of a strictly non-blocking network made by generalized vertical replication of reduced banyan networks.

splitters and combiners to the banyan networks. In fact a Shuffle/Exchange network with $s < n$ stages does not give full accessibility. So, if an EGS pattern connects the splitters to the Shuffle/Exchange networks, one inlet would be connected with the same inlet in all the planes, and then it would reach only a subset of the network outlets.

We define a new interconnection pattern of size NK called $\Sigma(N, K, n-s)$ such that a network inlet is connected with 2^{n-s} consecutive inlets of $K/(2^{n-s})$ consecutive planes that enable it to reach any network outlet. This kind of pattern is similar to an EGS one, with the difference that now 2^{n-s} adjacent outlets rather than one of the upstream stage are connected to 2^{n-s} adjacent inlets of the downstream stage. By labelling (i, j) and (i', j') the outlet j of the i-th splitter and the inlet j' of the i'-th Shuffle/Exchange network, respectively, the $\Sigma(N, K, n-s)$ pattern can be described as

$$\text{(10)} \qquad \Sigma(N, K, n-s): \begin{cases} j' = 2^{n-s} \left\lfloor \dfrac{i}{2^{n-s}} \right\rfloor + (j \bmod 2^{n-s}) \\ i' = \left\lfloor \dfrac{j}{2^{n-s}} \right\rfloor + \dfrac{K}{2^{n-s}} (i \bmod 2^{n-s}) \end{cases}$$

The vertical replication of Shuffle/Exchange networks with $s \leq n$ stages is made by a stage of N splitters $1 \times K$, K banyan networks $2^{n-m} \times 2^{n-m}$ and a stage of N combiners $K \times 1$. An EGS pattern connects the Shuffle/Exchange networks and the combiners, while a $\Sigma(N, K, n-s)$ pattern is placed between the splitters and the Shuffle/Exchange networks. An 8×8 network made by vertical replication of four Shuffle/Exchange networks 8×8 with $s = 2$ stages is shown in Figure 4.

When $s = n$ the pattern $\Sigma(N, K, 0)$ reduces to an EGS pattern and the vertical replication of Omega networks [**8**] is obtained.

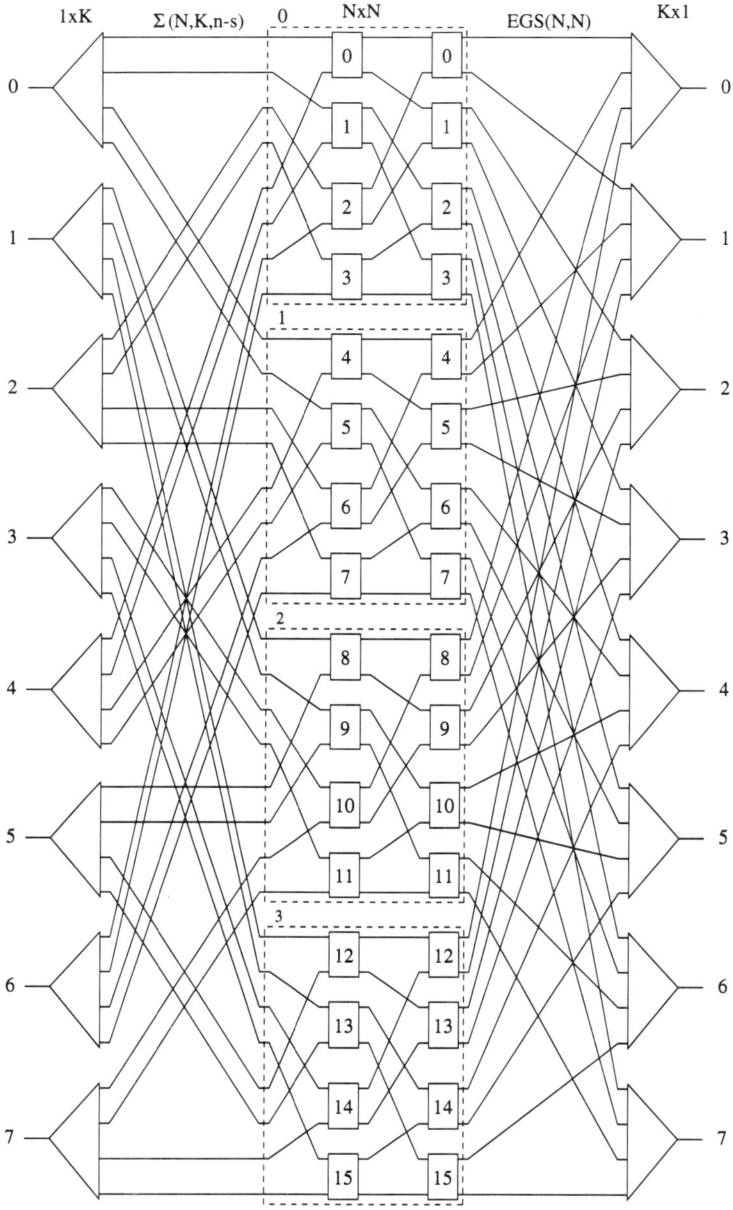

FIGURE 4. An 8×8 network made by vertical replication of four Shuffle/Exchange networks 8×8 with $s = 2$ stages

THEOREM 6. *By moving onto position* $x_{n-1} \cdots x_1$ *the switching element of stage* h $(0 \leq h \leq n-m-1)$ *that in the original network made by* 2^m *Omega networks* $2^{n-m} \times 2^{n-m}$ *has the index* $r_h(x_{n-1} \cdots x_1)$ *according to the rule*

$$
\begin{aligned}
&r_0\left(x_{n-1} \cdots x_1\right) = x_m \cdots x_1 x_{n-1} ... x_{m+1} \\
&r_h\left(x_{n-1} \cdots x_1\right) = x_{m+h} \cdots x_{h+1} x_{n-1} \cdots x_{m+h+1} x_h \cdots x_1 \qquad 0 < h < n-m-1 \\
&r_{n-m-1}\left(x_{n-1} \cdots x_1\right) = x_{n-1} \cdots x_{n-m} x_{n-m-1} \cdots x_1
\end{aligned}
$$

a Shuffle/Exchange network $N \times N$ with $n - m$ stages is obtained whose inlet permutation is

$$X_m (x_{n-1} \cdots x_1 x_0) = x_{n-m-1} \cdots x_0 x_{n-1} \cdots x_{n-m}$$

PROOF. The complete proof of this theorem can be found in Ref. [**10**]. A trace of this proof is also given in Appendix B. □

LEMMA 1. *If $\tilde{X}_m$ is the permutation made by $K = 2^m K'$ permutations X_m, then the following equivalence between the two permutations of size NK holds*

$$\Sigma (N, K, m) \equiv EGS(K, 2^m K) \cdot \tilde{X}_m$$

PROOF. The complete proof of this lemma can be found in Ref. [**10**]. A trace of this proof is also given in Appendix C. □

COROLLARY 1. *A network $N \times N$ made by generalized vertical replication of $2^m F$ reduced Omega networks with size $2^{n-m} \times 2^{n-m}$ is functionally equivalent to a network $N \times N$ made by vertical replication of $K = F$ Shuffle/Exchange networks with $s = n - m$ stages.*

PROOF. The generalized vertical replication of reduced Omega networks includes a stage of N splitters $1 \times F$, a central stage of $2^m F$ reduced Omega networks and a last stage of N combiners $F \times 1$. EGS patterns connect the splitters to the Omega networks and these networks to the combiners. By virtue of Theorem 6 the grouping of the $2^m F$ Omega networks into $K = F$ groups of 2^m networks gives K Shuffle/Exchange networks $N \times N$ with $s = n - m$ stages, with $\tilde{X}_m$ as inlet permutation. A permutation made by the cascade of an EGS pattern and the $\tilde{X}_m$ connects the splitters and the Shuffle/Exchange networks: based on Lemma 1 this is a $\Sigma(N, F, m)$ pattern. Thus we have obtained a network with a stage of N splitters $1 \times K$ connected through a $\Sigma(N, K, n-s)$ pattern to a stage of K Shuffle/Exchange networks $N \times N$ with $s = n - m$ stages which have access through an EGS pattern to a stage of N combiners $K \times 1$. □

By relying on the non-blocking conditions of the generalized vertical replication of reduced banyan networks found in the previous section and on the equivalence between these networks and vertical replication of Shuffle/Exchange networks with $s \leq n$ stages, rearrangeability and strict-sense non-blocking conditions for these latter networks can be found.

THEOREM 7. *A network $N \times N$ made by vertical replication of K Shuffle/Exchange networks with $s \leq n$ stages is rearrangeable if and only if*

$$K \geq 2^{\left\lfloor \frac{2n-s}{2} \right\rfloor} \tag{11}$$

PROOF. Corollary 1 states that the network made by generalized vertical replication of $2^m F$ Omega networks $2^{n-m} \times 2^{n-m}$ is functionally equivalent to a network

	N=8	16	32	64	128	256	512	1024
s=1	4	8	16	32	64	128	256	512
2	4	8	16	32	64	128	256	512
3	2	4	8	16	32	64	128	256
4		4	8	16	32	64	128	256
5			4	8	16	32	64	128
6				8	16	32	64	128
7					8	16	32	64
8						16	32	64
9							16	32
10								32

TABLE 5. Replication factor K of a rearrangeable vertically replicated Shuffle/Exchange network with $s \leq n$ stages.

made by vertical replication of K Shuffle/Exchange networks with $s = n - m$ stages. Based on Theorem 4, Eq. 8 has to be satisfied by the splitter fanout F in order to make the generalized vertical replication of reduced Omega networks rearrangeable (recall that an Omega network is just one type of banyan network). Therefore the replication factor K has to satisfy the same condition as F in Eq. 8 with $m = n - s$, by thus giving Eq. 11. □

THEOREM 8. *A network $N \times N$ made by vertical replication of K Shuffle/Exchange networks with $s \leq n$ stages is strictly non-blocking if and only if*

$$K \geq \begin{cases} 2^{n-s}\left(\frac{3}{2}2^{\frac{s}{2}} - 1\right) & s \text{ even} \\ 2^{n-s}\left(2^{\frac{s+1}{2}} - 1\right) & s \text{ odd} \end{cases} \tag{12}$$

PROOF. Theorem 5 states that the splitter fanout F has to satisfy Eq. 9 to make strictly non-blocking the generalized vertical replication of reduced Omega networks. As in the previous proof, because this network is functionally equivalent with the network made by vertical replication of K Shuffle/Exchange networks with $s \leq n$ stages, Eq. 9 has to be satisfied by the replication factor K, with $s = n - m$. The condition (12) is therefore found. □

The replication factors K of a rearrangeable and strictly non-blocking network made by vertical replication of Shuffle/Exchange networks with $s \leq n$ stages are shown in Tables 5 and 6, respectively. The diagonal $s = n$ gives the values found in Ref. [**2**] to make non-blocking (rearrangeable or strict-sense) the vertical replication of Omega networks. The row $s = 1$ gives $K = \frac{N}{2}$ because these are crossbar tree networks with one switching stage.

	N=8	16	32	64	128	256	512	1024
s=1	4	8	16	32	64	128	256	512
2	4	8	16	32	64	128	256	512
3	3	6	12	24	48	96	192	384
4		5	10	20	40	80	160	320
5			7	14	28	56	112	224
6				11	22	44	88	176
7					15	30	60	120
8						23	46	92
9							31	62
10								47

TABLE 6. Replication factor K of a strictly non blocking vertically replicated Shuffle/Exchange network with $s \leq n$ stages.

6. EGS networks

An EGS network of size $N \times N$ with 2×2 switching elements is now considered [**9**]. It is made by N splitters $1 \times F$, s stages of $\frac{NF}{2}$ switching elements 2×2 and N combiners $F \times 1$. Sufficient conditions on the splitter fanout F that make the EGS network strictly non-blocking are given in Ref. [**9**], by counting the maximum number of paths that can be blocked in the worst case. These non-blocking conditions differ in the two cases $s \leq n$ and $s > n$ stages, since the two kinds of networks have different properties. An EGS network with size 8×8 and $s = 2$ stages is shown in Figure 5.

THEOREM 9. *The position exchange of the switching elements in a network N_1 of size $NF \times NF$, made by $F = 2^m k$ (k integer) Shuffle/Exchange networks $N \times N$ with $s = n - m$ stages and a $\Sigma(N, F, m)$ inlet permutation according to*

$$r_h(x) = \left(x \bmod 2^{m+h}\right) + \frac{N}{2}\left(\left\lfloor \frac{x}{2^{m+h}} \right\rfloor \bmod F\right) + 2^{m+h}\left\lfloor \frac{x}{F 2^{m+h}} \right\rfloor \tag{13}$$

gives a network N_2 that is an EGS network $NF \times NF$.

PROOF. The complete proof of this theorem can be found in Ref. [**10**]. A trace of this proof is also given in Appendix D. □

COROLLARY 2. *An EGS network $N \times N$ with $s \leq n$ stages and splitter fanout F is functionally equivalent to a network made by vertical replication of $K = F$ Shuffle/Exchange networks with s stages, a $\Sigma(N, K, n - s)$ permutation after the splitters and an EGS permutation before the combiners*

PROOF. It follows directly from Theorem 9. □

As an application example of Corollary 2, Figure 5 is functionally equivalent to Figure 4.

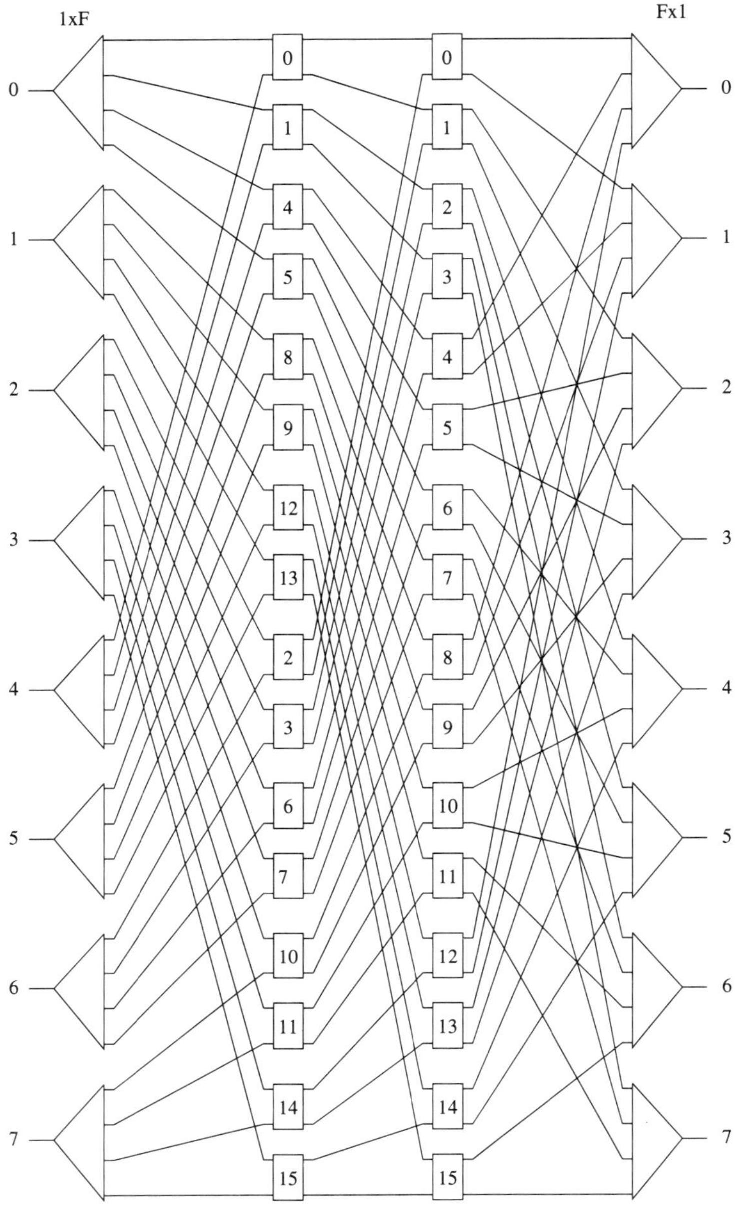

FIGURE 5. An 8×8 EGS network with $F = 4$ and $s = 2$.

Theorems 7 and 8 give the necessary and sufficient conditions to make non-blocking a network built by vertical replication of Shuffle/Exchange networks with $s \leq n$ stages. Because this network is functionally equivalent to an EGS network with the same number s of network stages (see Corollary 2), rearrangeable and strict-sense non-blocking conditions for these latter networks are easily found.

THEOREM 10. *An EGS network $N \times N$ with $s \leq n$ stages is rearrangeable if and only if the splitter fanout F satisfies the condition*

$$F \geq 2^{\left\lfloor \frac{2n-s}{2} \right\rfloor} \tag{14}$$

PROOF. An EGS network having a splitter fanout F and $s \leq n$ stages is functionally equivalent to a network made by vertical replication of $K = F$ Shuffle/Exchange networks with s stages (see Corollary 2). Therefore the rearrangeability condition on the splitter fanout F has to be the same found in Eq. 11. □

THEOREM 11. *An EGS network $N \times N$ with $s \leq n$ stages is strictly non-blocking if and only if the splitter fanout F satisfies the condition*

$$F \geq \begin{cases} 2^{n-s}\left(\dfrac{3}{2}2^{\frac{s}{2}} - 1\right) & s \text{ even} \\ \\ 2^{n-s}\left(2^{\frac{s+1}{2}} - 1\right) & s \text{ odd} \end{cases} \tag{15}$$

PROOF. An EGS network with splitter fanout F and $s \leq n$ stages is functionally equivalent to a network made by vertical replication of $K = F$ Shuffle/Exchange networks with s stages (see Corollary 2). Therefore the strictly non-blocking condition on the splitter fanout F has to be the same found in Eq. 12. □

It is worth noting that the previously known results about rearrangeable or strictly non-blocking EGS networks were only the sufficiency conditions of strict-sense non-blocking EGS network reported in Ref. [**9**]. Numerical values of the splitter fanout making an EGS network rearrangeable and strict-sense non-blocking are given by Tables 5 and 6, respectively.

7. Conclusions

A new wide class of interconnection networks has been defined, called Generalized Banyan Networks, built with banyan networks of reduced size in addition to splitters and combiners, with the aim of studying the non-blocking conditions of multistage interconnection networks with up to $\log_2 N$ stages. The classical banyan networks turn out to be just a subclass of GBNs. Rearrangeability and strict-sense non-blocking conditions for this new network class have been derived when the technique of vertical replication is used. The merging of adjacent stages of splitters and combiners has enabled us to derive the non-blocking conditions also for networks which replicate vertically banyan networks of reduced size. All the previous results known in the technical literature about non-blocking banyan networks with vertical replication have been confirmed as special cases of these new network classes. By means of topological equivalence between networks, new results on the non blocking conditions of Shuffle/Exchange networks with vertical replication and on EGS networks have been provided.

References

[1] L.R. Goke, G.J. Lipovski, "Banyan networks for partitioning multiprocessor systems", *Proc. of First Symp. on Computer Architecture*, Dec. 1973, pp. 21-30.

[2] C.-T. Lea, "Multi-log$_2$N networks and their applications in high-speed electronic and photonic switching systems", *IEEE Trans. on Communications*, Vol. 38, No. 10, Oct. 1990, pp. 1740-1749.

[3] C.-T. Lea, D.-J. Shyy, "Tradeoff of horizontal decomposition versus vertical stacking in rearrangeable nonblocking networks", *IEEE Trans. on Communications*, Vol. 39, No. 6, June 1991, pp. 899-904.

[4] D.-J. Shyy, C.-T. Lea, "Log$_2$(N,m,p) strictly nonblocking networks", *IEEE Trans. on Communications*, Vol. 39 No. 10, Oct. 1991, pp. 1502-1510.

[5] A. Jajszczyk, H.T. Mouftah, "Tree-type photonic switching networks", *IEEE Network*, Vol. 9, No. 1, Jan.-Feb. 1995, pp. 10-16.

[6] H.S. Stone, "Parallel processing with the perfect shuffle", *IEEE Trans. on Computers*, Vol. C-20, No. 4, Apr. 1972, pp. 358-366.

[7] T. Lang, H.S. Stone, "A shuffle-exchange network with simplified control", *IEEE Trans. on Computers*, Vol. C-25, No. 1, Jan. 1976, pp. 55-65.

[8] D.H. Lawrie, "Access and alignment of data in an array processor", *IEEE Trans. on Computers*, Vol. C-24 No. 12, Dec. 1975, pp. 1145-1155.

[9] G.W. Richards, "Theoretical aspects of multistage networks for broadband networks", Tutorial presentation at INFOCOM 93, San Francisco, Apr.-May 1993.

[10] I. Busi, "Rearrangeability and strict-sense non-blocking conditions in multistage interconnection networks for broadband communication networks", Thesis of Laurea, Politecnico di Milano, June 1997 (in italian).

Appendix

Appendix A. Proof of theorem 3

THEOREM 3. *A network N_1 with size $N \times N$ made by vertical replication of K generalized banyan networks with $n - m$ stages is functionally equivalent to a network N_2 of the same size made by generalized vertical replication of reduced banyan networks with a splitter fanout $F = 2^m K$ if the internal networks are moved according to*

$$i_l = i + 2^m(2^m - 1)\left\lfloor \frac{i}{2^m} \right\rfloor + 2^m(1 - F)\left\lfloor \frac{i}{F} \right\rfloor \tag{16}$$

PROOF. We need to prove that

1. moving the banyan networks in N_1 we obtain an EGS pattern between the splitters and the banyan networks as in N_2 ;
2. moving the banyan networks in N_1 we obtain an EGS pattern between the banyan networks and the combiners as in N_2.

The merging of two stages of splitters $1 \times K$ and 1×2^m in N_1 gives a splitter $1 \times F$ as in N_2.

The interconnection pattern between the splitters and the banyan networks in N_2 is an EGS permutation, such that the outlet (i, j) (outlet j of the splitter i) is connected to the inlet (i', j') (inlet j' of the banyan network i'). The banyan network i' has a logical label i'_l, such that the banyan network i'_l in N_1 has to be moved onto position i' to obtain N_2.

The outlet j of the splitter i in network N_2, whose size is $1 \times F$, is associated to the output j^* of the second-stage splitter l in N_1 connected through P_3 to the outlet $\tilde{j}$ of the first-stage splitter i. The splitter output (l, j^*) is connected through $\tilde{P}_1$ to the banyan inlet (i'_l, j'). We need to prove that the two expressions of (i'_l, j') are the same.

he splitter output (i, j) in the network N_2 is connected to the banyan inlet (i', j') by an EGS permutation

$$(17) \qquad \begin{cases} j' = \left\lfloor \dfrac{i}{2^m} \right\rfloor \\ i' = j + F\,(i \bmod 2^m) \end{cases}$$

The splitter outlet (i, j) in the network N_2 is connected to the banyan network i' , given by Eq. 17 whose logical label i'_l provided by Eq. 16

$$i'_l = j + F\,(i \bmod 2^m) + 2^m(2^m - 1)\left\lfloor \frac{i'}{2^m} \right\rfloor + 2^m(1 - F)\left\lfloor \frac{i'}{F} \right\rfloor$$

becomes

$$(18) \qquad i'_l = j^* + 2^{2m}\tilde{j} + 2^m\,(i \bmod 2^m)$$

using the equalities

$$j = 2^m\tilde{j} + j^* \qquad \tilde{j} = \left\lfloor \frac{j}{2^m} \right\rfloor \qquad j^* = j \bmod 2^m$$

$$\left\lfloor \frac{i'}{F} \right\rfloor = i \bmod 2^m$$

$$\left\lfloor \frac{i'}{2^m} \right\rfloor = \frac{F}{2^m}\,(i \bmod 2^m) + \tilde{j}$$

In the network N_1 the outlet $(i, \tilde{j})$ of the splitter in the first stage is connected to the splitter in the second stage labelled

$$(19) \qquad l = N\tilde{j} + i$$

and hence is also connected to the output (l, j^*) of the second stage of splitters. Therefore the permutation $\tilde{P}_1$ given in Eq. 3 becomes

$$(20) \qquad \begin{cases} j' = \left\lfloor \dfrac{l \bmod N}{2^m} \right\rfloor \\ i'_l = j^* + 2^m\,(l \bmod 2^m) + 2^{2m}\left\lfloor \dfrac{l}{N} \right\rfloor \end{cases}$$

Based on the relation given in Eq. 19 it can be shown that

$$l \bmod N = i$$

and hence

$$j' = \left\lfloor \frac{i}{2^m} \right\rfloor$$

which gives the first equation of (17).

Finally, using the following equality derived from Eq. 19

$$\left\lfloor \frac{l}{N} \right\rfloor = \tilde{j}$$

$$l \bmod 2^m = i \bmod 2^m$$

the second equation of (20) becomes

$$i'_l = j^* + 2^{2m}\tilde{j} + 2^m(i \bmod 2^m)$$

that is Eq. 18.

We have thus proven that the sequence of the permutation P_3 and $\tilde{P}_1$ becomes an EGS permutation, by properly moving the splitters and the banyan networks.

To prove that moving the banyan networks in N_1 we obtain an EGS pattern between the banyan networks and the combiners as in N_2, we associate to each banyan network i in N_2 a logical label i_l such that the banyan network i_l in N_1 has to be moved onto position i to obtain N_2.

In the network N_2 the combiner inlet (i', j') is backward connected to the banyan output (i, j), and the banyan network i has a logical label i_l given by Eq. 16.

In the network N_1 the combiner input (i', j') is associated to the splitter input $(l, \tilde{j})$ of the first stage, that is connected to the combiner input (i', j^*) of the last stage. The combiner inlet $(l, \tilde{j})$ is backward connected to the banyan outlet (i_l, j). We need to prove that the two expressions of (i_l, j) are the same.

In the network N_2 the combiner inlet (i', j') is backward connected to the banyan outlet (i, j) by an EGS pattern

$$\begin{cases} j = i' \bmod 2^{n-m} \\ i = 2^m j' + \left\lfloor \dfrac{i'}{2^{n-m}} \right\rfloor \end{cases} \tag{21}$$

The i-th banyan network has a logical label given by Eq. 16, which using Eq. 21 becomes

$$i_l = 2^m j' + \left\lfloor \frac{i'}{2^{n-m}} \right\rfloor + 2^m(2^m - 1)\left\lfloor \frac{i}{2^m} \right\rfloor + 2^m(1 - F)\left\lfloor \frac{i}{F} \right\rfloor$$

This expression becomes

$$i_l = 2^m\tilde{j} + 2^{2m}j^* + \left\lfloor \frac{i'}{2^{n-m}} \right\rfloor \tag{22}$$

using the following equations that can be derived from Eq. 21

$$j' = K\tilde{j} + j^* \qquad \tilde{j} = \left\lfloor \frac{j'}{K} \right\rfloor \qquad j^* = j' \bmod K$$

$$\left\lfloor \frac{i}{2^m} \right\rfloor = j'$$

$$2^m \left\lfloor \frac{i}{2^m} \right\rfloor = 2^m j' = F\tilde{j} + 2^m j^*$$

$$\left\lfloor \frac{i}{F} \right\rfloor = \tilde{j}$$

In the network N_1 made by vertical replication of GBNs the inlet (i', j^*) of the second stage of combiners is backward connected to the combiner of the first stage having a label

$$l = Nj^* + i'$$

So, using the definition (4), the inlet $(l, \tilde{j})$ is connected to the output (i_l, j)

$$\left\{ \begin{array}{l} j = l \bmod 2^{n-m} \\ i_l = 2^m \tilde{j} + 2^{2m} \left\lfloor \frac{l}{N} \right\rfloor + \left\lfloor \frac{l \bmod N}{2^{n-m}} \right\rfloor \end{array} \right. \tag{23}$$

Therefore using the definition of l we obtain from the first equation of (23)

$$j = i' \bmod 2^{n-m}$$

that is the first equation in (21).
Moreover, by means of the following relations based on the definition of l

$$l \bmod N = i'$$

$$\left\lfloor \frac{l}{N} \right\rfloor = j^*$$

the second equation of (23) gives

$$i_l = 2^m \tilde{j} + 2^{2m} j^* + \left\lfloor \frac{i'}{2^{n-m}} \right\rfloor$$

that is Eq. 22. □

Appendix B. Proof of theorem 6

THEOREM 6. *By moving onto position $x_{n-1} \cdots x_1$ the switching element of stage h $(0 \le h \le n-m-1)$ that in the original network made by 2^m Omega networks $2^{n-m} \times 2^{n-m}$ has the index $r_h(x_{n-1} \cdots x_1)$ according to the rule*

$$r_0\left(x_{n-1} \cdots x_1\right) = x_m \cdots x_1 x_{n-1} ... x_{m+1}$$
$$r_h\left(x_{n-1} \cdots x_1\right) = x_{m+h} \cdots x_{h+1} x_{n-1} \cdots x_{m+h+1} x_h \cdots x_1 \qquad 0 < h < n-m-1$$
$$r_{n-m-1}\left(x_{n-1} \cdots x_1\right) = x_{n-1} \cdots x_{n-m} x_{n-m-1} \cdots x_1$$

a Shuffle/Exchange network $N \times N$ with $n-m$ stages is obtained whose inlet permutation is

$$X_m\left(x_{n-1} \cdots x_1 x_0\right) = x_{n-m-1} \cdots x_0 x_{n-1} \cdots x_{n-m}$$

PROOF. We need to prove that

1. moving the SEs according to the specified rules r_h, all the interstage patterns become perfect shuffle permutations
2. the SEs in the last stage do not have to be moved
3. the inlet permutation is the cascade of $\tilde{X}_m$ and the perfect shuffle.

From Theorem 9 it is known that a $N \times N$ Shuffle/Exchange network with $n - m$ stages can be obtained from a network $N \times N$ made by 2^m Omega networks $2^{n-m} \times 2^{n-m}$.

By replacing in Eq. 13 N with 2^{n-m}, F with 2^m, m with 0, NF and n in Eq. 13 become now N and $n - m$, respectively, and Eq. 13 can be thus rewritten as

$$r_h(x) = \left(x \bmod 2^h\right) + 2^{n-m-1}\left(\left\lfloor \frac{x}{2^h} \right\rfloor \bmod 2^m\right) + 2^h \left\lfloor \frac{x}{2^{m+h}} \right\rfloor$$

We need to prove that this relation is equivalent, in binary form, to the r_h rules. In the case of $h = n - m - 1$, it can be shown that

$$r_{n-m-1}(x) = x$$

$$r_{n-m-1}(x_{n-1} \dots x_1) = x_{n-1} \dots x_{n-m} x_{n-m-1} \dots x_1$$

In the case of $h = 0$ we obtain

$$r_0(x) = 2^{n-m-1}\left(x \bmod 2^m\right) + \left\lfloor \frac{x}{2^m} \right\rfloor$$

In the binary representation of x, that is $x_{n-1} \dots x_{m+1} x_m \dots x_1$, the last m bits, $x_m \dots x_1$, are given by $(x \bmod 2^m)$, while the first $n - m - 1$ bits, $x_{n-1} \dots x_{m+1}$, are obtained as $\left\lfloor \frac{x}{2^m} \right\rfloor$, so that r_0 can be written as

$$r_0(x_{n-1} \dots x_1) = x_m \dots x_1 x_{n-1} \dots x_{m+1}$$

For $0 < h < n - m - 1$, $x = x_{n-1} \dots x_{m+h+1} x_{m+h} \dots x_{h+1} x_h \dots x_1$ can be obtained observing that $x_{n-1} \dots x_{m+h+1}$ is equal to $\left\lfloor \frac{x}{2^{m+h}} \right\rfloor$, $x_{m+h} \dots x_{h+1}$ is equal to $\left(\left\lfloor \frac{x}{2^h} \right\rfloor \bmod 2^m\right)$ and $x_h \dots x_1$ are given by $(x \bmod 2^h)$, so that we can write

$$r_h(x_{n-1} \dots x_1) = x_{m+h} \dots x_{h+1} x_{n-1} \dots x_{m+h+1} x_h \dots x_1$$

It has just been proven that moving the switching elements in a $N \times N$ network made by 2^m Omega networks of size $2^{n-m} \times 2^{n-m}$ gives a new network with all the interstage pattern of shuffle type. We still need to prove that the inlet permutation is the cascade of the X_m and the perfect shuffle.

The inlet permutation Y_m is such that the network inlet x is connected to the inlet z of the first switching stage. This inlet z has a logical label y such that $y_0 = z_0$ and $y_{n-1} \dots y_1 = r_0(z_{n-1} \dots z_1)$. It is known that x is connected to y through a shuffle permutation of size 2^{n-m} called σ_{n-m-1}, because it is the inlet permutation in a $2^{n-m} \times 2^{n-m}$ Omega network; so

$$\begin{aligned} y_{n-1} \dots y_{n-m} y_{n-m-1} \dots y_1 y_0 &= \sigma_{n-m-1}(x_{n-1} \dots x_0) \\ &= x_{n-1} \dots x_{n-m} x_{n-m-2} \dots x_0 x_{n-m-1} \end{aligned}$$

and since

$$y_{n-1} \dots y_{n-m} y_{n-m-1} \dots y_1 = r_0(z_{n-1} \dots z_1) = z_m \dots z_1 z_{n-1} \dots z_{m+1}$$

then Y_m is such that

$$\begin{aligned} z_{n-1}\dots z_{m+1}z_m\dots z_1z_0 &= y_{n-m-1}\dots y_1y_{n-1}\dots y_{n-m}y_0 \\ &= x_{n-m-2}\dots x_0x_{n-1}\dots x_{n-m}x_{n-m-1} \end{aligned}$$

Since Y_m is given by the cascade of X_m and σ_{n-1}, then

$$z_{n-1}\dots z_1z_0 = \sigma_{n-1}(y_{n-1}\dots y_0) = \sigma_{n-1}(X_m(x_{n-1}\dots x_0)) = Y_m(x_{n-1}\dots x_0)$$

$$z_{n-1}\dots z_{m+1}z_m\dots z_1z_0 = y_{n-2}\dots y_my_{m-1}\dots y_0y_{n-1} = x_{n-m-2}\dots x_0x_{n-1}\dots x_{n-m}x_{n-m-1}$$

and finally X_m is obtained

$$y_{n-1}y_{n-2}\dots y_my_{m-1}\dots y_0 = x_{n-m-1}x_{n-m-2}\dots x_0x_{n-1}\dots x_{n-m}$$

□

Appendix C. Proof of lemma 1

LEMMA 1. *If $\tilde{X}_m$ is the permutation made by $K = 2^mK'$ permutations X_m, then the following equivalence between the two permutations of size NK holds*

$$\Sigma(N, K, m) \equiv EGS(K, 2^mK) \cdot \tilde{X}_m$$

PROOF. We first describe formally the $\tilde{X}_m$ permutation and then we prove the thesis. The permutation X_m, whose size is N, can be defined in decimal notation as

$$y = X_m(x) = 2^m \left(x \bmod 2^{n-m}\right) + \left\lfloor \frac{x}{2^{n-m}} \right\rfloor$$

Therefore the permutation $\tilde{X}_m$, whose size is NK, is such that, by grouping its inlets and outlets in K groups of N elements each, the inlet (i,j) (inlet j of group i) is connected to outlet (i',j') according to

$$\begin{cases} i' = i \\ j' = 2^m\left(j \bmod 2^{n-m}\right) + \left\lfloor \dfrac{j}{2^{n-m}} \right\rfloor \end{cases} \tag{24}$$

The NK inlets of the EGS permutation can be divided into N sets of K elements and the NK outlets can be divided into 2^mK groups of $\frac{N}{2^m}$ elements each. The inlet (i,j) (inlet j of the group i) is connected to the outlet $(\tilde{i},\tilde{j})$ (outlet $\tilde{j}$ of the group $\tilde{i}$) by the EGS permutation. The NK outlets of the EGS permutation are also the inlets of $\tilde{X}_m$, and dividing this set of NK inlets into K sets of N elements, the outlet $(\tilde{i},\tilde{j})$ of the EGS permutation is the inlet (i'',j'') (outlet j'' of the group i'') of the $\tilde{X}_m$ permutation. By dividing the NK outlets of the $\tilde{X}_m$ permutation into K sets of N elements, the inlet (i'',j'') is connected to the outlet (i',j'). We need to prove that this outlet (i',j') connected to (i,j) is the same outlet (i',j') connected to (i,j) by the $\Sigma(N,K,m)$ permutation.

The inlet (i, j) is connected, by the EGS permutation, to the outlet $(\tilde{i}, \tilde{j})$ according to

$$\begin{cases} \tilde{i} = (Ki + j) \bmod 2^m K \\ \tilde{j} = \left\lfloor \dfrac{Ki + j}{2^m K} \right\rfloor \end{cases} \tag{25}$$

This outlet of the EGS permutation is equivalent to the inlet (i'', j'') of the $\tilde{X}_m$ permutation, given by

$$\begin{cases} i'' = \left\lfloor \dfrac{\tilde{i}}{2^m} \right\rfloor \\ j'' = 2^{n-m} \left(\tilde{i} \bmod 2^m\right) + \tilde{j} \end{cases} \tag{26}$$

The inlet (i'', j'') is connected to (i', j') given by the (24). Using Eq. 25 we obtain

$$\tilde{i} = j + K\,(i \bmod 2^m)$$

so that Eqs. (26) and (24) give

$$i' = i'' = \left\lfloor \frac{j}{2^m} \right\rfloor + \frac{K}{2^m}\,(i \bmod 2^m) \tag{27}$$

Using again Eqs. (25) and (26), we obtain

$$\tilde{j} = \left\lfloor \frac{i}{2^m} \right\rfloor$$

$$j'' = 2^{n-m}\left(\tilde{i} \bmod 2^m\right) + \left\lfloor \frac{i}{2^m} \right\rfloor = 2^{n-m}\,(j \bmod 2^m) + \left\lfloor \frac{i}{2^m} \right\rfloor$$

in which the last equality derives from the relation $K = 2^m K'$.
Since $\tilde{X}_m$ is such that

$$j' = 2^m \left(j'' \bmod 2^{n-m}\right) + \left\lfloor \frac{j''}{2^{n-m}} \right\rfloor$$

we can derive the final expression for j'

$$j' = 2^m \left\lfloor \frac{i}{2^m} \right\rfloor + (j \bmod 2^m) \tag{28}$$

using the following equalities

$$\left(j'' \bmod 2^{n-m}\right) = \left\lfloor \frac{i}{2^m} \right\rfloor$$

$$\left\lfloor \frac{j''}{2^{n-m}} \right\rfloor = j \bmod 2^m$$

Eqs. (27) and (28) prove that the EGS pattern cascaded to a $\tilde{X}_m$ permutation gives a $\Sigma(N, K, m)$ permutation. □

Appendix D. Proof of theorem 9

THEOREM 9. *The position exchange of the switching elements in a network N_1 of size $NF \times NF$, made by $F = 2^m k$ (k integer) Shuffle/Exchange networks $N \times N$ with $s = n - m$ stages and a $\Sigma(N, F, m)$ inlet permutation according to*

$$r_h(x) = \left(x \bmod 2^{m+h}\right) + \frac{N}{2}\left(\left\lfloor \frac{x}{2^{m+h}} \right\rfloor \bmod F\right) + 2^{m+h}\left\lfloor \frac{x}{F2^{m+h}} \right\rfloor \tag{29}$$

gives a network N_2 that is an EGS network $NF \times NF$.

PROOF. To prove this equivalence we need to prove that

1. moving the SEs of network N_1 in stage h and in stage $h+1$ ($h = 0, \cdots, s-1$) according to r_h we obtain a new network with all internal EGS interconnection pattern
2. the SEs in the last stage ($s-1$) do not have to be moved
3. moving the SEs in the first stage according to r_0 the inlet permutation of N_1, given by $\Sigma(N, F, n-s)$ followed by a perfect shuffle of size N, becomes an EGS permutation.

The outlet (i, j) (outlet j of switching element i) of stage h in N_2 is connected to the inlet (i', j') of stage $h+1$ ($h = 0, \cdots, s-1$). The SE i' has a logical label i'_l that is its position in N_1. The SE i_l of stage h in N_1, which has to be moved onto position i in N_2, is such that its outlet (i_l, j) is connected to inlet (i'_l, j') in stage $h+1$. We have to prove that the two expressions of (i'_l, j') are the same.

The outlet (i_l, j) of stage h in N_1 is connected to the inlet (i'_l, j') of the following stage

$$\begin{cases} i'_l = \dfrac{N}{2}\left\lfloor \dfrac{2i_l}{N} \right\rfloor + \left(2i_l \bmod \frac{N}{2}\right) + j \\ j' = \left\lfloor \dfrac{4\left(i_l \bmod \frac{N}{2}\right) + 2j}{N} \right\rfloor \end{cases} \tag{30}$$

This equation can be rewritten as

(31)

$$i'_l = \frac{N}{2}\left(\left\lfloor \frac{i}{2^{m+h}} \right\rfloor \bmod F\right) + \left(2i \bmod 2^{m+h+1}\right) + 2^{m+h+1}\left\lfloor \frac{i}{2^{m+h}F} \right\rfloor \bmod \frac{N}{2} + j$$

using the following equalities

$$\left(2i_l \bmod \frac{N}{2}\right) = \left(2i \bmod 2^{m+h+1}\right) + 2^{m+h+1}\left\lfloor \frac{i}{2^{m+h}F} \right\rfloor \bmod \frac{N}{2}$$

$$\left\lfloor \frac{2i_l}{N} \right\rfloor = \left\lfloor \frac{i}{2^{m+h}} \right\rfloor \bmod F$$

The outlet (i, j) in N_2 is connected to (i', j') by a permutation EGS, that is

$$\begin{cases} i' = \left(2i \bmod \frac{NF}{2}\right) + j \\ j' = \left\lfloor \dfrac{4i + 2j}{NF} \right\rfloor \end{cases} \tag{32}$$

According to Eq. 29 the i'-th switching element in the stage $h+1$ has a logical label

$$i'_l = \left(i' \bmod 2^{m+h+1}\right) + \frac{N}{2}\left(\left\lfloor \frac{i'}{2^{m+h+1}} \right\rfloor \bmod F\right) + 2^{m+h+1}\left\lfloor \frac{i'}{2^{m+h+1}F} \right\rfloor$$

that by means of the following relations based on Eq. 32

$$\left(i' \bmod 2^{m+h+1}\right) = \left(2i \bmod 2^{m+h+1}\right) + j$$

$$\left\lfloor \frac{i'}{2^{m+h+1}} \right\rfloor \bmod F = \left\lfloor \frac{i}{2^{m+h}} \right\rfloor \bmod F$$

$$\left\lfloor \frac{i'}{2^{m+h+1}F} \right\rfloor = \left\lfloor \frac{i}{2^{m+h}F} \right\rfloor \bmod 2^{n-m-h-2}$$

becomes

$$i'_l = \left(2i \bmod 2^{m+h+1}\right) + j + \frac{N}{2}\left(\left\lfloor \frac{i}{2^{m+h}} \right\rfloor \bmod F\right) + 2^{m+h+1}\left\lfloor \frac{i}{2^{m+h}F} \right\rfloor \bmod \frac{N}{2}$$

that is Eq. 31.

So we have proven that permuting the switching elements according to the rules r_h F Shuffle/Exchange networks N×N with $s = n - m$ stages are equivalent to a network with all EGS interstage patterns. This new network is N_2 if the output permutation is an identity and the inlet one is an EGS pattern.

The SEs in the last stage do not have to be moved and hence

$$r_{n-m-1}(x) = x$$

We now need to prove that the inlet permutation in N_2 is an EGS pattern. The two networks N_1 and N_2 have size NF. By grouping the NF inlets of each network in N sets of F elements, the inlet (i, j) (inlet j of the group i) of N_2 , is connected to the inlet (i', j') (inlet j' of SE i') of the first switching stage by an EGS permutation. The switching element i' in N_2 has a logical label i'_l given by Eq. 29.

The inlet (i, j) in N_1 is connected through the $\Sigma(N, F, m)$ permutation to $(\tilde{i}, \tilde{j})$, that is the inlet $\tilde{j}$ of the Shuffle/Exchange network $\tilde{i}$. The inlet $\tilde{j}$ is connected through a perfect shuffle permutation to the inlet j'' in the first switching stage. The inlet j'' of the Shuffle/Exchange network $\tilde{i}$ is the inlet j' of the switching element i'_l. We need to prove that the two expressions of (i'_l, j') are the same.

The EGS permutation in N_2 is such that (i, j) is connected to (i', j')

$$\text{(33)} \qquad \begin{cases} j' = \left\lfloor \dfrac{2i}{N} \right\rfloor \\ i' = j + F\left(i \bmod \frac{N}{2}\right) \end{cases}$$

The switching element i' of stage 0 has a logical label

$$i'_l = r_0(i') = (i' \bmod 2^m) + \frac{N}{2}\left(\left\lfloor \frac{i'}{2^m} \right\rfloor \bmod F\right) + 2^m \left\lfloor \frac{i'}{F2^m} \right\rfloor$$

that becomes

(34)

$$i'_l = (j \bmod 2^m) + \frac{N}{2}\left\lfloor \frac{j}{2^m} \right\rfloor + \frac{NF}{2^{m+1}}(i \bmod 2^m) + 2^m\left(\left\lfloor \frac{i}{2^m} \right\rfloor \bmod 2^{n-m-1}\right)$$

using the following equalities

$$(i' \bmod 2^m) = (j \bmod 2^m)$$

$$\left\lfloor \frac{i'}{2^m} \right\rfloor \bmod F = \left\lfloor \frac{j}{2^m} \right\rfloor + \frac{F}{2^m}(i \bmod 2^m)$$

$$\left\lfloor \frac{i'}{F2^m} \right\rfloor = \left\lfloor \frac{i}{2^m} \right\rfloor \bmod 2^{n-m-1}$$

derived using $F = 2^m k$ (k integer).

The inlet (i, j) in N_1 is connected to $(\tilde{i}, \tilde{j})$ by the $\Sigma(N, F, m)$ permutation

$$\begin{cases} \tilde{i} = \left\lfloor \frac{j}{2^m} \right\rfloor + \frac{F}{2^m}(i \bmod 2^m) \\ \tilde{j} = 2^m\left\lfloor \frac{i}{2^m} \right\rfloor + (j \bmod 2^m) \end{cases}$$

and the inlet $\tilde{j}$ is connected to j'', by a perfect shuffle, so that

$$j'' = (2\tilde{j} \bmod N) + \left\lfloor \frac{2\tilde{j}}{N} \right\rfloor$$

This last expression becomes

$$j'' = \left\lfloor \frac{2i}{N} \right\rfloor + 2\left\{2^m\left(\left\lfloor \frac{i}{2^m} \right\rfloor \bmod 2^{n-m-1}\right) + (j \bmod 2^m)\right\}$$

using the following equalities

$$\left\lfloor \frac{2\tilde{j}}{N} \right\rfloor = \left\lfloor \frac{2i}{N} \right\rfloor$$

$$(2\tilde{j} \bmod N) = 2^{m+1}\left(\left\lfloor \frac{i}{2^m} \right\rfloor \bmod 2^{n-m-1}\right) + (2j \bmod 2^{m+1})$$

The inlet j'' of the first switching stage in the Shuffle/Exchange network $\tilde{i}$ is the inlet (i'_l, j')

$$\begin{cases} i'_l = \frac{N}{2}\tilde{i} + \left\lfloor \frac{j''}{2} \right\rfloor \\ j' = j'' \bmod 2 \end{cases}$$

By means of the expression of j'' it can be found that

$$(j'' \bmod 2) = \left\lfloor \frac{2i}{N} \right\rfloor$$

$$\left\lfloor \frac{j''}{2} \right\rfloor = 2^m\left(\left\lfloor \frac{i}{2^m} \right\rfloor \bmod 2^{n-m-1}\right) + (j \bmod 2^m)$$

So we obtain

$$j' = \left\lfloor \frac{2i}{N} \right\rfloor$$

$$i'_l = \frac{N}{2}\left\lfloor \frac{j}{2^m} \right\rfloor + \frac{NF}{2^{m+1}}(i \bmod 2^m) + 2^m\left(\left\lfloor \frac{i}{2^m} \right\rfloor \bmod 2^{n-m-1}\right) + (j \bmod 2^m)$$

that is Eqs. (33) and (34). □

DEPT. OF ELECTRONICS AND INFORMATION, POLITECNICO DI MILANO, PIAZZA LEONARDO DA VINCI 32, 20133 MILANO (ITALY)
E-mail address: achille@elet.polimi.it

DIMACS Series in Discrete Mathematics
and Theoretical Computer Science
Volume 42, 1998

Isomorphism of Classical Rearrangeable Networks

M. Douglas McIlroy and Joseph P. Savicki

ABSTRACT. The isomorphism of multistage connecting networks based on square switches wired in Benes's iterative pattern, Clos's recursively expanded pattern, and other derived patterns is proved for general switch sizes as well as 2×2's. Proofs based on mixed-radix numberings of the switches in each stage are exhibited succinctly in commuting diagrams. The renumberings that embody the isomorphism are summarized in simple graphics.

Introduction

Good topologies for rearrangeable networks are well known. Clos described a class of multistage topologies recursively [**C**], and Benes described a class iteratively [**B**]. Benes topologies are isomorphic to Clos topologies built of square switches, a fact that is often cited but seldom proved in the literature, except for the special case of 2×2 primitive switches [**A**][**W**]. We exhibit the isomorphism as a rearrangement of the digits of mixed-radix numbers that identify the primitive switches within each stage. The well known 2×2 case is taken as a warm-up, then the general case is treated.

An $N \times N$ *rearrangeable network* allows any permutation of its N inputs to be connected one-to-one to its N outputs. Benes, by a group-theoretic argument, justified a general construction of such a network as a progression of similarly constructed *stages,* where each stage consists of a column of *square switches,* whose inputs and outputs are wired to adjacent stages in a standard pattern [**B**]. An $n \times n$ square switch, though often thought of concretely as a crossbar, is itself an $n \times n$ rearrangeable connecting network. Hereafter a ''network'' will always be understood to be rearrangeable, a ''switch'' will always be square, and ''size-n'' will be used interchangeably with ''$n \times n$''.

1. Binary case

1.1 Binary case, Clos network. The Clos design[1] has a structure that can be understood by inspection. It uses $N\log N - N/2$ size-2 switches and can be tweaked down to $N\log N - N + 1$, close to the information-theoretic lower bound of $\log(N!) \approx N\log N - 1.44N$. (Logarithms base 2.) The network, sketched in Figure 1, has outside stages of $N/2$ size-2 switches and a middle stage of two size-$N/2$ switches. The middle

Subject classification: 94C10, switching theory.

[1]We use the name ''Clos'' for convenience and familiarity. This class of networks had been described before [**K**] and is more restrictive than Clos's full class, in which primitive switches need not be square.

switches, if larger than 2×2, are recursively constructed the same way. Input and output terminals are numbered $0 \,..\, N-1$. Terminals on the same outside switch i are said to be *mates* of each other. Two wires from each outside switch i go respectively *up* and *down* to terminals i on the near side of middle switches 0 and 1.

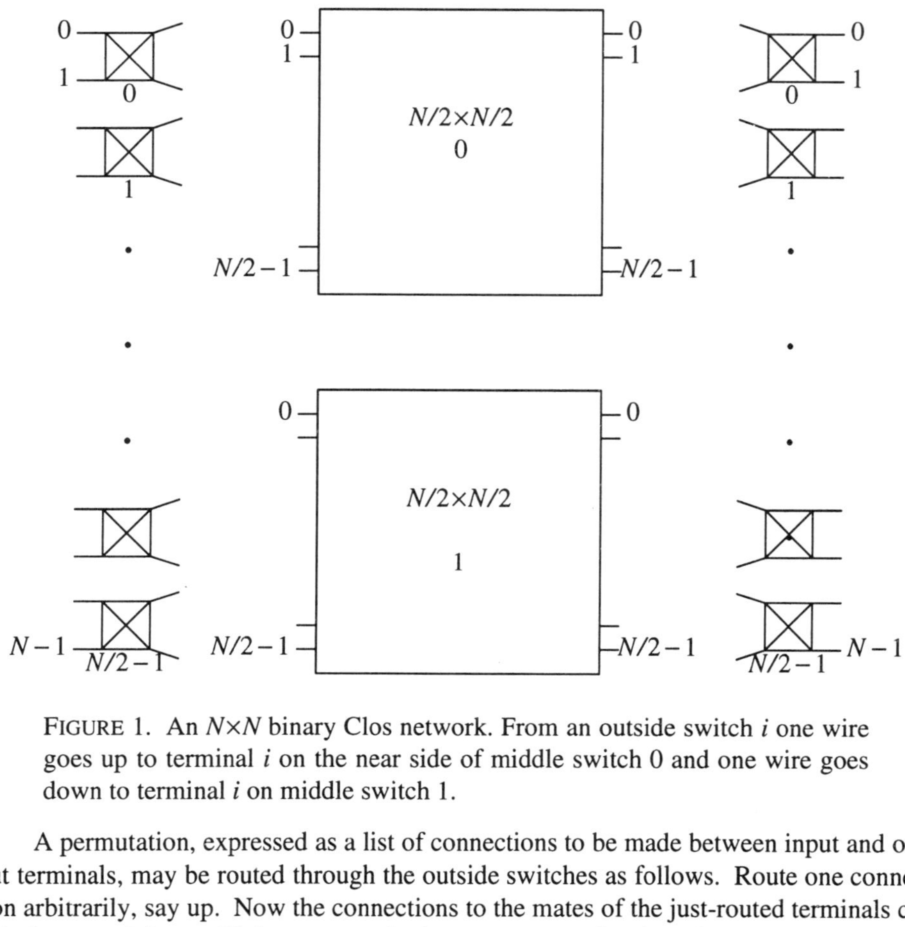

FIGURE 1. An $N \times N$ binary Clos network. From an outside switch i one wire goes up to terminal i on the near side of middle switch 0 and one wire goes down to terminal i on middle switch 1.

A permutation, expressed as a list of connections to be made between input and output terminals, may be routed through the outside switches as follows. Route one connection arbitrarily, say up. Now the connections to the mates of the just-routed terminals can only be routed down. If those mates don't connect to each other, the mates of the terminals they connect to must be routed up, and so on down and up in alternation until two terminals connect to each other. If any connections remain unrouted, route one arbitrarily and continue as before.

The routing of connections through the outside switches determines permutations to be realized by the middle switches. When the outside routing is complete, connections can be routed through the middle switches recursively until the 2×2 base case is reached. An arbitrary permutation has been realized, so the network is indeed rearrangeable.

The first arbitrary choice of input terminal and routing direction may be fixed once and for all, which means that one outside switch, say input switch 0, can be wired straight through. By similarly eliminating one 2×2 switch from each decomposable nested switch, one saves $2^0 + 2^1 + \cdots + 2^{k-2}$ size-2 switches. The modified design uses $N lg_2 N - N + 1$ switches, within $O(N)$ of the information-theoretic minimum, $lg N!$. It was described in different terms by A. E. Joel {**J**}.

1.2. Binary case, Benes network. Benes's design for an $N \times N$ binary network, where $N = 2^k$, has $2k-1$ columns of $N/2$ switches each. The wiring, symmetric about the middle column, connects adjacent columns in a perfect-shuffle pattern. Number the inputs and outputs of each column sequentially from top to bottom with k-bit binary numbers (Figure 2). Wire output i in column j, for $0 \le j < k$, to input $L(i)$ of column $j+1$, where $L(i)$ rotates i one bit to the left.

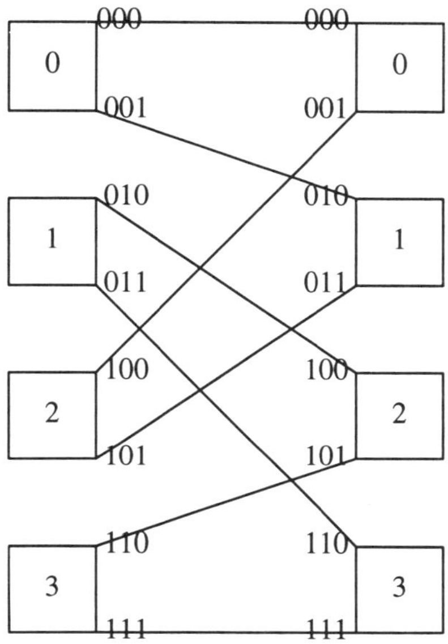

FIGURE 2. Two of the five columns of an 8×8 binary Benes network. For the full network, see Figure 4.

1.3. Binary case, isomorphism. A recursively expanded Clos network has the same number of rows and columns as the Benes network of the same size (Figure 4). The wiring between columns 0 and 1 of the expanded Clos network is the inverse of the wiring between two columns of the Benes network (Figure 3). Moreover, the two patterns are isomorphic up to renumbering of switches.

THEOREM 1. *The Benes construction of a binary network is isomorphic to the expanded Clos construction of a binary network of the same size.*

PROOF. We shall display the isomorphism explicitly. Let $s.t$ denote the catenation of bit strings s and t, $\bar{s}$ the reversal of s, and $|s|$ the number of bits in s.

Write terminal number i as $h.m.p$, where h is the high bit, m the middle $k-2$ bits, and p the low bit. $h.m$ designates a switch and p a port on that switch. In a Benes network, terminal $i = h.m.p$ in column j is wired to $L(i) = L(h.m.p) = m.p.h$ in column $j+1$. The map L rotates i one bit left.

In an expanded Clos network, terminal $i = h.m.p$ in column 0 is wired to terminal $p.h.m = R(i)$ in column 1, where $R(i)$ rotates i right one bit. In general, connections between columns j and $j+1$ are local to a nested switch of size $N/2^j$. Thus terminal i in column j is wired to terminal $R_j(i)$ in column $j+1$, where R_j rotates the low $k-j$ bits right 1 bit.

To map a Benes network into an expanded Clos network, apply a renumbering B_j to

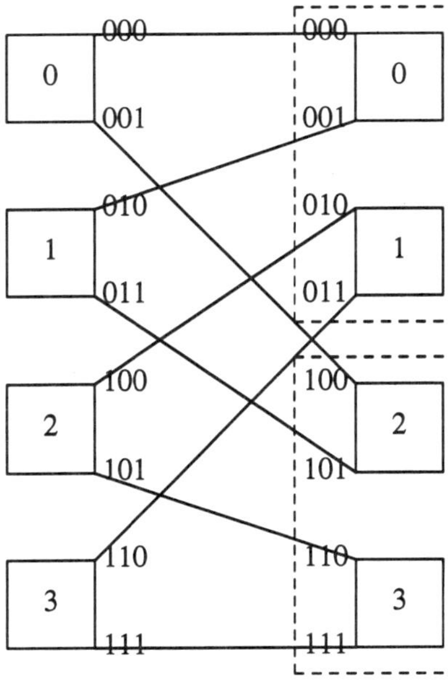

FIGURE 3. Columns 0 and 1 of a 5-stage 8×8 recursively expanded Clos network. Dashed lines outline the first stage of two contained size-4 networks, which are themselves 3-stage Clos networks.

column j, where B_j splits a k-bit number into three parts of sizes $k-1-j$, j, and 1 bits, reverses the first part, and exchanges the first part with the second:

$$B_j(a.b.c) = b.\bar{a}.c, \quad |a.b.c|=k,\ |b|=j,\ |c|=1,\ 0 \le j \le k-1$$

Since B_j leaves the last bit of a terminal number unchanged, it preserves matings. Isomorphism may be confirmed by checking that the operation on each arrow in the commutative diagram, Figure 5, maps the terminal number at one end of the arrow to the number at the other end. The numbers at the ends of horizontal arrows are the numbers of two terminals in one network that are joined by a wire. The numbers at the ends of vertical arrows are the numbers of corresponding terminals in the two networks. In the terminal number in the upper left corner p is the within-switch port number, h the high bit of the switch number, l_j the low j bits of the switch number, and m_j the remaining "middle" $k-j-2$ bits of the switch number.

In Figure 5, the action of arrow B_j follows from the identifications $a=h.m_j$, $b=l_j$, $c=p$; and the action of B_{j+1} follows from $a=m_j$, $b=l_j.p$, $c=h$.

Some incidental, but interesting, properties of the mappings B_j may be observed:

(i) B_0 and B_1 are the same. They reverse the bits of the switch number: $B_0(s.p) = B_1(s.p) = \bar{s}.p$.

(ii) The first and last switches in each column are fixed: $B_j(i) = i$ for $i \in \{0,1,N-2,N-1\}$ and all j.

(iii) The center column is fixed: $B_{k-1}(i) = i$ for all i.

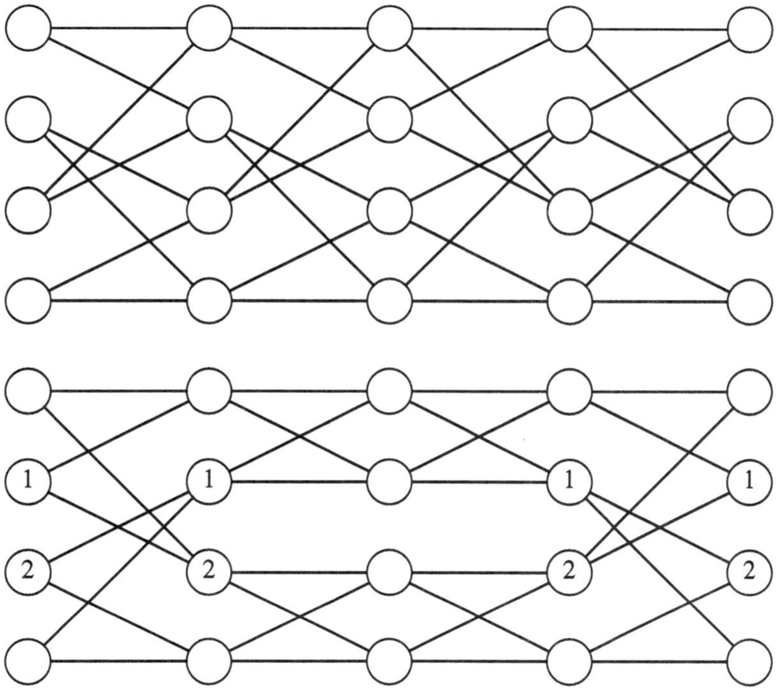

FIGURE 4. Wiring patterns for a $k=3$ (i.e. 8×8) binary Benes network above and a recursively expanded Clos network below. The patterns have left-right and top-bottom mirror symmetry. The patterns are isomorphic: exchanging positions of switches labeled 1 and 2 within their columns converts the lower diagram to the upper one.

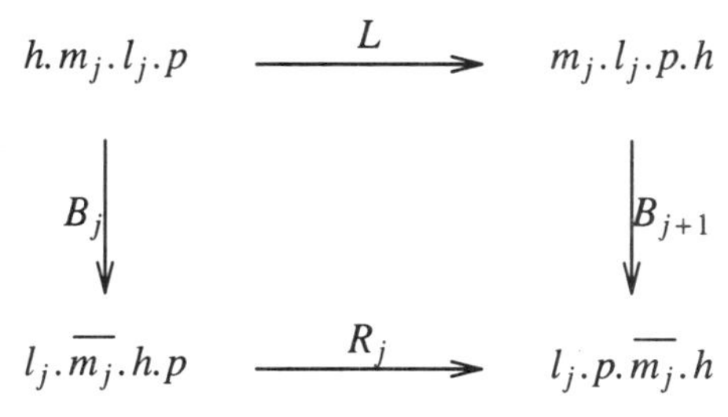

FIGURE 5. Correspondence between wiring patterns of a binary Benes network above and a recursively expanded Clos network below. Horizontal arrows are functions for wiring column j to column $j+1$, where $0\leq j<k-1$. Columns with $2k-2\geq j>k-1$ are wired symmetrically.

2. General case

A general size-N Benes network has $2k-1$ columns with mirror symmetry about the middle column. Column j, for $0\leq j<k$, consists of N/N_j size-N_j switches, as does column $2k-2-j$. The product of sizes is N:

$$\prod_{0 \le j < k} N_j = N.$$

The previous reasoning, carried through in base b instead of base 2, confirms that a Benes network of size b^k made of $b \times b$ switches is isomorphic to an expanded Clos network in which the size of middle switches shrinks by a factor of b at each nesting level. More generally, we have

THEOREM 2. *A $2k-1$-column Benes network with switches of size N_j in columns j and $2k-2-j$, $0 \le j < k$, is isomorphic to a Clos network expanded to k levels, with outside switches at the jth nesting level being size N_j.*

PROOF. We reason with mixed-radix number representations. To keep track of radixes, a number may be tagged with a list of its radixes. For example, a two-digit number with digit a_1 in base b_1 and digit a_2 in base b_2 is written $a_1.a_2[b_1,b_2]$. From the definition of radix representation, we have

$$a_1.a_2[b_1,b_2] = (a_1b_2+a_2)[b_1b_2].$$

Let $M_j = N/N_j$ be the number of switches in column j of the Benes network. Also let $N_{ab} = \prod_{a \le j < b} N_j$ and $M_{ab} = N/N_{a,b+1}$. These identities hold:

$$N = N_{0k} = N_{0j}N_{jk}, \quad 0 \le j \le k$$

As before, number switches in column j from 0 at the top to $M_j - 1$ at the bottom and number terminals from 0 to $N-1$.

2.1. Wiring pattern, Benes network. Let terminal i, also identified as port p of switch s in column j, where $0 \le j < k-1$, be wired to terminal i' (switch s', port p') in column $j+1$. For simplicity, we restrict attention to columns 0 and 1. Trivially,

$$i = s.p[M_0,N_0] \tag{1}$$

$$i' = s'.p'[M_1,N_1]. \tag{2}$$

Successive wires from column 0 are connected to successive switches cyclically in column 1. Thus terminal i connects to switch $s' = i \bmod M_1$. Ports on a column-1 switch are used in order, so port p' is used on the p'-th cycle (counting from 0) through the switches of column 1. Hence $p' = i \operatorname{div} M_1$, where div is the integer-quotient operator. Observe that

$$p'.s'[N_1,M_1] = M_1(i \operatorname{div} M_1) + (i \bmod M_1) = i. \tag{3}$$

Substituting $M_0 = N_1M_{01}$ and $M_1 = N_0M_{01}$ into (1) and (3) gives

$$i = p'.s'[N_1,N_0M_{01}] = s.p[N_1M_{01},N_0]$$

Expand $s'[M_{01}N_0]$ into $u.v[M_{01},N_0]$ and $s[N_1M_{01}]$ into $h.w[N_1,M_{01}]$:

$$p'.u.v[N_1,M_{01},N_0] = h.w.p[N_1,M_{01},N_0],$$

It follows that $p'=h$, $u=w$, and $v=p$. Substituting into (1) and (2), we determine a mapping $i \to i'$ that tells how wires join terminals in columns 0 and 1:

$$i = h.w.p[N_1,M_{01},N_0] \to w.p.h[M_{01},N_0,N_1] = i'$$

This wiring function rotates the (mixed radix) terminal number left, as did the wiring function for a binary Benes network. Since we have appealed to no peculiar properties of columns 0 and 1, we may replace 0 and 1 by j and $j+1$ to get the wiring function L between columns j and $j+1$.

$$L(h.w.p[N_{j+1}, M_{j,j+1}, N_j]) = w.p.h[M_{j,j+1}, N_j, N_{j+1}] \tag{4}$$

2.2. Wiring pattern, Clos network. In the corresponding 3-stage Clos network, column 1 consists of N_0 size-M_0 switches, so terminal $i = s.p[M_0, N_0]$ in column 0 is wired to terminal $i' = p.s[N_0, M_0] = p.s[N_0, M_{01}N_1]$ in column 1. Nested inside each column-1 switch are M_{01} size-N_1 switches. Rewriting $s[M_{01}N_1]$ as $u.v[M_{01}, N_1]$, we find that the wiring function from column 0 to column 1 is given by

$$i[N] = u.v.p[M_{01}, N_1, N_0] \to p.u.v[N_0, M_{01}, N_1] = i'[N]$$

The wiring function rotates the (mixed-radix) terminal number right, as did that for the expanded binary Clos network. At nesting level j, the wiring pattern is that of N_{0j} independent Clos networks of size N_{jk}. Split the terminal numbers i and i' into high- and low-order parts:

$$i[N] = i^+.i^-[N_{0j}, N_{jk}]$$
$$i'[N] = i'^+.i'^-[N_{0j}, N_{jk}].$$

The wiring function affects only the low-order part, so $i^+ = i'^+$. The effect on the low-order part is like wiring columns 0 and 1 of a Clos network recursively expanded to $k-j$ levels with switch sizes N_j through N_{k-1}. Hence the general wiring function, R_j, satisfies

$$R_j(i^+.u.v.p[N_{0j}, N_{j+2,k}, N_{j+1}, N_j]) = i^+.p.u.v[N_{0j}, N_j, N_{j+2,k}, N_{j+1}]. \tag{5}$$

2.3. General case, isomorphism. Collecting the wiring functions (4) and (5), we get a tentative diagram, which lacks appropriate functions on the vertical arrows.

$$\begin{array}{ccc}
h.w.p[N_{j+1}, M_{j,j+1}, N_j] & \xrightarrow{\quad L \quad} & w.p.h[M_{j,j+1}, N_j, N_{j+1}] \\
\Big\downarrow & & \Big\downarrow \\
i^+.u.v.p[N_{0j}, N_{j+2,k}, N_{j+1}, N_j] & \xrightarrow{\quad R_j \quad} & i^+.p.u.v[N_{0j}, N_j, N_{j+2,k}, N_{j+1}]
\end{array}$$

When every N_j is 2, the diagram should reduce to Figure 5. Following this lead, notice that $M_{j,j+1} = N_{0j}N_{j+2,k}$, and replace $w[M_{j,j+1}]$ by $m.l[N_{j+2,k}, N_{0j}]$ to make the top lines similar. In the bottom line substitute $v = h$, $i^+ = l$ and $u = \overline{m}$. The resulting diagram commutes, provided

$$B_j(a.b.c[N_{j+1,k}, N_{0j}, N_j]) = b.\overline{a}.c[N_{0j}, N_{j+1,k}, N_j],$$

where a number u expressed in a compound base N_{mn} is understood as shorthand for its multidigit representation:

$$u[N_{mn}] = u_m.u_{m+1}. \cdots .u_{n-1}[N_m, N_{m+1}, \cdots, N_{n-1}]$$
$$\bar{u}[N_{mn}] = u_{n-1}.u_{n-2}. \cdots .u_m[N_{n-1}, N_{n-2}, \cdots, N_m].$$

These identifications lead to a commuting diagram, Figure 6, the same as Figure 5 except for the mixed-radix representation. This completes the proof of Theorem 2.

$$\begin{array}{ccc}
h.m.l.p[N_{j+1}, N_{j+2,k}, N_{0j}, N_j] & \xrightarrow{L} & m.l.p.h[N_{j+2,k}, N_{0j}, N_j, N_{j+1}] \\
\Big\downarrow B_j & & \Big\downarrow B_{j+1} \\
l.\bar{m}.h.p[N_{0j}, N_{j+2,k}, N_{j+1}, N_j] & \xrightarrow{R_j} & l.p.\bar{m}.h[N_{0j}, N_j, N_{j+2,k}, N_{j+1}]
\end{array}$$

FIGURE 6a. Correspondence between wiring patterns of a general Benes network (upper arrow) and a recursively expanded Clos network (lower arrow).

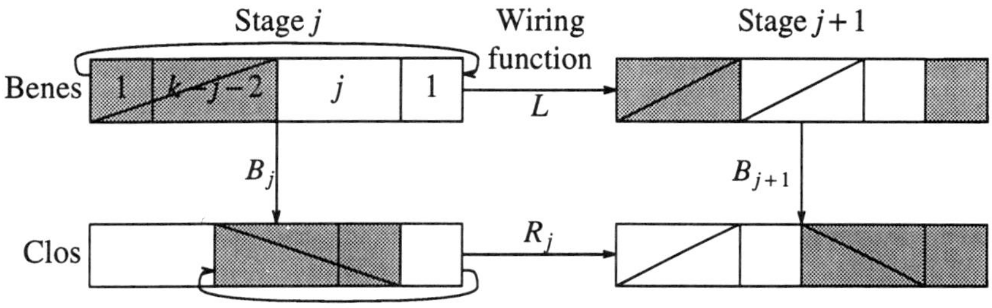

FIGURE 6b. Graphic representation of the correspondence. Numbers are shown as boxes representing strings of k mixed-radix digits. Boxes are partitioned to show substrings of the indicated lengths. Circling arrows show the digit rotations of the wiring functions L and R_j. Diagonals suggest digit reversals.

2.4. More isomorphic models. From the arguments above follows the further isomorphism of some other network models. The networks are tripartite – a middle column flanked by left and right "half-networks". In the two models already considered, the half networks are severally isomorphic and the middle columns are invariant under the mapping B_{k-1}. Thus a hybrid network with one Benes half and one expanded-Clos half is also isomorphic to the two original models.

Another isomorph is a Benes network with the two halves reversed. To see this recall that the wiring of columns 0 and 1 in a Benes network is isomorphic to that of columns 0 and 1 in an expanded Clos network, which is in turn the reversal of the wiring of columns 0 and 1 of a Benes network. Since all columns of a Benes half network are wired by the same rule, each pair of a pair of columns extends to a half network. Then isomorphism of the whole network follows from symmetry: the isomorphisms of the left and right halves extend consistently to the middle column.

The isomorphism of Benes and reverse Benes is proved without words in Figure 7a, which also portrays a fact familiar from FFT's: digit reversal inverts a shuffle.

The isomorphism of Clos, Benes, and reverse Benes networks implies that reverse Clos networks also belong to the same isomorphism class. Still more ways to draw the same network can be invented by ringing changes on Figure 6b, as for example Figure 7b.

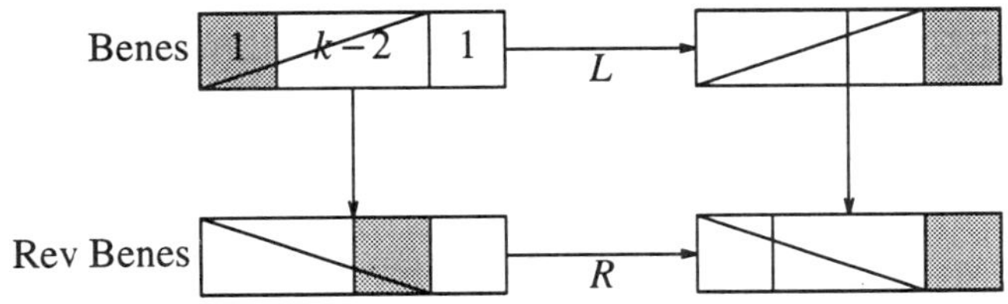

FIGURE 7a. Isomorphism between Benes and reverse Benes networks.

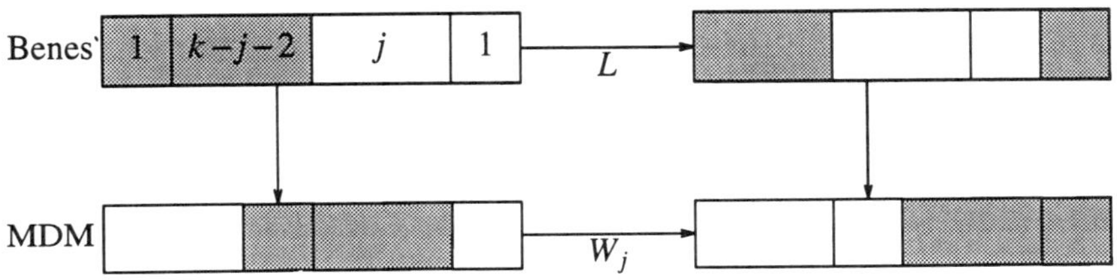

FIGURE 7b. Isomorphism of a Benes network and a "modified data manipulator" [**W**]. Transformation W_j swaps two digits.

Summary

We have given no strictly new results. The binary case has often been demonstrated in the literature and Andrzej Jajszczyk presented a proof of the general case (in Polish) some thirty years ago, using a different argument. Our treatment, however, is particularly simple. The proof is captured algebraically in a single commutative diagram, Figure 6a, and graphically in Figure 6b.

References

[A] D. P. Agrawal, *Graph theoretical analysis and design of multistage interconnection networks,* IEEE Trans. on Computers **C-32** (1983) 637-648.

[B] V. E. Benes, *Mathematical Theory of Connecting Networks and Telephone Traffic,* Academic Press, New York, 1965.

[C] C. Clos, *A study of non-blocking switching networks,* Bell System Tech. J. **32** (1953) 406-424.

[J] A. E. Joel Jr., *On permutation switching networks,* Bell System Tech. J. **47** (1968) 813-825.

[K] W. Keister, A. E. Ritchie and S. H. Washburn, *Design of Switching Circuits,* Van Nostrand, New York, 1951.

[W] C. Wu and T. Feng, *On a class of multistage interconnection networks,* IEEE Trans. on Computers **C-29** (1980) 694-702.

BELL LABORATORIES, MURRAY HILL, NJ 07974
Current address: Department of Computer Science, Dartmouth College, Hanover, NH 03755
E-mail address: doug@cs.dartmouth.edu

LUCENT TECHNOLOGIES, 67 WHIPPANY ROAD, WHIPPANY, NJ 07981
E-mail address: jsavicki@lucent.com

DIMACS Series in Discrete Mathematics
and Theoretical Computer Science
Volume **42**, 1998

Characterizing Bit Permutation Networks

Gerard J. Chang* Frank K. Hwang Li-Da Tong*
Department of Applied Mathematics
National Chiao Tung University
Hsinchu 30050, Taiwan

Abstract

In recent years, many multistage interconnection networks using 2×2 switching elements have been proposed for parallel architectures. Typical examples are baseline networks, banyan networks, shuffle-exchange networks and their inverses. As these networks are blocking, such networks with extra stages have also been studied extensively. These include Benes networks and $\Delta \oplus \Delta'$ networks. Recently, Hwang, Liaw and Yeh studied k-extra-stage networks which is a generalization of the above networks. They also investigated the equivalence issue among some of these networks. In this paper, we study a more general class of networks, which we call $(m + 1)$-stage d-nary bit permutation networks. We characterize the equivalence of such networks by sequence of positive integers.

Index terms. Multistage interconnection network, switching network, permutation routing, Sterling number, rearrangeably nonblocking.

1 Introduction

Consider a multistage interconnection network $\mathcal{V}$ with $N = d^{n+1}$ inputs and outputs, and has $m + 1$ stages of N/d crossbars of size $d \times d$.

*Supported in part by the National Science Council under grant NSC86-2115-M009-002.

1991 *Mathematics Subject Classification.* Primary 94C15.

Let the jth crossbar in a stage be labeled by j in the d-nary number (with n bits). A *bit-i group* consists of those crossbars whose labels are identical except the ith bit. Such a group will be labeled by a d-nary number x with n bits which is identical to any member in the group except bit i is replaced by the symbol x_0, which stands for the set $\{0,1,...,d-1\}$. $\mathcal{V}$ will be called a $(m+1)$*-stage d-nary bit permutation network* if the linking between stage k to stage $k+1$ is always from a bit-i_k group G to a bit-j_k group G', where G' is a permutation of G, for $k = 0,1,...,m-1$. For detail description and notation of bit permutation networks, see Section 2.

Note that $(n+1)$-stage binary bit permutation networks include all self-routing networks like Omega, banyan, baseline, and their inverse networks. Binary bit permutation networks have been hotly studied in the literature [1] [3] [6] [8] [11] for their topological equivalence. Bermond, Fourneau and Jean-Marie [3] characterized the Omega-equivalent class by the $P(i,j)$ property. An $(n+1)$-stage network satisfies the $P(i,j)$ property if the subnetwork from stage i to stage j has exactly 2^{n-j+i} components. Then an $(n+1)$-stage network with the unique path property is in the Omega-equivalent class if and only if it satisfies the $P(i,j)$ property for all $0 \le i \le j \le n$.

Another special class of bit permutation networks consists of $(n+1)$-stage networks with extra stages. A k-extra-stage network is a cascade of a $(n+1)$-stage network with k extra stages also satisfying the bit permutation linking pattern. Lea and Shyy [7] [9] proposed adding extra stages to a binary inverse banyan network while the k extra stages are added by pattern F^{-1} (see below). Hwang, Liaw, and Yeh [5] generalized the study of equivalence by adding extra stages to a binary Omega-equivalent network with the following patterns for extra stages:

(i) F: they are identical to the first k stages of the network,
(ii) F^{-1}: identical to the mirror image of the first k stages,
(iii) L: identical to the last k stages,
(iv) L^{-1}: identical to the mirror image of the last k stages.

In this paper we determine the equivalence classes among all $(m+1)$-stage d-nary bit permutation networks. We characterize such a network by an m-sequence over $\{1,2,\cdots,n\}$. Namely, every $(m+1)$-stage d-nary bit permutation network is reduced to an m-sequence over $\{1,2,\cdots,n\}$ and equivalence is determined by some easily computable sequence statistics. Note that the sequence is independent of d. For $m=n$, this characterization of course corresponds to the $P(i,j)$ characterization. But the sequence-graph correspondence is not in an obvious way. With the power and convenience of the sequence characterization, we easily give an explicit solution of the size of the s-stage bit permutation class. Recently, Hu, Shen and Yang [4] gave an $O(N^4 \log N)$-time algorithm to check the equivalence of combined $(2n-1)$-stage networks, which are obtained by cascading two Omega-equivalent networks. In this paper, we give an mn-time algorithm for checking the equivalence of two $(m+1)$-stage bit permutation networks. In particular, the running time of this algorithm is $O(\log^2 N)$ when the network has $2n-1$ stages.

2 Networks

We start the discussion of bit permutation networks by examining the following classical example.

A typical Omega-equivalent network consists of N input terminals, N output terminals, and $\log_2 N$ columns (stages) of 2×2 switching elements in which each column has $N/2$ switching elements. Figure 1 shows a 2×2 switching element; and Figure 2 shows a baseline network with $N = 16$, in which a terminal i is represented by its binary number representation (x_3, x_2, x_1, x_0) and is adjacent to a switching element named by (x_3, x_2, x_1).

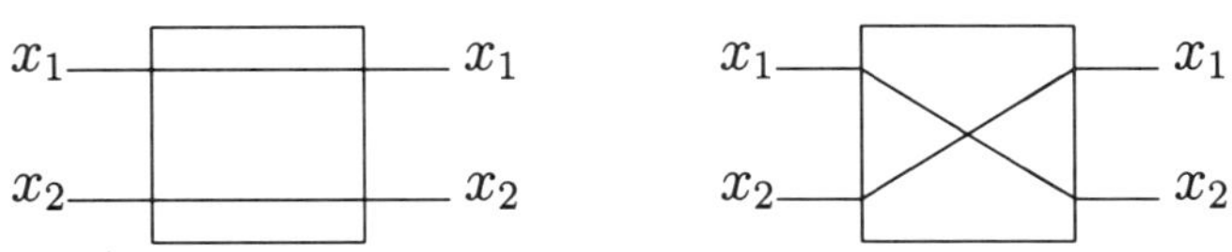

(a) Straight connection (b) Crossed connection

Figure 1. A 2×2 switching element.

One can view the baseline network in Figure 2 as a graph whose vertices are those 32 switching elements named by $(x_3, x_2, x_1)_i$, where $0 \le i \le 3$ and $x_1, x_2, x_3 \in \{0, 1\}$, and there are links

from $(x_3, x_2, x_1)_0$ to $(x_0, x_3, x_2)_1$,
from $(x_3, x_2, x_1)_1$ to $(x_3, x_0, x_2)_2$, and
from $(x_3, x_2, x_1)_2$ to $(x_3, x_2, x_0)_3$,

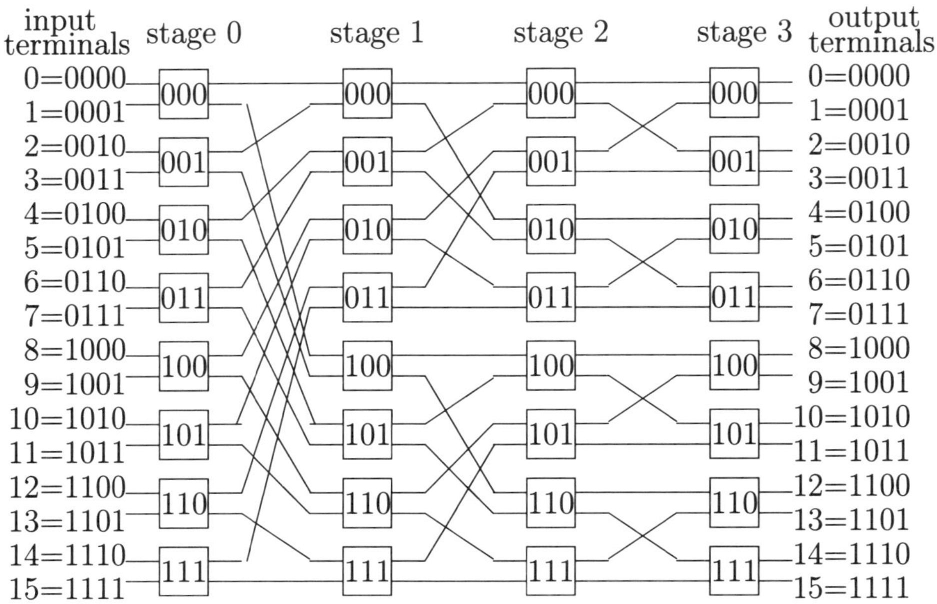

Figure 2. A baseline network with $N = 16$.

where $x_0 \in \{0, 1\}$, meaning there are two links, one with $x_0 = 0$ and the other with $x_0 = 1$. A link from a switching element x at stage i to a switching element y at stage $i+1$ exists if the bits of y can be obtained from the bits of x by a permutation depending only on i. For instance, we can represent the links from stage 0 to stage 1 by a permutation f_1 on $\{0, 1, 2, 3\}$ with

$$f_1(0) = 1,\ f_1(1) = 2,\ f_1(2) = 3,\ f_1(3) = 0.$$

In this way, the links from stage 0 to stage 1 are those from $(x_3, x_2, x_1)_0$ to $(x_{f_1(3)}, x_{f_1(2)}, x_{f_1(1)})_1$. We can also say that for a link from $x = (x_3, x_2, x_1)_0$ to $y = (x_0, x_3, x_1)_1$, a coordinate x_j at jth position of x moves to $f_1^{-1}(j)$th position of y; where the coordinate x_0 moves from "outside" into y means there are two such links. Similarly, the following permutations f_2 and f_3 represent links from stage 1 to stage 2 and stage 2 to stage 3, respectively:

$$f_2(0) = 1,\ f_2(1) = 2,\ f_2(2) = 0,\ f_2(3) = 3;$$

$$f_3(0) = 1,\ f_3(1) = 0,\ f_3(2) = 2,\ f_3(3) = 3.$$

Throughout this paper, we shall use the *cycle* notation for permutations. That is, the cycle $(i_1, i_2, \cdots, i_n)$ represents the permutation f with

$$f(i_1) = i_2, f(i_2) = i_3, \cdots, f(i_{n-1}) = i_n, f(i_n) = i_1,$$
$$\text{and } f(j) = j \text{ for all other } j.$$

Then f_1 can be represented by $(0,1,2,3)$, f_2 by $(0,1,2)$, and f_3 by $(0,1)$.

Not only the baseline network in Figure 2 can be represented by the permutations f_i, all bit permutation networks can be represented in this way. We can also use a general $d \times d$ switching elements instead of 2×2 switching elements. A general setting is as follows.

Definition. Suppose n is a positive integer and $f_1, f_2, ..., f_m$ are $m \geq 0$ permutations on $\{0, 1, ..., n\}$ such that $f_i(0) \neq 0$ for $1 \leq i \leq m$. The $(m+1)$*-stage d-nary bit permutation network* $N_d(n, f_1, f_2, ..., f_m)$ is the network whose vertices are those $(x_n, x_{n-1}, ..., x_1)_i$ with $0 \leq i \leq m$ and $x_j \in \{0, 1, ..., d-1\}$ for $1 \leq j \leq n$; and each $(x_n, x_{n-1}, ..., x_1)_{i-1}$ is adjacent to $(x_{f_i(n)}, x_{f_i(n-1)}, ..., x_{f_i(1)})_i$ where $x_0 \in \{0, 1, ..., d-1\}$. In other words, there is an edge from $(x_n, x_{n-1}, ..., x_1)_{i-1}$ to $(y_n, y_{n-1}, ..., y_1)_i$ whenever $y_j = x_{f_i(j)}$ for $1 \leq j \leq n$.

The following example shows a $(m+1)$-stage binary bit permutation network with $N = 2^{n+1}$ input and output terminals.

Example 1. Interchange the role of the coordinates x_1 and x_2 of (x_3, x_2, x_1) in stages 2 and 3 of the baseline network in Figure 2. The resulting network, which is topologically equivalent to BL, is $N_2(3, f_1, f_2, f_3)$ with

$$f_1 = (0,1,2,3), \ f_2 = (0,2,1) \text{ and } f_3 = (0,2),$$

see Figure 3. Note that if one interchanges the switching elements $(x_3, 0, 1)_i$ with $(x_3, 1, 0)_i$ for $i = 2, 3$, the resulting network is the baseline network.

3 Topological equivalence

The main effort of this paper is to establish methods for determining if two bit permutation networks $N_d(n, f_1, f_2, ..., f_m)$ and $N_d(n, f'_1, f'_2, ..., f'_m)$ are topologically equivalent or graphically isomorphic . The following theorem is the foundation of our theory.

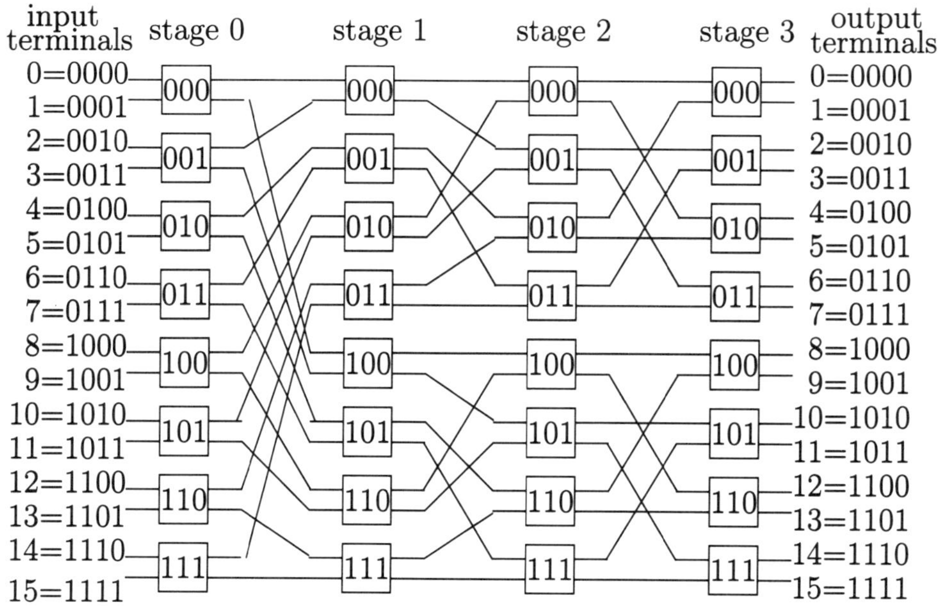

Figure 3. The network in Example 1.

Theorem 1 *If there exist permutations* $g_0, g_1, ..., g_m$ *on* $\{0, 1, ..., n\}$ *such that* $g_i(0) = 0$ *for* $0 \leq i \leq m$ *and* $f'_i = g_{i-1}^{-1} \circ f_i \circ g_i$ *for* $1 \leq i \leq m$, *then* $N_d(n, f_1, f_2, ..., f_m)$ *is isomorphic to* $N_d(n, f'_1, f'_2, ..., f'_m)$.

Proof. Consider the bijection g from the vertex set of $N_d(n, f_1, f_2, ..., f_m)$ to the vertex set of $N_d(n, f'_1, f'_2, ..., f'_m)$ defined by

$$(x_n, x_{n-1}, ..., x_1)_i \xrightarrow{g} (x_{g_i(n)}, x_{g_i(n-1)}, ..., x_{g_i(1)})_i \text{ for } 0 \leq i \leq m.$$

In other words, $g((x_n, x_{n-1}, ..., x_1)_i) = (x'_n, x'_{n-1}, ..., x'_1)_i$ whenever $x'_j = x_{g_i(j)}$ for $1 \leq j \leq n$.

In order to see these two networks are isomorphic, we only need to check that g is edge preserving. Suppose e is an edge from $(x_n, x_{n-1}, ..., x_1)_{i-1}$ to $(y_n, y_{n-1}, ..., y_1)_i$ in $N_d(n, f_1, f_2, ..., f_m)$, i.e., $y_j = x_{f_i(j)}$ for $1 \leq j \leq n$. Let

$$g((x_n, x_{n-1}, ..., x_1)_{i-1}) = (x'_n, x'_{n-1}, ..., x'_1)_{i-1}, \text{ i.e., } x'_j = x_{g_{i-1}(j)}$$

$$\text{and} \quad g((y_n, y_{n-1}, ..., y_1)_i) = (y'_n, y'_{n-1}, ..., y'_1)_i, \text{ i.e., } y'_j = y_{g_i(j)}$$

for $1 \le j \le n$. Then,

$$y'_j = y_{g_i(j)} = x_{f_i \circ g_i(j)} = x_{g_{i-1} \circ f'_i(j)} = x'_{f'_i(j)} \text{ for } 1 \le j \le n.$$

Thus, there exists an edge from $(x'_n, x'_{n-1}, ..., x'_1)_{i-1}$ to $(y'_n, y'_{n-1}, ..., y'_1)_i$. Conversely, an edge in $N_d(n, f'_1, f'_2, ..., f'_m)$ also corresponds to an edge in $N_d(n, f_1, f_2, ..., f_m)$.

Q.E.D.

As a quick application of Theorem 1, consider BL in Figure 2 as $N_2(3, f_1, f_2, f_3)$ with $f_1 = (0,1,2,3)$, $f_2 = (0,1,2)$, $f_3 = (0,1)$; and the network in Example 1 as $N_2(3, f'_1, f'_2, f'_3)$ with $f'_1 = (0,1,2,3)$, $f'_2 = (0,2,1)$, $f'_3 = (0,2)$. Then, $g_0 = (2,3)$ and $g_1 = g_2 = g_3 = (1,2)$ show that these two networks are isomorphic.

Theorem 2 *Every bit permutation network $N_d(n, f_1, f_2, ..., f_m)$ is isomorphic to some bit permutation network $N_d(n, f'_1, f'_2, ..., f'_m)$, where $f'_i = (0, k_i)$ with $k_i \in \{1, 2, ..., n\}$ for $1 \le i \le m$.*

Proof. We shall prove by induction on j the following claim which implies the theorem.

Claim(j): $N_d(n, f_1, f_2, ..., f_m)$ is isomorphic to some $N_d(n, f'_1, f'_2, ..., f'_m)$, where $f'_i = (0, k_i)$ with $k_i \in \{1, 2, ..., n\}$ for $1 \le i \le j$.

Claim(0) is clearly true. Suppose Claim($j-1$) holds. Consider a general $j \ge 1$. Let $k_j = f'_j(0)$. Then $k_j \in \{1, 2, ..., n\}$. Let $g_j = (f'_j)^{-1} \circ (0, k_j)$ and all other g_i are identity permutations. It is easy to check that $g_i(0) = 0$ for $0 \le i \le m$. According to Theorem 1, $N_d(n, f'_1, f'_2, ..., f'_m)$, and therefore $N_d(n, f_1, f_2, ..., f_m)$, is isomorphic to $N_d(n, f''_1, f''_2, ..., f''_m)$, where $f''_i = g_{i-1}^{-1} \circ f'_i \circ g_i$ for $1 \le i \le m$. In particular, $f''_i = f'_i = (0, k_i)$ for $1 \le i \le j-1$, and $f''_j = f'_j \circ g_j = f'_j \circ (f'_j)^{-1} \circ (0, k_j) = (0, k_j)$. This gives Claim(j).

Q.E.D.

So, basically we only need to consider the networks with $f_i = (0, k_i)$. For convenience, we shall use $N_d(n, k_1, k_2, ..., k_m)$ as a short notation for the network $N_d(n, f_1, f_2, ..., f_m)$ with $f_i = (0, k_i)$ and $k_i \in \{1, 2, ..., n\}$ for $1 \le i \le m$.

Our next step is to determine when are two networks isomorphic.

Theorem 3 *If g is a permutation on $\{0, 1, 2, ..., m\}$ with $g(0) = 0$, then the network $N_d(n, k_1, k_2, ..., k_m)$ is isomorphic to $N_d(n, g(k_1), g(k_2), ..., g(k_m))$.*

Proof. The theorem following from Theorem 1 and the fact that $g \circ (0, k_i) \circ g^{-1} = (0, g(k_i))$. Note that we apply Theorem 1 by choosing all g_i as g. Q.E.D.

For any sequence $(k_1, k_2, ..., k_m)$ over $\{1, 2, ..., n\}$ with $a = |\{k_1, k_2, ..., k_m\}|$, let $1 = i_1 < i_2 < ... < i_a$ are those indices such that $k_{i_r} \notin \{k_1, k_2, ..., k_{i_r-1}\}$ for $1 \le r \le a$. Choose any permutation g on $\{1, 2, ..., n\}$ with the property that $g(k_{i_r}) = r$ for $1 \le r \le a$. Then $(k_1^*, k_2^*, ..., k_m^*) = (g(k_1), g(k_2), ..., g(k_m))$ is a sequence with the property that $\{k_1^*, k_2^*, ...k_m^*\} = \{1, 2, ..., a\}$ and for any $1 \le r \le a$, $\{k_1^*, k_2^*, ..., k_{i_r}^*\} = \{1, 2, ..., r\}$ when i_r is the minimum index with $k_{i_r}^* = r$. Such a sequence is called a *canonical sequence* over $\{1, 2, ..., n\}$.

Corollary 4 *Any $N_d(n, f_1, f_2, ..., f_m)$ is isomorphic to $N_d(n, k_1, k_2, ..., k_m)$ for some canonical sequence over $\{1, 2, ..., n\}$.*

Theorem 5 *If $a = |\{k_1, k_2, ..., k_m\}|$, then $N_d(n, k_1, k_2, ..., k_m)$ has d^{n-a} connected components.*

Proof. First we claim that for any two vertices $x = (x_n, x_{n-1}, ..., x_1)_i$ and $y = (y_n, y_{n-1}, ..., y_1)_j$ such that $x_k = y_k$ for all $k \notin \{k_1, k_2, ..., k_m\}$, there exists a path joining them.

For the case in which $i = 0$ and $j = m$, $x = x^{(0)} \to x^{(1)} \to \cdots \to x^{(m)} = y$ is a desired path, where $x^{(i)}$ is a vertex in stage i that is obtained from $x^{(i-1)}$ by replacing its k_i-th coordinate with y_{k_i}. For the general case, consider the vertices $x^* = (x_n, x_{n-1}, ..., x_1)_0$ and $y^* = (y_n, y_{n-1}, ..., y_1)_m$. By the above construction, there exists an x^*-y^* path. It is clear that $(x_n, x_{n-1}, ..., x_1)_i \to (x_n, x_{n-1}, ..., x_1)_{i-1} \to \cdots \to (x_n, x_{n-1}, ..., x_1)_0 = x^*$ is an x-x^* path. Similarly, there exists a y^*-y path. Thus x and y are joined by a path.

On the other hand, a move from any vertex to its neighbor never changes its k-th coordinate for $k \notin \{k_1, k_2, ..., k_m\}$. Thus two vertices with different values at the k-th coordinate for some $k \notin \{k_1, k_2, ...k_m\}$ are not in a same connected component of the network. Therefore, the network has d^{n-a} connected components. Q.E.D.

Theorem 6 *For any two canonical sequences $(k_1, k_2, ..., k_m)$ and $(k'_1, k'_2, ..., k'_m)$ over $\{1, 2, ..., n\}$, the network $N_d(n, k_1, k_2, ..., k_m)$ is isomorphic to $N_d(n, k'_1, k'_2, ..., k'_m)$ if and only if $(k_1, k_2, ..., k_m) = (k'_1, k'_2, ..., k'_m)$.*

Proof. The two networks are clearly isomorphic if the two sequences are equal.

Conversely, suppose $(k_1, k_2, ..., k_m) \neq (k'_1, k'_2, ..., k'_m)$. Let i be the minimum index such that $k_i \neq k'_i$, say $k_i < k'_i$ and $k_r = k'_r$ for $1 \leq r \leq i-1$. Since the two sequences are canonical, $i \geq 2$. By the definition, there exists some $1 \leq r \leq i-1$ such that $k_r = k'_r \in \{k_i, k'_i\}$. Choose a maximum such r. Then $|\{k_r, k_{r+1}, ..., k_i\}| = a \neq a' = \{k'_r, k'_{r+1}, ..., k'_i\}$. Note that $N_d(n, k_r, k_{r+1}, ..., k_i)$ has d^{n-a} connected components while $N_d(n, k'_r, k'_{r+1}, ..., k'_i)$ has $d^{n-a'}$ connected components. Hence the two networks are not isomorphic. Q.E.D.

Remark. By Theorem 6, the $N_d(n, k_1, k_2, ..., k_m)$ in Corollary 4 is unique. Such a network is called the *canonical representation* of $N_d(n, f_1, f_2, ..., f_m)$.

A condition slightly weaker than the $P(i, j)$ characterization follows from Theorem 6.

Corollary 7 *An $(n+1)$-stage network with the unique path property and satisfying the $P(i, i+1)$ property for $i = 0, 1, ..., n$, is in the Omega-equivalent class if and only if it also satisfying the $P(0, j)$ property for $j = 1, 2, ..., n$.*

Proof. By satisfying the $P(i, i+1)$ property, the network is a bit permutation network. By satisfying the $P(0, j)$ property, the number of components is increasing in j, which implies $k_j \neq k_i$ for all $i < j$. It follows $(k_1, k_2, ..., k_n)$ are all distinct; but there is only one such canonical sequence. Corollary 7 follows from Theorem 6 immediately. Q.E.D.

Theorem 8 *The number of equivalent classes among $(m+1)$-stage d-nary bit permutation networks is $\sum_{t=1}^{n} \sum_{i=0}^{t} \frac{1}{t!}(-1)^{t-i}\binom{t}{i} i^m$.*

Proof. By Corollary 4 and Theorem 6, we only need to count the set $C(m, t)$ of canonical sequences of length m with t distinct elements for $1 \leq t \leq n$. Denote by $P(m, t)$ the set of all partitions of $\{1, 2, ..., m\}$ into t nonempty subsets. Define a mapping h from $C(m, t)$ to $P(m, t)$ by $h(k_1, k_2, ..., k_m) = \{C_1, C_2, ..., C_t\}$, where $C_i = \{j : k_j = i\}$. It is easy to see that h is one-to-one. On the other hand, for any partition $\{C_1, C_2, ..., C_t\}$ in $P(m, t)$, we may assume that $\min C_1 \leq \min C_2 \leq$

$\cdots \leq \min C_t$. Let $k = (k_1, k_2, ..., k_m)$, where $k_i = j$ for $i \in C_j$. Then $h(k) = \{C_1, C_2, ..., C_t\}$. So h is onto. Therefore $|C(m,t)| = |P(m,t)|$, which by definition is the Sterling number of second kind $S(m,t)$, see [10]. It is well-known that

$$S(m,t) = \frac{1}{t!}\sum_{i=0}^{t}(-1)^{t-i}\tbinom{t}{i}i^m.$$

Summing over t, we obtain Theorem 8. Q.E.D.

By the theorems in this Section, a bit permutation network $N_d(n, f_1, f_2, ..., f_m)$ is topologically equivalent to its canonical representation $N_d(n, k_1, k_2, ..., k_m)$, whose topology is determined by the canonical sequence $(k_1, k_2, ..., k_m)$. We summarize an efficient algorithm from the proofs of Theorems 2 and 3.

Algorithm. Find the canonical representation of a bit permutation network.
Input. A bit permutation network $N_d(n, f_1, f_2, ..., f_m)$.
Output. The canonical sequence $(k_1^*, k_2^*, ..., k_m^*)$ for $N_d(n, f_1, f_2, ..., f_m)$.
Method.

```
f̄_0 = (0); f̄_{m+1} = (0);
for j = 1 to m do f̄_j = f_j;
for j = 1 to m do
    k_j = f̄_j(0);
    f̄_{j+1} = (0, k_j) ∘ f̄_j ∘ f̄_{j+1};
    end for;
for j = 1 to n do mark[j] = 0;
a = 0;
for j = 1 to m do
    if mark[k_j] = 0 then
        a = a + 1;
        mark[k_j] = a;
        k*_j =mark[k_j];
        end then;
    end for;
```

The time complexity of this algorithm is $O(mn)$.

References

[1] D. P. Agrawal, Graph theoretical analysis and design of multistage interconnection networks, *IEEE Trans. Computers*, vol. C-32, no. 7, pp. 637-648, July 1983.

[2] V. E. Benes, On rearrangeable three-stage connecting networks, *Bell System Technical J.*, vol. XLI, no. 5, pp. 1481-1492, Sept. 1962.

[3] J.-C. Bermond, J. M. Fourneau, and A. Jean-Marie, Equivalence of multistage interconnection networks, *Inform. Proc. Lett.* vol. 26, no. 1, pp. 45-50, 1987.

[4] Q. Hu, X. Shen, and J. Yang, Topologies of combined $(2\log N-1)$-stage interconnection networks, *IEEE Trans. Computers*, vol. C-46, no. 1, pp. 118-124, Jan. 1997.

[5] F. K. Hwang, S. C. Liaw, and H. G. Yeh, Equivalence classes for extra-stage networks, manuscript, 1997.

[6] C. P. Kruskal and M. Snir, A unified theory of interconnection network structure, Theo. Comput. Sci. 48 (1986) 75-94.

[7] C.-T. Lea and D.-J. Shyy, Tradeoff of horizontal decomposition versus vertical stacking in rearrangeable nonblocking networks, *IEEE Trans. Commun.*, COM-37, no. 6, pp. 879-904, June 1991.

[8] D. S. Parker, Notes on shuffle exchange-type switching networks, *IEEE Trans Computers*, vol. C-29, no. 3, pp. 213-222, March 1980.

[9] D.-J. Shyy and C.-T. Lea, $\text{Log}_2(N, m, p)$ strictly nonblocking networks, *IEEE Trans. Commun.*, COM-39, no. 10, pp. 1502-1510, Oct. 1991.

[10] R. P. Stanley, *Enumerative Combinatorics, Volume I,* Wadsworth, Inc., Belmont, California, 1986.

[11] C. -L. Wu and T. Y. Feng, On a class of multistage interconnection networks, *IEEE Trans. Computers*, vol. C-29, no. 8, pp. 694-702, Aug. 1980.

DIMACS Series in Discrete Mathematics
and Theoretical Computer Science
Volume **42**, 1998

Structure and Density of Sparse Crossbar Concentrators

Emre Gündüzhan and A. Yavuz Oruç

ABSTRACT. A sparse crossbar (n, m)-concentrator is a bipartite graph with n source and m sink vertices, $m \leq n$, in which there exists a matching between every m source vertices and the m sink vertices. In this paper, we investigate the structure, and the density of sparse crossbar (n, m)-concentrators among all 2^{nm} bipartite graphs. We establish that the density of sparse crossbar concentrators is bounded from below by 0.2887 when $m = n$, from above by $1/e$ when $m = n/2$, and it tends to 0 when $m = 1$, as $n \to \infty$. We also derive upper and lower bounds on the density of sparse crossbar (n, m)-concentrators for an arbitrary $m \leq n$. The lower bounds provide an insight into the structure of sparse crossbar concentrators, while the upper bounds give a partial characterization of bipartite graphs which fail to have a concentrator property.

1. Introduction

Consider a device with n distinguished terminals, called *inputs*, and m distinguished terminals, called *outputs*, where each input is connected directly to some of the outputs by switches, called *crosspoints*. Such a device is called a sparse crossbar (n, m, c)-concentrator if there exists a set of crosspoints by which every $c \leq m$ inputs can be connected to some c outputs. The parameter c is called the *capacity* of the concentrator. When $c = m$, the concentrator is said to have a full capacity, and we drop c from the notation.

Sparse crossbar concentrators have been investigated in the literature in connection with efficient constructions of more powerful networks [**Pip77, Bas81, JO93, MO94**]. They have also been studied independently for their crosspoint complexity [**Mas77, NM82**] and structural properties [**OH96, GO96**]. It has been established that an (n, m, c)-concentrator requires at least $(n-c+1)m/(m-c+1)$ crosspoints, and this bound is tight when $c = m$ [**OH96**]. More specifically, it was shown that a sparse crossbar (n, m)-concentrator can be constructed with $(n - m + 1)m$ crosspoints using a bipartite graph, called a *fat-and-slim crossbar* for any integers n and $m \leq n$. More recently, it was shown that sparse crossbar (n, m)-concentrators can be constructed with a minimum crosspoint complexity, and using nearly the

1991 *Mathematics Subject Classification.* Primary 94C15; Secondary 05A05.

This work is supported in part by the National Science Foundation under grant No. NCR-9405539.

same number of crosspoints per each input for arbitrary values of n and $m \leq n$ [**GO96**].

In this paper, we consider another interesting problem about sparse crossbar concentrators. The problem is to determine how little or abundantly the sparse crossbar full capacity concentrators are contained among all 2^{nm} sparse crossbars with n inputs and m outputs. This is a question of density, and as such, it is a measure of resilience of sparse crossbar concentrators among all bipartite graphs. In particular, the density of sparse crossbar (n, m)-concentrators among all 2^{nm} sparse crossbars represents the probability that a randomly constructed sparse crossbar is a concentrator. Aside from this motivation, the density problem leads to a more complete understanding of the structure of sparse crossbar concentrators as we will see. In particular, the lower bounds on the density of sparse crossbar concentrators shed light on how such graphs are structured, while the upper bounds expose the reasons why many bipartite graphs fail to be a concentrator.

The rest of the paper is organized as follows. In the next section, we present basic mathematical facts that will be needed in establishing our results. In Section 3, we summarize some of the recently constructed sparse crossbar concentrators. In Section 4, we formalize the density concept, and examine the density of sparse crossbar concentrators for a few small values of n and m. In Section 5, we present upper and lower bounds on the density of sparse crossbar concentrators. The paper is concluded in Section 6 with the analysis and comparisons of the lower and upper bounds, and suggestions for future research.

2. Definitions and Preliminary Facts

We begin with some definitions.

DEFINITION 1. A sparse crossbar (n, m)-concentrator is a bipartite graph with n inputs and m outputs such that there exists a matching between every m of the inputs and the m outputs.

DEFINITION 2. The incidence matrix of an (n, m)-concentrator is an $m \times n$ binary matrix $B = [b_{i,j}]_{m \times n}$ such that $b_{i,j} = 1$ if and only if there exists an edge between input j and output i.

A key result which is central to the characterization of sparse concentrators is the bipartite graph version of Hall's theorem:

THEOREM 1. *Let G be a bipartite graph with n inputs and $m \leq n$ outputs. Every subset of m inputs has a matching in G iff every j inputs are connected to at least j outputs, $1 \leq j \leq m$.*

The next two results follow from Hall's theorem.

THEOREM 2. *A bipartite graph with n inputs and m outputs is a sparse concentrator if and only if its incidence matrix does not contain an $(m - k + 1) \times k$ zero submatrix, $k = 1, 2, \ldots, m$.*

PROOF. Let G be a bipartite graph, and B_G be its incidence matrix. If B_G contains an $(m - k + 1) \times k$ zero submatrix then the k inputs associated with this zero submatrix could at most be connected to $m - (m - k + 1) = k - 1$ outputs, and any m inputs that include those k inputs cannot be matched with the m outputs. On the other hand, suppose that B_G does not contain any $(m - k + 1) \times k$ zero

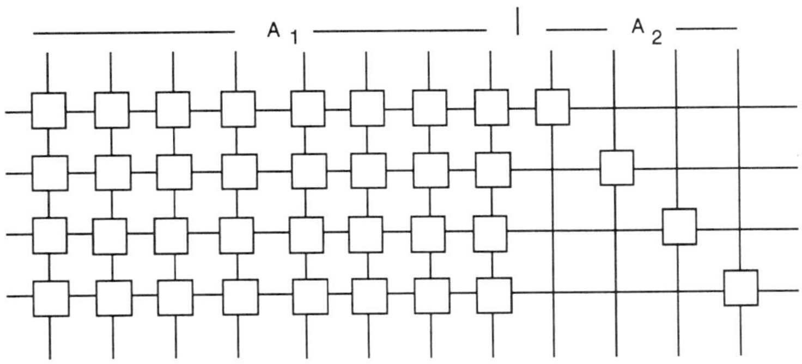

FIGURE 1. A (12, 4)-fat-and-slim crossbar.

submatrices. This implies that every k inputs in G are connected to at least k outputs. Therefore, by Hall's theorem there must exist a matching between every m-inputs and the m outputs. □

THEOREM 3. *In a sparse crossbar (n, m)-concentrator, every k of the m outputs must be connected*[1] *to at least $n - m + k$ inputs.*

PROOF. If some k outputs are connected only to $n-m+k-1$ or fewer inputs, then some $m - k + 1$ or more inputs will be connected to only $m - k$ outputs, contradicting that the graph in question is a concentrator. □

3. Fat-and-slim and Banded Sparse Crossbars

One way to determine whether or not sparse crossbar concentrators occur abundantly among the set of all bipartite graphs is to construct families of sparse concentrators, and then compute their cardinalities. Two relatively large families of sparse crossbar concentrators have recently been introduced in the literature [**OH96, GO96**].

DEFINITION 3. Let $\mathcal{G} = (I, O)$ be a bipartite graph with n inputs and m outputs. Suppose that I is partitioned into two sets A_1 and A_2, where $|A_1| = n-m$ and $|A_2| = m$. $\mathcal{G}$ is called an (n, m)-fat-and-slim crossbar if each of the $n-m$ inputs in A_1 is connected to all the m outputs (fat section), and if each of the m inputs in A_2 is connected to a single but distinct output (slim section).

A (12, 4)-fat-and-slim crossbar is depicted in Figure 1.

THEOREM 4 ([**OH96**]). *Every (n, m)-fat-and-slim crossbar is an (n, m)- concentrator, and it uses the minimum number of crosspoints possible, i.e. $(n-m+1)m$.*

COROLLARY 1. *There exist at least*

$$n \times (n-1) \times (n-2) \times \cdots \times (n-m+1) \tag{1}$$

distinct sparse crossbar (n, m)-concentrators.

[1]We say that k outputs are connected to r inputs if each of the r inputs is connected to at least one of the k outputs.

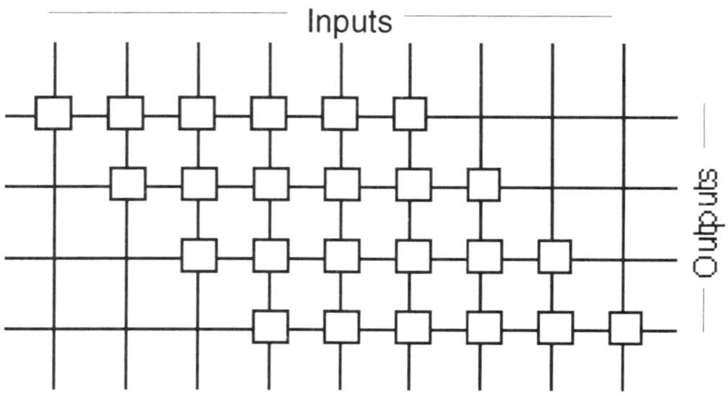

FIGURE 2. A banded sparse crossbar concentrator.

PROOF. The $(n-m)$-input fat section can be fixed in $\binom{n}{m}$ ways, and for each such fat section, the m crosspoints can be distributed into the m rows in $m!$ ways.□

By Stirling's approximation, the cardinality, $|FS(n,m)|$, of the family of fat-and-slim crossbar concentrators is

$$|FS(n,m)| \approx 2^{m \log_2 \frac{n-m}{e} + (n+0.5)\log_2 \frac{n}{n-m}}. \quad (2)$$

Dividing this by 2^{nm} shows that, while fat-and-slim crossbars form a large family of sparse crossbar (n,m)-concentrators, their density tends to 0 as $n \to \infty$.

Another family of sparse crossbar concentrators, called *banded sparse crossbars* can be obtained by placing blocks of $n-m+1$ "1"s into the m rows of an $m \times n$ incidence matrix.

DEFINITION 4. An (n,m)-sparse crossbar G is called (column) banded if its incidence matrix $B_G = [b_{i,j}]$ is given by

$$b_{i,j} = \begin{cases} 1, & \text{if } i \le j \le i+n-m \\ 0, & \text{if } j < i \text{ or } j > i+n-m \end{cases}$$

for $i = 1, 2, \ldots, m$.

A $(9,4)$-banded crossbar is depicted in Figure 2.

THEOREM 5. *Every banded (n,m)-sparse crossbar is a concentrator.*

The theorem can be proved using the following lemma [**GO96**] (which will also be needed later in Section 5.2).

LEMMA 1 (Guo-Oruç). *Let X and Y be any two columns of an $n \times m$ binary matrix A. Suppose X covers[2] Y, and let $a_{i_1,x}, \ldots, a_{i_r,x}$ be any r rows of "1"s in X. Let B be a matrix obtained from A by exchanging $a_{i_l,x}$ with $a_{i_l,y}$, $1 \le l \le r$. If A does not have an $(m-k+1) \times k$ zero submatrix, for any $k, 1 \le k \le m$, then neither does B.*

DEFINITION 5. Two sparse crossbar concentrators G_1 and G_2 are said to be topologically equivalent if G_1 can be obtained from G_2 by permuting the inputs and/or outputs of G_2, and conversely.

[2]We say that column X covers column Y if whenever a row in Y contains 1 so does X.

DEFINITION 6. Any sparse crossbar concentrator which is topologically equivalent to a banded sparse crossbar is also called banded.

COROLLARY 2. *For $2 \leq m \leq n/2$, there exist at least*

$$\binom{n}{n-2m+2}(2m-2)! = n \times (n-1) \times (n-2) \cdots \times (n-2m+3) \tag{3}$$

distinct banded sparse crossbar concentrators.

PROOF. The $n-2m+2$ full columns (i.e., those containing only "1" entries) of a banded sparse crossbar concentrator can be fixed in $\binom{n}{n-2m+2}$ ways, and any permutation of the remaining $2m-2$ columns gives a distinct banded sparse crossbar concentrator. □

Comparing (3) with (1) shows that the set of banded sparse crossbars is larger than the set of fat-and-slim crossbars when $n-2m+3 < n-m+1$, or $m > 2$. However, it can be shown that the density of banded sparse crossbar concentrators also tends to 0 as $n \to \infty$.

We can construct a larger family of sparse crossbar concentrators, by increasing the number of full columns to $n-2m+2i, 1 \leq i \leq m$. It is obvious that the resulting sparse crossbar for each $i, 1 \leq i \leq m$ is a concentrator, and as in the proof of the corollary, each permutation of the remaining $2m-2i$ columns will result in a distinct concentrator after the $n-2m+2i$ full columns are fixed. Hence we have the following theorem:

THEOREM 6. *For $2 \leq m \leq n/2$, there exists a family of sparse crossbar concentrators with no less than*

$$\sum_{i=1}^{m} \binom{n}{n-2m+2i}(2m-2i)!$$

elements.

It is easy to verify that this set of sparse crossbar concentrators does not overlap with the set of fat-and-slim sparse crossbars.

4. The Density of Sparse Crossbar Concentrators

We have described two families of sparse crossbar concentrators and the two together form a set of

$$(\prod_{j=1}^{m} n-j+1) + (\sum_{i=1}^{m} \prod_{j=n-2m+2i+1}^{n} j)$$

elements. The main problem we now wish to deal with is "what portion of all the 2^{nm} bipartite graphs are actually sparse crossbar (n,m)-concentrators?". Formally, we are interested in the density of sparse crossbar concentrators among the 2^{nm} sparse crossbars with n inputs and m outputs, where:

DEFINITION 7. The density $d_{C(n,m)}$ of the set of sparse crossbar (n,m) - concentrators, $C(n,m)$, among all sparse crossbars with n inputs and m outputs is given by:

$$d_{C(n,m)} = \lim_{n\to\infty} |C(n,m)| \times 2^{-nm}$$

It is easy to see that when $m = 1$, there is only one sparse crossbar concentrator for all $n \geq 1$, and hence

$$d_{C(n,1)} =^{\lim}_{n\to\infty} 2^{-n} = 0.$$

For other values of m, the problem is not as trivial. For example, suppose $m = n$. In this case, we are dealing with square matrices, and to count all $n \times n$ matrices in which there is a matching of each column with a distinct row, we must have an actual algorithm to construct all such matrices. For $n = 2$, it is easy to see that of the sixteen 2×2 binary matrices, exactly seven correspond to sparse crossbar concentrators. For $n = 3$ and $n = 4$, we can count them by examining the structure of the matrices that correspond to sparse crossbar concentrators. In this way, it can be verified that there are 247 sparse crossbar $(3, 3)$-concentrators, and 37811 sparse crossbar $(4, 4)$-concentrators. These numbers suggest that $d_{C(n,n)} = 1$, but the most we can show is $d_{C(n,n)} \geq 0.288$.

THEOREM 7. *The density of sparse crossbar (n, n)-concentrators is at least* 0.288.

PROOF. The proof will follow from a more generalized statement in the next section (See Remark 1).

5. Density Bounds

In this section, we present two upper bounds and two lower bounds on the density of sparse crossbar concentrators.

5.1. Upper Bounds. The most obvious upper bound can be derived using Theorem 3 with $k = 1$. The theorem states that every sparse crossbar (n, m)-concentrator must have at least $(n - m + 1)m$ crosspoints. It follows that all $m \times n$ matrices with fewer than $(n - m + 1)m$ "1"s should correspond to bipartite graphs which are not concentrators. Hence,

$$d_{C(n,m)} \leq \lim_{n\to\infty} \frac{2^{nm} - \sum_{i=0}^{(n-m+1)m-1} \binom{nm}{i}}{2^{nm}}, \tag{4}$$

or

$$d_{C(n,m)} \leq \lim_{n\to\infty} \frac{\sum_{i=(n-m+1)m}^{nm} \binom{nm}{i}}{2^{nm}}. \tag{5}$$

Using Moon's inequality [**EJ71**],

$$\sum_{i>r/2+\lambda}^{r} \binom{r}{i} < 2^r e^{-2\lambda^2/r},$$

in the interval $0 \leq \lambda \leq r/2$, with $r = nm, \lambda = nm/2 - m(m-1)$, yields the upper bound

$$d_{C(n,m)} \leq \lim_{n\to\infty} e^{-2((nm/2-m(m-1))^2/nm}, \tag{6}$$

or

$$d_{C(n,m)} \leq \lim_{n\to\infty} e^{-2m((n/2-(m-1))^2/n} \tag{7}$$

in the interval $1 \leq m \leq n/2 + 1$. It follows that

COROLLARY 3.

$$d_{C(n,n/2)} \leq 1/e \approx 0.37.$$

We can derive another upper bound on $d_{C(n,m)}$ in the same interval by lower bounding the number of bipartite graphs which are not concentrators, as above. This time, we count those bipartite graphs in any row of which there are m or more zeros. Again, by Theorem 3, these correspond to bipartite graphs which are not concentrators.

Let R_k denote the set of $m \times n$ matrices with m or more zeros in the kth row. Let $|R_k|$ denote the number of elements in R_k. Then,

$$|R_{j_1}| = \sum_{i=m}^{n} \binom{n}{i} 2^{n(m-1)},$$

$$|R_{j_1} \cap R_{j_2}| = \left(\sum_{i=m}^{n} \binom{n}{i} \right)^2 2^{n(m-2)},$$

$$\vdots$$

$$|R_{j_1} \cap \cdots \cap R_{j_m}| = \left(\sum_{i=m}^{n} \binom{n}{i} \right)^m,$$

for any distinct indices $j_1, \ldots, j_m$.

We can use these equalities to derive a lower bound on the number, $x_{C(n,m)}$, of bipartite graphs which are not concentrators, by the inclusion exclusion principle.

$$\begin{aligned}
x_{C(n,m)} &\geq |R_1 \cup R_2 \cup \cdots \cup R_m| \\
&= \sum_i |R_i| - \sum_{i,j} |R_i \cap R_j| + \cdots + (-1)^{m+1} |R_1 \cap \cdots \cap R_m| \\
&= \sum_{j=1}^{m} \binom{m}{j} (-1)^{j+1} \left(\sum_{i=m}^{n} \binom{n}{i} \right)^j 2^{n(m-j)} \\
&= -2^{nm} \sum_{j=1}^{m} \binom{m}{j} \left(\sum_{i=m}^{n} \binom{n}{i} (-2^{-n}) \right)^j .
\end{aligned}$$

Using the equality

$$\sum_{j=1}^{m} \binom{m}{j} x^j = (x+1)^m - 1,$$

we get

$$\begin{aligned}
x_{C(n,m)} &\geq -2^{nm} \left(\left(\sum_{i=m}^{n} \binom{n}{i} (-2^{-n}) + 1\right)^m - 1 \right), \\
&= 2^{nm} - \left(2^n - \sum_{i=m}^{n} \binom{n}{i}\right)^m \\
&= 2^{nm} - \left(\sum_{i=0}^{m-1} \binom{n}{i}\right)^m.
\end{aligned}$$

Since $d_{C(n,m)} = 1 - \lim_{n\to\infty}(x_{C(n,m)}/2^{nm})$, we get

$$(8) \qquad d_{C(n,m)} \leq \lim_{n\to\infty} \left(\frac{1}{2^n}\sum_{i=0}^{m-1}\binom{n}{i}\right)^m .$$

Rewriting the inequality as

$$(9) \qquad d_{C(n,m)} \leq \lim_{n\to\infty} \left(\frac{1}{2^n}\sum_{i=n-m+1}^{n}\binom{n}{i}\right)^m ,$$

and once again, using Moon's inequality with $\lambda = (n-m+1) - n/2$, we obtain the same upper bound

$$(10) \qquad d_{C(n,m)} \leq \lim_{n\to\infty} e^{-2m(n/2-m+1)^2/n},$$

in the interval $1 \leq m \leq n/2+1$.

As we will see in the next section, if the second bound is computed directly using the expression in (9), it provides a tighter upper bound on $d_{C(n,m)}$. This bound can be improved by considering larger values of k in Theorem 2. However, the counting gets more complicated, and lest we count these non-concentrator graphs due to the higher order terms precisely, their contribution is negligible on $x_{C(n,m)}$.

5.2. Lower Bounds. To find a lower bound on $d_{C(n,m)}$, we count the number of $m \times n$ matrices that correspond to an (n,m)-concentrator, and can be generated from a fat-and-slim concentrator. We use the transformation lemma (Lemma 1) that has been stated before.

First assume that $m \leq n-m$. This makes the fat region of the concentrator at least as large as the slim region. Suppose we choose any i columns from the fat region and any i columns from the slim region, and swap some entries between the first columns selected from each group, then between second columns selected from each group, and so on. For each pair of columns, there can be at most $m-1$ swaps, giving $2^{m-1}-1$ different arrangements of the two columns (we exclude the case of no swaps). This gives $(2^{m-1}-1)^i$ different concentrators for each selection of the groups. Summing this up for all the possible selections, we get

$$|C(n,m)| \geq \sum_{i=0}^{m}\binom{m}{i}\binom{n-m}{i}(2^{m-1}-1)^i.$$

Now, suppose we add k columns to the fat region, and remove k columns from the slim region. For each k, we will get a different concentrator since the total number of crosspoints is changed. We can use the above generation process on the concentrators obtained in this way, which modifies the bound as

$$|C(n,m)| \geq \sum_{k=0}^{m}\sum_{i=0}^{m-k}\binom{m-k}{i}\binom{n-m+k}{i}(2^{m-1}-1)^i.$$

We can use the same counting technique for the $m > n-m$ case. In this case, the summation limits will be different. Also, increasing the fat region will eventually bring the problem to the previous case, so there will be two summation terms. The general bound for $|C(n,m)|$ is given by

$$|C(n,m)| \geq \begin{cases} \sum_{k=0}^{m}\sum_{i=0}^{m-k}\binom{m-k}{i}\binom{n-m+k}{i}(2^{m-1}-1)^i & , \quad m \leq n-m \\ \sum_{k=0}^{m-n/2}\sum_{i=0}^{n-m+k}\binom{m-k}{i}\binom{n-m+k}{i}(2^{m-1}-1)^i & \\ \quad +\sum_{k=1}^{n/2}\sum_{i=0}^{n/2-k}\binom{n/2-k}{i}\binom{n/2+k}{i}(2^{m-1}-1)^i, & m > n-m \end{cases}$$

and

$$d_{C(n,m)} \geq \lim_{n\to\infty}\frac{|C(n,m)|}{2^{nm}}.$$

This lower bound can be improved using another construction. The key idea is to ensure that each input has a distinct output to be matched with. This is guaranteed if we enter a "1" in a different column in each row as we prove in the next theorem.

THEOREM 8. *There exist at least*

$$(11) \qquad (\prod_{i=1}^{m-1} 2^i(2^{m-i}-1)) \times (\sum_{i=n-m+1}^{n}\binom{n}{i}).$$

sparse crossbar concentrators.

PROOF. We know from Theorem 3 that every row in the incidence matrix of a sparse crossbar concentrator must have at least $n-m+1$ "1"s. Hence, the first row can be fixed in any one of

$$\sum_{i=n-m+1}^{n}\binom{n}{i}$$

ways. Once the first row is fixed this way, we fill some $n-m$ columns which have "1" entries in the first row with "1"s, and mark a "1" among the remaining "1"s in the first row. This insures that the input which corresponds to the column with the marked "1" can be matched with an output. This process can be iterated by allowing the remaining $m-1$ columns in the second row to assume any one of $2^{m-1}-1$ patterns of "1's and "0"s (we only need to exclude the all "0" pattern). Furthermore, the entry at the intersection of the second row and the column in which the 1 has been marked can be either 0 or 1. Hence, the unspecified m entries in the second row can be fixed in any one of $2(2^{m-1}-1)$ ways. Iterating this argument for the next row, and so on, we see that the total number of sparse crossbar (n,m)-concentrators is not less than the expression given in (11). □

REMARK 1. When $n=m$, the lower bound in (11) reduces to

$$(12) \qquad \prod_{i=0}^{n-1} 2^i(2^{n-i}-1) = 2^{n^2}\prod_{i=1}^{n}(1-2^{-i}).$$

Dividing this expression by 2^{n^2}, we find that the density of a sparse crossbar (n, n)-concentrator satisfies the inequality

$$d_{C(n,n)} \geq \lim_{n\to\infty} \prod_{i=1}^{n} (1 - 2^{-i}). \tag{13}$$

Using the Euler expansion [**Bro42**]

$$\prod_{i=1}^{\infty} (1 - q^i) = 1 - (q + q^2) + (q^5 + q^7) - (q^{12} + q^{15}) + \ldots,$$

with $q = 1/2$, and the indices being alternately $n(3n \pm 1)/2$, we find that

$$d_{C(n,m)} \geq 1 - (1/2 + 1/4) + (1/32 + 1/128) - (1/4096 + 1/32768) \geq 0.288.$$

The exact values of the lower and upper bounds are plotted in Figures 3, 4, and 5. It is seen that the second upper bound is always tighter than the first upper bound. On the other hand, the first lower bound is slightly tighter than the second lower bound when $m = 0.2n$, but the second lower bound gets tighter as m approaches n. In fact, when $m = n$, it tends to 0.288, whereas the first lower bound tends to 0.

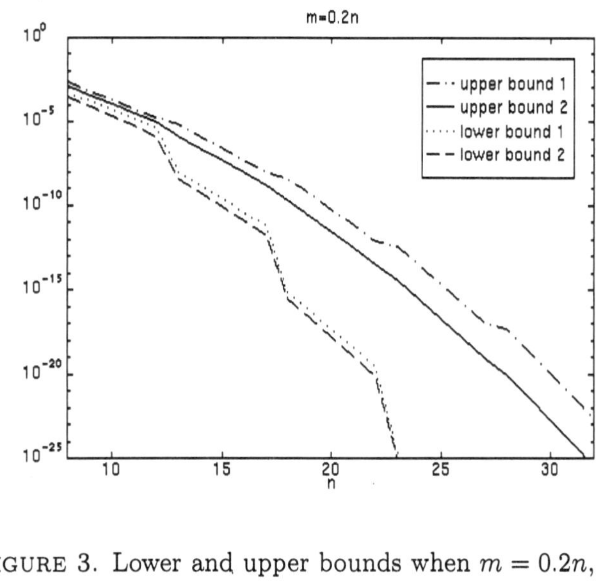

FIGURE 3. Lower and upper bounds when $m = 0.2n, 1 \leq n \leq 40$.

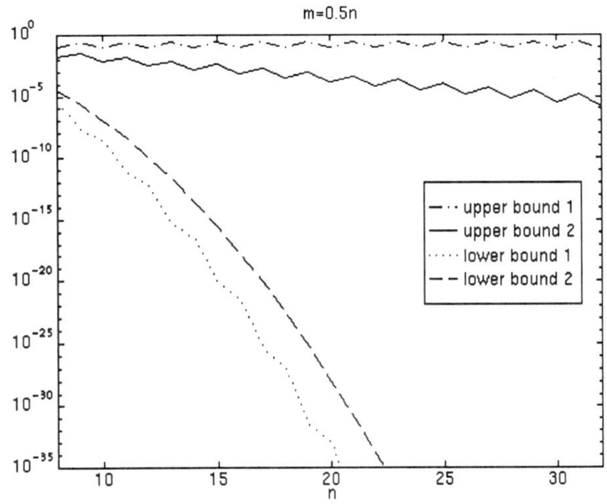

FIGURE 4. Lower and upper bounds when $m = 0.5n, 1 \leq n \leq 40$.

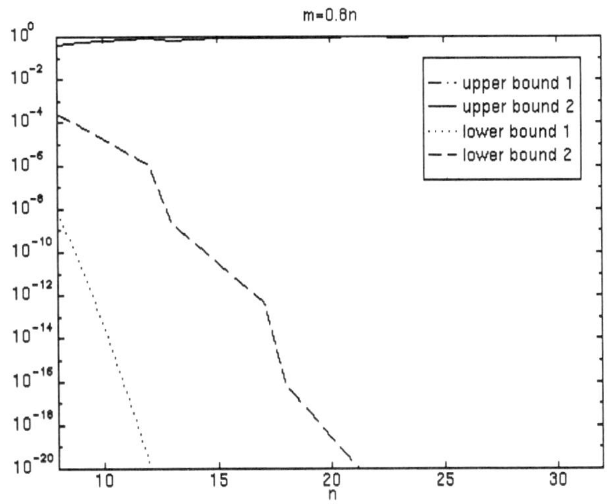

FIGURE 5. Lower and upper bounds when $m = 0.8n, 1 \leq n \leq 40$.

6. Concluding Remarks

In this paper, we have investigated the density of sparse crossbar concentrators within the set of all bipartite graphs with n inputs and m outputs. The results we presented here offer some clues on how extensively such graphs are found among bipartite graphs. It has been shown that the density of sparse crossbar concentrators is strictly greater than 0 (in fact, it is not less than 0.288 when the number of inputs equals the number of outputs). In the other direction, it has been shown that the density of sparse crossbar $(n, n/2)$-concentrators is less than $1/e$.

Added in the Proof: Recently, we proved by a probabilistic argument that the density of sparse crossbar (n, m)-concentrators tends to 1 when $m/n > 1/2$, and it tends to 0 when $m/n < 1/2$. A detailed account of this result will be presented elsewhere.

References

[Bas81] L. A. Bassalygo, *Asymptotically optimal switching circuits*, Prob. of Inform. Trans. **17** (1981), 206–211.

[Bro42] T. J. A. Bromwich, *An introduction to the theory of infinite series*, MacMillan and Co., Limited, London, 1942, pp. 116–117.

[EJ71] P. Erdös and J. Spencer, *Probabilistic Methods in Combinatorics*, Academic Press, 1974, p. 18.

[GO96] W. Guo and A. Y. Oruç, *Regular sparse crossbar concentrators*, IEEE Trans. Comput. (to appear).

[JO93] C. Jan and A. Y. Oruç, *Fast self-routing permutation switching on an asymptotically minimum cost network*, IEEE Trans. on Comput. (Dec. 1981), 1469–1479.

[LO95] C. Y. Lee and A. Y. Oruç, *Design of efficient and easily routable generalized connectors*, IEEE Trans. on Commun. (Feb. 1995), 646–650.

[Mas77] G. M. Masson, *Binomial switching networks for concentration and distribution*, IEEE Trans. on Commun. **COM-25** (Sept. 1977), 873–883.

[NM82] S. Nakamura and G. M. Masson, *Lower bounds on crosspoint in concentrators*, IEEE Trans. on Comp. (Dec. 1982), 1173–1178.

[MO94] M. V. Chien and A. Y. Oruç, *Adaptive binary sorting schemes and associated interconnection networks*, IEEE Trans. on Par. and Distr. Systems **5** (June 1994), 561–571.

[OH96] A. Y. Oruç and H. M. Huang, *Crosspoint complexity of sparse crossbar concentrators*, IEEE Trans. on Information Theory (Sept. 1996), 1466–1471.

[Pip77] N. Pippenger, *Superconcentrators*, SIAM Jour. on Comput. **6** (1977), 298–304.

Electrical Engineering Department, University of Maryland, College Park, MD 20742

E-mail address: emre@eng.umd.edu

Electrical Engineering Department, University of Maryland, College Park, MD 20742

E-mail address: yavuz@eng.umd.edu

DIMACS Series in Discrete Mathematics
and Theoretical Computer Science
Volume 42, 1998

Non-blocking Routing Properties of Clos Networks

Tony T. Lee and Philip P. To

ABSTRACT. We study the routing algorithms of the Clos network and a large class of multistage interconnection networks (MIN) based on an address numbering scheme of these interconnection networks. It is well-known that if the sèt of connection requests is ordered, then the routing within a banyan network is non-blocking. We show that a whole family of MINs as well as the Clos networks possess the same non-blocking properties. In this context, the Clos network can be considered as the cascade combination of the omega network and the reverse omega network as far as routing is concerned. Furthermore, we can extend this numbering scheme and the associated routing properties to construct broadcast and rearrangebly non-blocking multiconnection networks.

1. Introduction

The most important class of interconnection networks ever studied is the Clos networks, first proposed by C. Clos [**1**] in 1953. A three-stage Clos network consists of three stages of switch modules. Each module in one stage is connected to every modules in the next stage. The three-stage Clos network is rearrangeably non-blocking [**2**]. It can realize any permutation of inputs and outputs by rearranging existing connections. Route assignment within the network is performed by means of a control algorithm. One well-known example is the looping algorithm proposed in [**3**] to control the Benes network, which is a Clos network composed entirely of 2×2 switching elements. Work has also been done to design parallel algorithms for route assignment (for example, [**4, 5**]). These routing algorithms, however, are not feasible in packet switching systems since they require a relatively long period of time to execute. In packet switching system, a self-routing algorithm is required so that the packets are routed based only on locally available information such as their destination addresses. The advantage of the self-routing approach is that when the network becomes large, on-the-fly route assignment for incoming packets in every time slot is either impossible or not feasible. In [**6**], Douglass and Oruc showed that self-routing in Clos network is impossible in general. As a result, the Clos network is believed to be not suitable for packet switching. Although some work (for instance, [**7, 8**]) has indicated that certain particular permutations can

1991 *Mathematics Subject Classification.* Primary 94A99.

This work was supported by the Research Grant Council of Hong Kong under Earmarked Grant CUHK276/94E.

be self-routed in the Clos network, these permutations are not general enough to allow the use of Clos network as a packet switch.

It is well-known that a seemingly unrelated network known as the banyan network possesses the non-blocking and self-routing properties and can be used effectively in high-speed packet switches. In particular, if the input packets to the banyan network are concentrated and their corresponding output addresses are monotonic, the network is non-blocking and each input packet can be self-routed to its destination by means of the output address. The reverse banyan network, which is the mirror image of the banyan network, also possesses the non-blocking and self-routing properties. In [**9**], a non-blocking and self-routing switch is proposed which is essentially the cascade combination of a sorting network, such as the Batcher sorting network [**10**], and a banyan network. The sorting network sorts the packets according to their output destinations and resolves any output port contention to make the set of packets satisfy the non-blocking conditions of the banyan network. When the packets enters the banyan network, they are self-routed to their destinations in a non-blocking manner. This way of switch construction is generally referred to as the *sort-banyan principle* and is the basis of a lot of existing non-blocking switch architectures.

In this paper, we show by a precise mathematical proof that if the connection requests to the Clos network are ordered, then they can also be self-routed to their destinations in a non-blocking manner. Moreover, the routing tag can be completely determined by the rank of each request and its output address. As we will show, the non-blocking and self-routing properties of a network are tied to the numbering scheme, which is in turn inherent from the topological structure of the network. In this context, we will show that the Clos network is actually the cascade combination of the reverse omega network and the omega network. The non-blocking self-routing algorithm of the Clos network can be further generalized to the General Interval Splitting Algorithm. Thus broadcasting within the Clos network without blocking becomes possible. With these findings, the realization of a multicast packet switch and a five-stage rearrangeably non-blocking multiconnection network will be discussed.

The remainder of this paper is organized as follows. In Section 2, the numbering scheme of the three-stage Clos network is introduced and the non-blocking property of the self-routing algorithm based on this numbering scheme is investigated. We will show that the routing tag of a packet can be completely determined from its rank and the destination address. We will also show that the non-blocking and self-routing properties of the omega network and the reverse omega network are actually inherent from the Clos network. In Section 3, we generalize the routing algorithm to a replication algorithm for the broadcast Clos network. In Section 4, the construction of multicast networks based on the Clos network is investigated and we will discuss the construction of a multicast packet switch and a five-stage rearrangeably non-blocking multiconnection network. Conclusions will be given in Section 5.

2. The Clos Network

An $N \times N$ Clos network is characterized by two integer parameters p and q, where p is the number of input-stage and output-stage modules and q is the number of middle-stage modules. As shown in Fig. 1, apart from numbering the inputs and

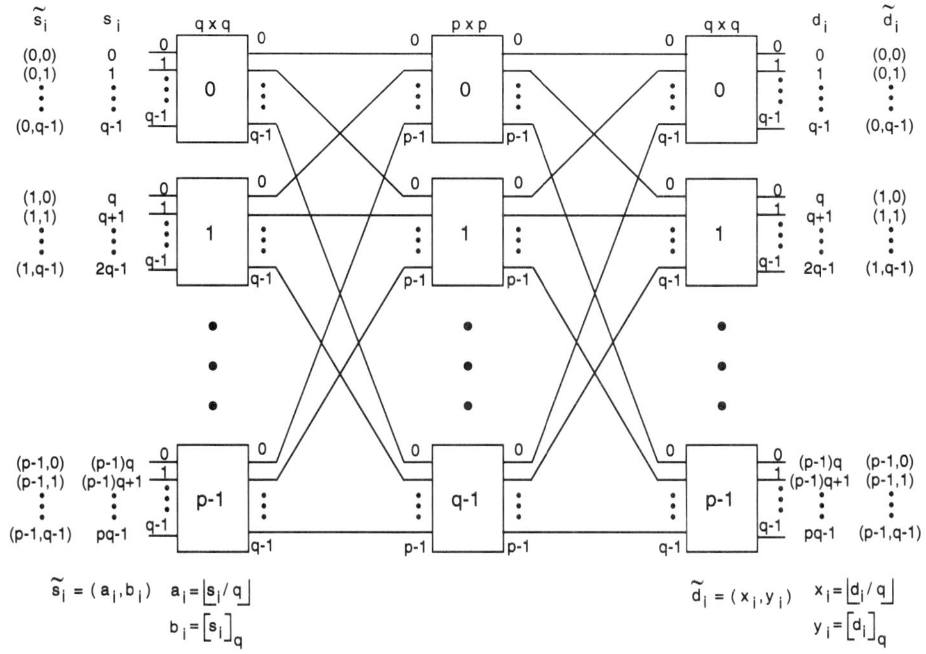

FIGURE 1. A three-stage Clos network.

outputs consecutively from 0 to $N-1$, a two-tuple addressing scheme may be used: an input is numbered (a_i, b_i) if it is the b_i'th input/output port on input/output module a_i. It can be easily seen that a_i is of base p $(0 \leq a_i \leq p-1)$ and b_i is of base q $(0 \leq b_i \leq q-1)$. The two-tuple address $\tilde{s}_i$ and $\tilde{d}_i$ can be obtained from their corresponding consecutive addresses s_i and d_i by

$$(2.1) \qquad \tilde{s}_i = (\lfloor \frac{s_i}{q} \rfloor, [s_i]_q) \qquad \text{and} \qquad \tilde{d}_i = (\lfloor \frac{d_i}{q} \rfloor, [d_i]_q)$$

where $\lfloor \frac{a}{b} \rfloor$ is the largest integer smaller than a/b and $[a]_b$ is the reminder of a/b. In the rest of the paper, we will denote the address (a_i, b_i) by $a_i b_i$ whenever there is no ambiguity. The two-tuple addresses are essential to self-routing in the Clos network. In particular, to go from any middle-stage module to an output $x_i y_i$, a packet is routed to the x_i'th output in the middle-stage module and to the y_i'th output in the output-stage module. Using this numbering scheme, self-routing is possible starting from any of the middle-stage modules. The remaining question is how to route to packets from the input-stage modules to the middle-stage modules in a distributed manner without causing internal path conflicts.

2.1. Non-blocking Route assignment. Let $C = \{0, 1, \ldots, q-1\}$ be the set of middle-stage modules and let $\pi = \{(s_0, d_0), \ldots, (s_{n-1}, d_{n-1})\}$ be the input-output permutation (which may be partial). A sufficient condition for the assignment $f : \pi \rightarrow C$ to be non-blocking is

$$(2.2) \qquad f(s_i, d_i) = f(s_j, d_j) \;\Rightarrow\; |\, s_i - s_j \,| \geq q \text{ and } |\, d_i - d_j \,| \geq q$$

where $d_i = \pi(s_i)$ and $d_j = \pi(s_j)$. This condition guarantees that no more than one input or output on the same module will be assigned to the same middle-stage module so that the assignment is non-blocking. For an assignment f, the routing tag for the connection (s_i, d_i) is therefore given by

$$(f(s_i, d_i), \lfloor \frac{d_i}{q} \rfloor, [d_i]_q). \tag{2.3}$$

Due to the symmetric structure of the Clos network, the routing tag for the reverse path from d_i to s_i is

$$(f(s_i, d_i), \lfloor \frac{s_i}{q} \rfloor, [s_i]_q). \tag{2.4}$$

A non-blocking route assignment algorithm was discussed in [**11**] in which monotonic connection requests are assigned to the middle-stage modules in a sequential manner. In the following, we show that the basis of the algorithm in fact stems from a fundamental lemma in number theory and that the route assignment can be obtained simply from the rank of each connection request.

The set of connection requests π is monotonic if the active inputs $s_0, s_1, \dots, s_{n-1}$ and their corresponding outputs $d_0, d_1, \dots, d_{n-1}$ satisfy either one of the following conditions:

$$\begin{array}{lcll} s_i < s_j & \Rightarrow & d_i < d_j & \text{for all } i, j; \text{ or} \\ s_i < s_j & \Rightarrow & d_i > d_j & \text{for all } i, j. \end{array}$$

Monotonic connection requests are considered since they exhibit a simple non-blocking scheme due to the following lemma.

LEMMA 2.1. *Let $x_0, x_1, ..., x_{n-1}$ be a strict monotonic sequence of integers and for all i, define $g(x_i)$ as follows*

$$g(x_i) = [m + i]_q \qquad \textit{for all } i.$$

where m and q are constant integers. Then $g(x_i) = g(x_j)$ implies $\mid x_i - x_j \mid \geq q$ for $i \neq j$.

PROOF. Without loss of generality, assume that the sequence is increasing and let $i < j$. It follows that $g(x_i) = g(x_j)$ implies that $i + lq = j$, where $l \geq 1$. Now

$$\begin{array}{rcl} x_j - x_i & = & (x_j - x_{j-1}) + (x_{j-1} - x_{j-2}) + ... + (x_{i+1} - x_i) \\ & \geq & \underbrace{1 + ... + 1}_{j-i} = j - i = lq \geq q. \end{array}$$

□

Lemma 2.1 can be applied to devise a route assignment algorithm for ordered connection requests. It is called the Rank-based Assignment Algorithm and it performs route assignment according to the "rank" of each connection request.

THEOREM 2.2 (Rank-based Assignment Algorithm). *Let the set of connection requests* $\pi = \{(s_0, d_0), \ldots, (s_{n-1}, d_{n-1})\}$ *be monotonic. The assignment*

$$f(s_i, d_i) = [m + i]_q, \tag{2.5}$$

where m *is a constant integer,* q *is the number of middle-stage modules and* i *is the rank of the connection request* $(s_i, d_i = \pi(s_i))$*, is non-blocking.*

Proof. Since the set of connection requests $\pi = \{(s_0, d_0), \ldots, (s_{n-1}, d_{n-1})\}$ is a monotonic input-output permutation, both sequences $(s_0, \ldots, s_{n-1})$ and $(d_0, \ldots, d_{n-1})$ are monotonic. Let

$$\begin{aligned} g_1(s_i) &= f(s_i, \pi(s_i)) = f(s_i, d_i) = [m+i]_q \qquad \text{and} \\ g_2(d_i) &= f(\pi^{-1}(d_i), d_i) = f(s_i, d_i) = [m+i]_q \end{aligned}$$

for all i. Thus, $f(s_i, d_i) = f(s_j, d_j)$ implies $g_1(s_i) = g_1(s_j)$ and $g_2(d_i) = g_2(d_j)$. It follows from Lemma 1 that $\mid s_i - s_j \mid \geq q$ and $\mid d_i - d_j \mid \geq q$. Hence the assignment is non-blocking according to condition given in (2.2). □

2.2. Recursiveness. With the Rank-based Assignment Algorithm, each middle-stage module in the Clos network is itself a Clos network with monotonic connection requests. The route assignment algorithm can therefore be applied recursively. To see this, consider a middle-stage module M and consider two connection requests (s_i, d_i) and (s_j, d_j) assigned to M. Assume that $s_i < s_j$ and $d_i < d_j$. Due to the Rank-based Assignment Algorithm, we have $s_j - s_i \geq q$ and $d_j - d_i \geq q$. Let the ports on M to which s_i and s_j are connected be m_i and m_j respectively and let the ports to which d_i and d_j are connected be n_i and n_j respectively. Now, from (2.3) and (2.4), we have

$$\begin{aligned} m_j &= \lfloor \frac{s_j}{q} \rfloor \geq \lfloor \frac{s_i + q}{q} \rfloor = \lfloor \frac{s_i}{q} \rfloor + 1 = m_i + 1 > m_i; \quad \text{and} \\ n_j &= \lfloor \frac{d_j}{q} \rfloor \geq \lfloor \frac{d_i + q}{q} \rfloor = \lfloor \frac{d_i}{q} \rfloor + 1 = n_i + 1 > n_i. \end{aligned}$$

The rank order is preserved in each middle-stage module. The active inputs and their corresponding outputs on M are therefore monotonic.

In general, if N can be factorized into $N = q_0 \cdot q_1 \cdot \ldots \cdot q_{n-1}$, we can recursively generate the Clos network into a $(2n-1)$-stage network. Let the stages be numbered from 0 to $2n-2$. The size of each stage-i module is q_i for $0 \leq i \leq n-1$ and is equal to q_{2n-2-i} for $n-1 \leq i \leq 2n-2$. Each input and output address can then be numbered by an n-tuple address. For an input address $S = s_0 s_1 \ldots s_{n-1}$ and an output address $D = d_0 d_1 \ldots d_{n-1}$, the bases of both s_i and d_i are equal to q_{n-1-i} for $0 \leq i \leq n-1$. In the following, we derive a general representation of the routing tag. First consider a 3-stage Clos network with parameters p_0 and q_0. Suppose a packet enters from input s_i and is destined for output d_i. Let the rank of the packet be $r_0(s_i)$. By the Rank-based Assignment Algorithm, the packet is routed to the middle-stage module M given by $[r_0(s_i)]_{q_0}$. Now M is further factorized into a smaller Clos network with parameters p_1 and q_1. The rank of the packet in M can be computed from $r_0(s_i)$ and is given by $r_1(s_i) = \lfloor r_0(s_i)/q_0 \rfloor$. In fact, we can obtain these information directly by dividing $r_0(s_i)$ by q_0:

TABLE 1. The set of connection requests to a 16 × 16 Benes network.

s	0	1	3	4	7	10	12
d	1	2	4	6	9	12	14
r	0	1	2	3	4	6	7
$\tilde{r} = \alpha_3\alpha_2\alpha_1\alpha_0$	0000	0001	0010	0011	0100	0110	0111
$\tilde{d} = \beta_3\beta_2\beta_1\beta_0$	0001	0010	0100	0110	1001	1100	1110
$R = \alpha_0\alpha_1\alpha_2\beta_3\beta_2\beta_1\beta_0$	0000001	1000010	0100100	1100110	0011001	0111100	1111110

$$r_0(s_i) = \alpha_1 q_0 + \alpha_0.$$

The remainder $\alpha_0 = [r_0(s_i)]_{q_0}$ represents the middle-stage module assigned to the packet and the quotient $\alpha_1 = \lfloor r_0(s_i)/q_0 \rfloor$ represents the rank of the packet in M. Within M, the same procedure is repeated and we can obtain the route information by dividing $r_1(s_i)$ by q_1. In general, if $N = q_0 \cdot q_1 \cdot \ldots \cdot q_{n-1}$, we can express $r_0(s_i)$ as

$$\begin{aligned} r_0(s_i) &= \alpha_1' q_0 + \alpha_0 \\ &= (\alpha_2' q_1 + \alpha_1) q_0 + \alpha_0 = \alpha_2' q_1 q_0 + \alpha_1 q_0 + \alpha_0 \\ &= ((\alpha_3' q_2 + \alpha_2) q_1 + \alpha_1) q_0 + \alpha_0 = \alpha_3' q_2 q_1 q_0 + \alpha_2 q_1 q_0 + \alpha_1 q_0 + \alpha_0 \\ &= \ldots \end{aligned}$$

and in general,

$$r_0(s_i) = \sum_{i=1}^{n-1} \alpha_i \prod_{j=0}^{i-1} q_j + \alpha_0. \tag{2.6}$$

In a similar way, the destination address can be written as

$$d_i = \sum_{i=1}^{n-1} \beta_i \prod_{j=0}^{i-1} q_j + \beta_0. \tag{2.7}$$

The routing tag from s_i to d_i is therefore given by

$$(\alpha_0, \alpha_1, \alpha_2, \ldots, \alpha_{n-2}, \beta_{n-1}, \beta_{n-2}, \ldots, \beta_1, \beta_0). \tag{2.8}$$

As an example, consider the special case where $q_i = 2$ for all i. In that case, a Benes network is formed and the input and output addresses are represented by binary strings. Fig. 2 shows a 16 × 16 Benes network with the connection requests shown in Table 1. Upon entering the network, each packet can determine its routing tag from the binary representation of its rank and the destination address. The rank of the packets may be computed distributedly by an adder network cascaded in front of the Benes network.

2.3. General Clos-type Networks. In the following, we show that the Clos network is in fact the cascade combination of the omega and the reverse omega network. We will show that this is precisely the reason why route assignment in the Clos network is particularly easy for monotonic connection requests. The

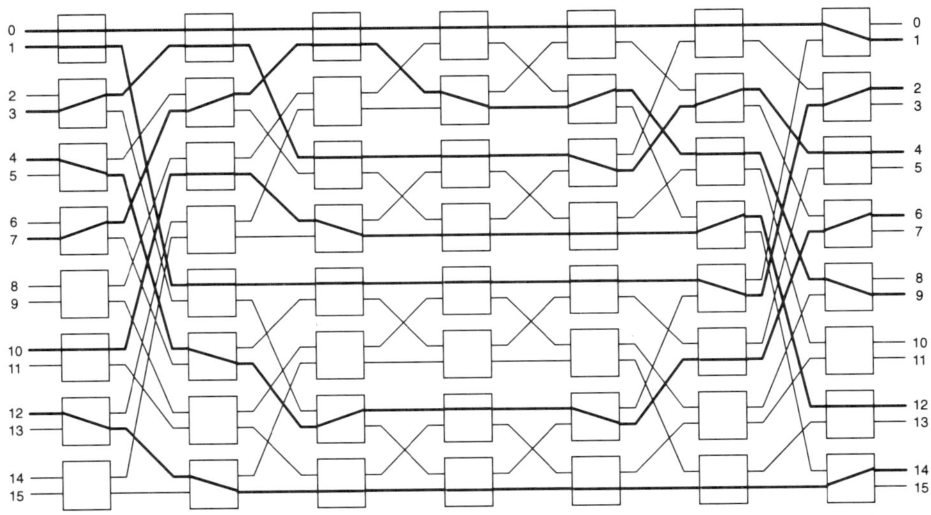

FIGURE 2. A 16 × 16 Benes network.

principle behind the Rank-based Assignment Algorithm can also be derived from the address numbering schemes of the omega and the reverse omega network.

It is well-known that the omega network is non-blocking and self-routing if the set of input packets is concentrated and their output destinations are monotonic [**12, 13**]. Similarly, the reverse omega network, which is merely the mirror image of the reverse omega network, is non-blocking if the set of input packets is concentrated and their outputs are monotonic. Both the omega and the reverse omega networks can be recursively constructed. In fact, their 2 × 2 realizations, namely the banyan network and the reverse banyan network, have been used extensively in building high-speed ATM switches. The former is usually used with a batcher sorting network to build a point-to-point switch based on the sort-banyan principle while the latter usually performs the function of a concentrator.

Figure 3 shows the network formed from the cascade combination of a 12 × 12 reverse omega network and an omega network of the same size. The inputs and outputs of network are numbered by the two-tuple addressing scheme. Suppose this five-stage network is given a set of monotonic connection requests. To satisfy the non-blocking condition of the reverse omega network, we can deliberately route the input packets to a set of concentrated ports in stage 3. In this case, any sets of concentrated ports, including those which are cylically concentrated, can be used. Once the packets enter the the omega network, since they are concentrated and their outputs are monotonic, they can be routed to their destinations in a distributed and non-blocking manner. This five-stage network is therefore non-blocking given monotonic connection requests. Since no switching is performed in stage 3, we can eliminate it and combine the modules in stage 2 and stage 4. The resulting network is the three-stage Clos network. Therefore the three-stage Clos network is non-blocking to monotonic connection requests.

Note that the port addresses y_ix_i's in stage 3 are obtained from the port addresses x_iy_i's in the previous stage through a shuffle connection pattern. Therefore, assigning the packets to a set of concentrated ports in stage 3 (in increasing x_i's) is actually equivalent to assigning the packets to the middle-stage modules in the

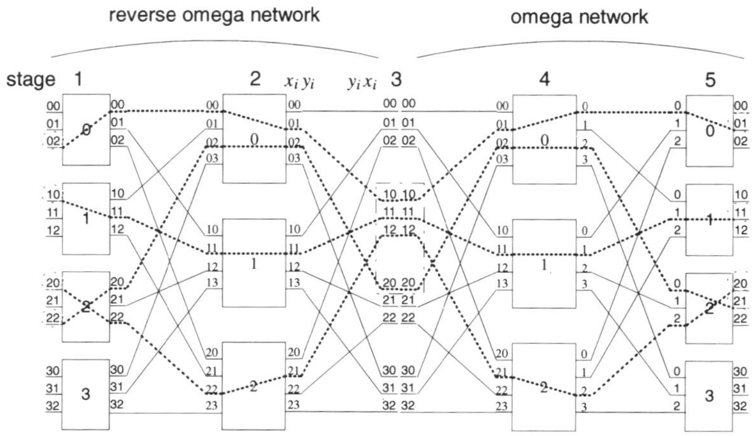

FIGURE 3. The cascade combination of a reverse omega network and an omega network.

Clos network by their ranks. This forms the basis of the Rank-based Assignment Algorithm.

In general, any multistage interconnection network (MIN) can be combined with its reverse network to form a general Clos-type network which possesses the same non-blocking and self-routing properties. In a MIN, there is a unique path for each input-output pair. If the MIN is $N \times N$ and is composed of 2×2 switching elements, the number of alternate paths for each input-output pair in the combined network will be equal to $N/2$. These alternate paths are formed by the product of the paths in the MINs.

3. Broadcast Clos Network

In this section, we extend the non-blocking properties of the Clos network to construct a non-blocking broadcast Clos network [**14**]. A broadcast Clos network is a Clos network with switching nodes which can perform packet replication. Let the active inputs be $s_0, s_1, \ldots, s_{n-1}$ ($s_j > s_i$ if $j > i$). Each active input may go to a number of outputs. Let their corresponding *sets of outputs* be $D_0, D_1, \ldots, D_{n-1}$ respectively. The set of connection requests is monotonic if

$$\begin{aligned} s_i < s_j \quad &\Rightarrow \quad d_i < d_j \qquad \forall d_i \in D_i \text{ and } \forall d_j \in D_j \qquad \text{or;} \\ s_i < s_j \quad &\Rightarrow \quad d_i > d_j \qquad \forall d_i \in D_i \text{ and } \forall d_j \in D_j; \end{aligned}$$

If the set of connection requests is monotonic, route assignment can be done by the Rank-based Assignment Algorithm for broadcast Clos networks.

THEOREM 3.1 (Rank-based Assignment Alg. for Broadcast Clos Net.). *Let the set of connection requests $\pi = \{(s_0, D_0), \ldots, (s_{n-1}, D_{n-1})\}$ be monotonic. The assignment*

$$f(s_i, D_i) = [m + i]_q, \tag{3.1}$$

where q is the number of middle-stage modules, m is a constant integer and i is the rank of the connection request, is non-blocking.

Proof. Construct the sequence $(d_0, d_1, \ldots, d_{n-1})$ by arbitrary choosing a d_i from D_i for $0 \leq i \leq n-1$. Since π is monotonic, the sequences $(s_0, s_1, \ldots, s_{n-1})$ and $(d_0, d_1, \ldots, d_{n-1})$ are monotonic. By Theorem 2.2, the network is non-blocking for this choice of d_i's. Since the choice of d_i from D_i is arbitrary, the broadcast Clos network is non-blocking. □

3.1. Routing and Replication. The replication process in the broadcast Clos network consists of two parts. The input packets are first routed to the middle-stage modules by the Rank-based Assignment Algorithm. Packet replication is then performed from the middle-stage modules to the output-stage modules. Recall that the Rank-based Assignment Algorithm preserves the order of the connection requests. Using the same notation as in Section 2.2, the rank $r(s_i)$ of a connection request (s_i, D_i) can be written as

$$r(s_i) = \sum_{i=1}^{n-1} \alpha_i \prod_{j=0}^{i-1} q_j + \alpha_0.$$

The routing tag for the packet to go from input to the middle-stage module assigned to it is given by $(\alpha_0, \alpha_1, \alpha_2, \ldots, \alpha_{n-2})$. Once the packets reach the middle-stage modules, packet replications and routing is controlled by the General Interval Splitting Algorithm which is discussed in the following.

We denote the switch modules from the middle stage to the output stage as broadcast switch modules due to their capability in performing replication. A packet arriving at a broadcast switch module can be routed to any one of the output links, or it can be replicated if necessary and sent out to a subset of the output links. For the non-blocking operations of the broadcast Clos network, assume that the set of connection requests is monotonic and the set of output destinations assigned to each input packet is consecutive. This assumption is justified because in a copy network, the replicated packets do not need to be routed to some specific output ports. Routing is done by a point-to-point switching network which is cascaded behind the copy network. We can therefore deliberately assign the sets of output destinations to the input packets such that the connection requests are monotonic. Under such assumption, the set of output destinations of each packet consists of an address interval which can be represented by two numbers, the *minimum* and *maximum*.

Let $N = q_0 \cdot q_1 \cdot \ldots \cdot q_{n-1}$. The broadcast Clos network consists of $2n-1$ stages numbered by $0, 1, \ldots, 2n-2$. An address interval is a set of contiguous n-tuple. Suppose a node at stage i $(n-1 \leq i \leq 2n-2)$ receives a packet containing an address interval specified by: $\min(i-1) = m_{n-1} \ldots m_{2n-2}$ and $\max(i-1) = M_{n-1} \ldots M_{2n-2}$, where the argument $(i-1)$ indicates that the packet comes from a node in stage $(i-1)$. Since the size of the modules in different stages may be different, each tuple in an address can be of different bases. As discussed in Section 2.2, m_j and M_j, where $n-1 \leq j \leq 2n-2$, are of base q_{2n-2-j}. When a node in stage i receives a packet, the following procedure is performed:

1. If $m_i = M_i$, then send the packet out on link m_i.
2. If $m_i \neq M_i$, then $(M_i - m_i + 1)$ copies of the packet are required. Replicate the packet, modify the headers and send the packets out on links m_i to M_i.

The purpose of header modification is to split the original address interval into several sub-intervals, each contained in a replicated packet [**15**]. It is easy to see

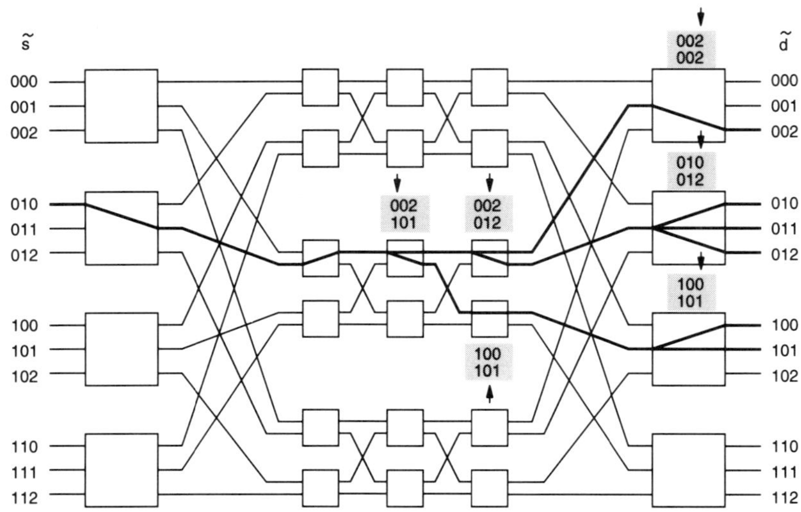

FIGURE 4. Packet replication process in a five-stage Clos network.

that $m_j = M_j$ for $j = n-1, \ldots, i-1$ holds for every packet which arrives at stage i, where $n-1 \leq i \leq 2n-2$. The event $m_i > M_i$ is impossible. For the packet sent out on link m_i, the header becomes:

$$
\begin{aligned}
\min(i) &= \min(i-1) = (m_{n-1}, \ldots, m_i, m_{i+1}, \ldots, m_{2n-2}) \\
\max(i) &= (m_{n-1}, \ldots, m_i, q_{(2n-2)-(i+1)} - 1, \ldots, q_0 - 1).
\end{aligned} \tag{3.2}
$$

This is due to the fact that the bases of $m_{i+1}, m_{i+2}, \ldots, m_{2n-2}$ are $q_{(2n-2)-(i+1)}$, $q_{(2n-2)-(i+2)}, \ldots, q_0$ respectively. For the packet sent out on link M_i, the header becomes:

$$
\begin{aligned}
\min(i) &= (M_{n-1}, \ldots, M_i, 0, \ldots, 0) \\
\max(i) &= \max(i-1) = (M_{n-1}, \ldots, M_i, M_{i+1}, \ldots, M_{2n-2}).
\end{aligned} \tag{3.3}
$$

For those packets going to output j where $m_i < j < M_i$, the header becomes:

$$
\begin{aligned}
\min(i) &= (m_{n-1}, \ldots, m_{i-1}, j, 0, \ldots, 0) \\
\max(i) &= (m_{n-1}, \ldots, m_{i-1}, j, q_{(2n-2)-(i+1)} - 1, \ldots, q_0 - 1)
\end{aligned} \tag{3.4}
$$

Figure 4 shows the operations of the General Interval Splitting Algorithm on a five-stage broadcast Clos network. An input or output address is represented by a three-tuple. The first and the second components are of base 2 while the third one is of base 3. By recursively applying the Rank-based Assignment algorithm, the packet is routed to one of the middle-stage modules. The packets are then subsequently replicated and routed to the destinations using the General Interval Splitting Algorithm. The address intervals contained in the header of each packet are shown in the shaded boxes.

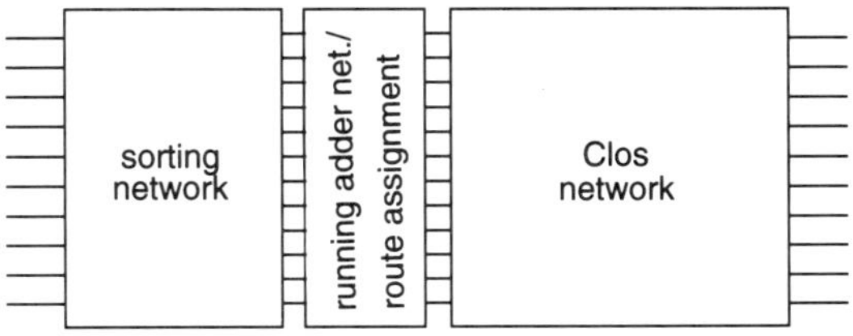

FIGURE 5. The sort-Clos network architecture.

4. Multicasting in Clos Networks

With the non-blocking routing algorithm and the numbering scheme, the Clos network can be used to build a non-blocking packet switch by cascading a sorting network in front of the Clos network. We denote this as the *sort-Clos principle.* The architecture is shown in Fig. 5. The sorting network sorts the input packets according to their destinations (but without the need to concentrate them.) An algorithm, for instance the ring reservation algorithm [**16**], is executed at the inputs of the running adder network to resolve output port contention. The running adder network computes the rank of each winning packet for route assignment by the Rank-based Assignment Algorithm. Once the route assignment is completed, each packet will be given a routing tag with which it can self-route to its destination in a non-blocking manner.

If the Clos network in Fig.5 is recursively factorized by 2×2 switching elements, it becomes the cascade combination of the reverse banyan network and the banyan network. The sort-Clos architecture therefore degenerates to the sort-banyan network with the two-phase contention resolution algorithm [**17**]. The sort-Clos principle is therefore the generalization of the sort-banyan principle.

In the following, we show that the non-blocking and self-routing properties of the Clos network are equally applicable to multicast networks in both packet switching and circuit switching networks. Two applications are discussed, namely a multicast packet switch and a five-stage rearrangeably non-blocking multiconnection network.

4.1. Multicast Packet Switch. In [**15**], Lee proposed a non-blocking copy network architecture based on the broadcast banyan network. It received a lot of attention due to its non-blocking and self-routing properties. The network structure is shown in Fig. 6. The copy network consists of a Running Adder Network (RAN), a set of Dummy Address Encoders (DAE's), a concentrator network, a Broadcast Banyan Network (BBN) and a group of Trunk Number Translators (TNT's). Each packet entering the copy network contains a Broadcast Channel Number (BCN), which identifies the broadcast connection, and a Copy Number (CN), which specifies the number of copies requested. Packet replication is performed by means of an encoding process and a decoding process. During the encoding process, the RAN and DAE's compute a dummy address interval for each broadcast packet based on its copy number. The dummy address interval is used in the BBN to ensure non-blocking operations. During the decoding process, the concentrator concentrates the input packets based on the running sum of their activity bits. The concentrator network is essentially the cascade combination of a running adder network and a reverse banyan network. Non-blocking packet replications are then performed in

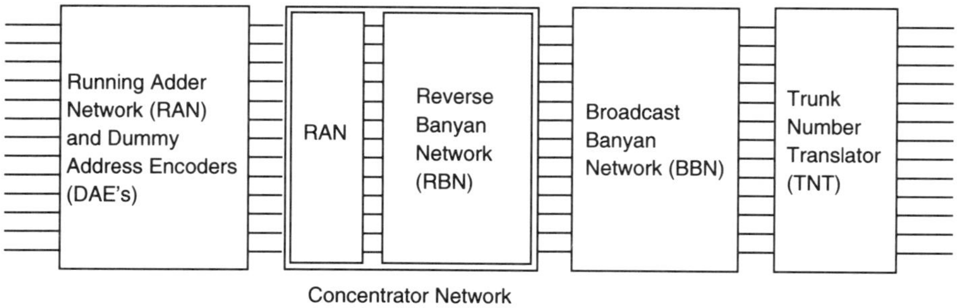

FIGURE 6. Non-blocking copy network architecture.

the BBN by using the Boolean Interval Splitting Algorithm. By means of the BCN and the copy index attached to each packet during the replication process, the TNT attaches the destination address to each resulting packet through the use of table-lookup. The packets can then be routed to their destinations in the subsequent point-to-point switch.

Using a similar observation as discussed in Section 2.3, we can conclude that the cascade combination of the concentrator and the banyan network in Fig. 6 is in fact a broadcast Clos network. In fact, the broadcast Clos network is the cascade combination of the reverse omega network and the broadcast omega network. Therefore, Lee's non-blocking copy network architecture in fact stems from the basic properties of the Clos network.

4.2. Rearrangeably Non-blocking Multiconnection Networks. In circuit switching, a lot of work has been done to construct non-blocking multiconnection networks. In a multiconnection network, each output may be connected to any one of the inputs for a total of N^N different connection patterns. In other words, a mapping $\pi = \{(s_0, \pi(s_0)), \ldots, (s_{n-1}, \pi(s_{n-1}))\}$, where $\pi(i)$ is a set of output destinations (not necessarily continuous), can be realized if

$$\pi(i) \cap \pi(j) = \emptyset \qquad \text{for} \quad i \neq j \quad \text{and} \quad 0 \leq i, j \leq n-1.$$

Most of the literature (for example, [**18, 19, 20, 21**]) focused on determining the number of middle-stage modules required for a three-stage Clos network to be non-blocking to multiconnection requests, while some (for example, [**22**]) considered two-stage network constructions. It can be shown, however, that a rearrangeably non-blocking multiconnection network can easily be constructed by the cascade combination of the broadcast Clos network and the Clos network. The cascade Benes architecture proposed in [**11**] for $N : 1$ connections basically demonstrates the same principle. In the following, we will illustrate the operations of the five-stage rearrangeably non-blocking multiconnection network by means of an example. Consider the network formed by cascading a broadcast Clos network and a Clos network as shown in Fig. 7. Suppose the following set of connection requests is to be satisfied:

$$\pi = \begin{pmatrix} 0 & 2 & 5 & 10 \\ \{6,9\} & \{0,4,7\} & \{1,5,10\} & \{3\} \end{pmatrix}.$$

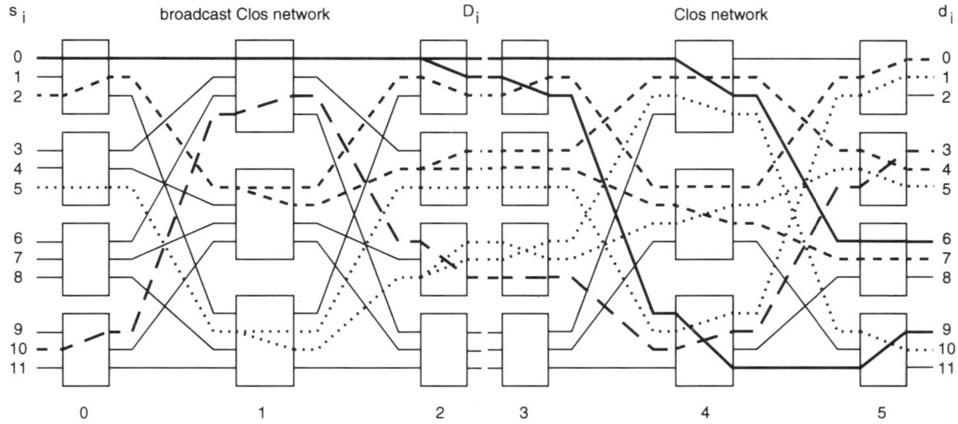

FIGURE 7. Cascade combination of a broadcast Clos network and a Clos network.

TABLE 2. Connection requests to the broadcast Clos network.

s_i	0	2	5	10
D_i	0,1	2,3,4	5,6,7	8
$m+i$	0	1	2	3
$f(s_i, D_i) = [m+i]_q$	0	1	2	0

Using the previous notations, we have $\pi(0) = \{6,9\}$, $\pi(2) = \{0,4,7\}$, $\pi(5) = \{1,5,10\}$ and $\pi(10) = \{3\}$. The broadcast Clos network is responsible for broadcasting the inputs based on the number of output destinations of each input. Given π, the number of output destinations of each input is determined and π is transformed to a set of connection requests for the broadcast Clos network. Let $\mid \pi(s_i) \mid$ denotes the number of elements in $\pi(s_i)$. In this example, input s_i is assigned a set of consecutive outputs D_i in stage 2 based on $\mid \pi(s_i) \mid$ (Table 2). By using the Rank-based Assignment Algorithm and the General Interval Splitting Algorithm, the inputs are broadcasted to the outputs in stage 2 in a non-blocking manner.

When the calls emerge from the broadcast Clos network, they subsequently enter the Clos network. Since the calls appears consecutively at the output of stage 2, the set of connection requests for the Clos network can be easily determined. In our example, let $\pi_j(s_i)$ denotes the j'th element of the set $\pi(s_i)$ and let π' be the set of connection requests in the Clos network. We can obtain π' as follows:

$$\pi' = \begin{pmatrix} 0 & 1 & 2 & 3 & 4 & 5 & 6 & 7 & 8 \\ \pi_0(0) & \pi_1(0) & \pi_0(2) & \pi_1(2) & \pi_2(2) & \pi_0(5) & \pi_1(5) & \pi_2(5) & \pi_0(10) \end{pmatrix}$$

$$= \begin{pmatrix} 0 & 1 & 2 & 3 & 4 & 5 & 6 & 7 & 8 \\ 6 & 9 & 0 & 4 & 7 & 1 & 5 & 10 & 3 \end{pmatrix}$$

This is because there are 2 output destinations from input 0, 3 from inputs 2 and 5, and 1 from input 10 respectively. Based on π', a non-blocking route assignment can be determined and the connection requests can be established in a non-blocking manner.

Obviously, no switching is performed between modules in stage 2 and modules in stage 3. We can therefore combine them together to form a five-stage rearrangeably non-blocking multiconnection network. Note, however, that this network is only suitable for circuit switching applications since centralized route assignment is required in the Clos network.

5. Conclusions

In this paper, we have studied the numbering scheme and the routing algorithms of the Clos network and a large class of multistage interconnection networks. We showed that there is a strong relationship between the numbering scheme, the non-blocking self-routing algorithm and the topological structure of a network. In this regard, we have shown that the Clos network possesses the non-blocking and self-routing properties with a proper numbering scheme and that the Clos network is actually the cascade combination of the omega and the reverse omega network. With the non-blocking properties of the Clos network, we have generalized and unified the sort-banyan principle, the construction of multicast packet switch based on broadcast banyan network, as well as the construction of rearrangeably non-blocking multiconnection networks.

References

[1] C. Clos, "A Study of Non-Blocking Switching Networks," *Bell Sys. Tech. Jour.*, Mar. 1953, pp. 406–424.

[2] V.E. Benes, "On Rearrangeable Three-Stage Connecting Networks," *Bell Sys. Tech. Jour.*, Sept. 1962, pp. 1481–1492.

[3] D.C. Opferman and N.T. Tsao-Wu, "On a Class of Rearrangeable Switching Networks Part I: Control Algorithm", *Bell Sys. Tech. Jour.*, May-Jun. 1971, pp. 1579–1600.

[4] D. Nassimi and S. Sahni, "Parallel Algorithms to Setup Up the Benes Permutation Network," *IEEE Trans. Comput.*, Vol. C-31, Feb. 1982, pp. 148–154.

[5] T.T. Lee and S.Y. Liew, "Parallel Routing Algorithms for Clos Network," *Proc. IEEE INFOCOM'96*, 1996, pp. 279–286.

[6] B.G. Douglass and A.Y. Oruc, "On Self-routing in Clos Connection Networks," *IEEE Trans. on Commun.*, Vol. 41, No. 1, Jan. 1993, pp. 121–124.

[7] D. Nassimi and S. Sahni, "A Self-Routing Benes Network, and Parallel Permutation Algorithms," *IEEE Trans. Comput.*, Vol. C-30, May 1981, pp. 157–161.

[8] R. Boppana and C.S. Raghevandra, "On Self-routing in Benes and Shuffle-exchange Networks," *Proceedings of ICC'88*, 1988, pp. 196–200.

[9] A. Huang and S. Knauer, "Starlite: A Wideband Digital Switch," *Proc. GLOBECOM '84*, Nov. 1984, pp.121–125.

[10] K.E. Batcher, "Sorting Networks and Their Applications," *The Proceedings of AFIPS 1968 SJCC,* 1968, pp.307–314.

[11] K. Sezaki, Y. Tanaka and M. Akiyama, "N:1 Connection Switching Networks Suited for Time Division Switching," *Computer Networks and ISDN Systems*, 20, 1990, pp. 383–389.

[12] D.H. Lawrie, "Access and Alignment of Data in An Array Processor," *IEEE Trans. on Comput.*, Vol. C-24, No. 12, Dec. 1975, pp. 99–109.

[13] T.T Lee and P.P. To, "Non-blocking and Self-routing Properties of Sort-Clos Networks," *Int. Jour. of Comput. Sys. Science and Engineering*, Vol. 12, No. 2, Mar. 1997, pp. 159–169.

[14] P.P. To and T.T. Lee, "Generalized Non-blocking Copy Networks," *Proceedings of IEEE ICC'97*, Jun. 1997.

[15] T.T. Lee, "Non-blocking copy networks for multicast packet switching," *IEEE J. Select. Areas Commun.*, Vol. 6, No. 9, Dec. 1988, pp. 1455–1467.

[16] B. Bingham and H. Bussey, "Reservation-based contention resolution mechanism for Batcher-banyan packet switches," *Electron. Lett.*, 23rd, Vol. 24, No. 13, Jun. 1988, pp. 772–773.

[17] J.Y. Hui and E. Arthurs, "A broadband switch for integrated transport," *IEEE J. Select. Areas Commun.,* Vol. SAC-5, No. 8, Oct. 1987, pp. 1264–1273.

[18] F.K. Hwang, "Rearrangeability of Multiconnection Three-stage Networks," *Networks*, Vol. 2, 1972, pp. 301–306.
[19] G.M. Masson and B.W. Jordan, "Generalized Multi-stage Connection Networks," *Networks*, Vol. 2, 1972, pp. 191–209.
[20] C.D. Thompson, "Generalized Connection Networks for Parallel Processor Intercommunication," *IEEE Trans. on Comp.*, Vol. C-27, No. 12, Dec. 1978, pp. 1119–1125.
[21] F.K. Hwang and A. Jajszczyk, "On Nonblocking Multiconnection Networks," *IEEE Trans. on Comm.*, Vol. COM-34, No. 10, Oct. 1986, pp. 1038–1041.
[22] G.W. Richards and F.K. Hwang, "A Two-stage Rearrangeable Broadcast Switching Network," *IEEE Trans. on Comm.*, Vol. COM-33, No. 10, Oct. 1985, pp. 1025–1035.

Department of Information Engineering, The Chinese University of Hong Kong, Shatin, N.T., Hong Kong.

Current address: Department of Information Engineering, The Chinese University of Hong Kong, Shatin, N.T., Hong Kong.

E-mail address: `ttlee@ie.cuhk.edu.hk`

Department of Information Engineering, The Chinese University of Hong Kong, Shatin, N.T., Hong Kong.

Current address: Department of Information Engineering, The Chinese University of Hong Kong, Shatin, N.T., Hong Kong.

E-mail address: `ptto@ie.cuhk.edu.hk`

DIMACS Series in Discrete Mathematics
and Theoretical Computer Science
Volume 42, 1998

An algorithm for the construction of concentrators from 2×2 sorters

Shou-Yen Robert Li*, Gar Man Koo, and Hui Li
Department of Information Engineering
The Chinese University of Hong Kong
e-mail:bobli@ie.cuhk.edu.hk
gmkoo4@ie.cuhk.edu.hk,hli6@ie.cuhk.edu.hk

June 20,1997

Abstract

This paper presents polynomial-time algorithms for the construction of deterministic and internally non-blocking concentrators from multi-stage parallel cascades of 2×2 sorters. The approach is a generalization of "Fast Knockout" and "Sortout" techniques. The goal is to construct best known m-to-n networks of 2×2 sorters for m-to-n concentration, where m and n are within the practical range. The main criterion on the complexity of concentration is the number of stages in the cascade.

*Research by the first author was supportedin part by UPGc Earmarked Grants 275/94E and 480/95E.Mathematics subject classification:94C10 Switching theory,application of Boolean algebra; 68Q99,68R99

1 Introduction

Concentration is a commonly used technique in resolving contention in switching (See, for examples,[2] [4] [5] [10] [11]). One way to construct an m-to-n concentrator in packet switching is by sorting the m inputs to the extent that the largest n among them can be separated from the remaining m-n. Such a sorter can be constructed as a multi-stage network of parallel sorting cells, where outputs of cells at one stage are inputs to cells at a later stage. The building block of the network is the sorting cell, which is typically a 2×2 sorter as depicted by Fig. 1.1.

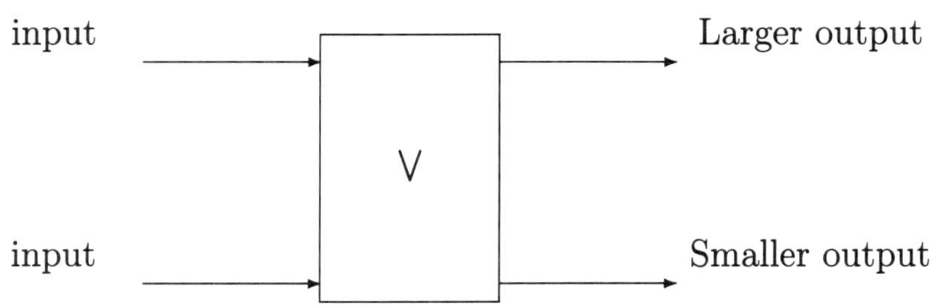

Figure 1.1. A $2\times$ 2 sorter.

It is straightforward to operate the 2×2 sorters in a *self-route* fashion. Various types of building blocks in the construction of self-route switching fabrics can be regarded as special cases of sorting devices. For example, consider the 4×4 routing cells with two 0-bound outputs and two 1-bound outputs as depicted in Fig.1.2. This routing cell can be regarded as a sorter according to the order of

$$\text{'0-bound'} > \text{'idle'} > \text{'1-bound'}$$

among inputs.

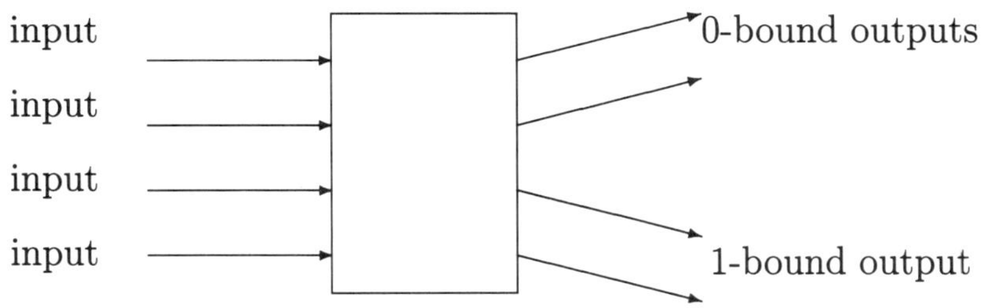

Figure 1.2. Various type of routing cells are special cases of sorter.

Thus concentration by a network of 2×2 sorters is especially suitable for self-route ATM switching. In a multi-stage parallel packet switching fabric, concentration performed at a certain stage reduces the number of packets that advance to later stages and thereby reduces the blocking probability in later stages. When properly designed, such reduction in the blocking probability can more than offset the blocking rate incurred inside the concentrator.

Unlike like software algorithms for sorting, such as "quicksort" [6] and hashing[3], concentration algorithms are hardware algorithms that do not have a centralized flow control and require no external storage. A great deal of literature has been devoted to sorting networks and architectures of concentrators, but relatively little is known about the complexity until recently. The time complexity is the packet delay through the process of concentration, which can be measured by the number of stages, also called the *depth*. The space complexity of a concentration network can be measured either by the depth or by the number of sorting cells, which is also called the *cost*. In [9], a concentrator of O(N) cost has been constructed, where the coefficient of the linear order is formidably large.

This paper is aiming at the construction of efficient concentrators with parameters in the practical range rather than the asymptotic order of the complexity. The constructions in this paper will typically incur low cost with respect to the depth. For this reason, this paper shall concentrate on low-depth constructions. Besides, there are the following reasons to emphasize the depth. Since there can be at most m/2 sorting cells at each stage, the two measures, depth and cost, are proportional to each other to a certain extent. This proportion is not exact since some of the outputs from a stage may skip the next stage. However, in many applications, inputs to sorting cells at each stage need to be synchronized in order to facilitate the detailed design. Thus, when a stage is skipped, a *delay element* needs to be introduced in order to maintain the synchronization. When the extra cost of delay elements is added on top of sorting cells, the proportion between cost and depth is enhanced. Moreover, even when synchronization is not required, often the VLSI layout aligns sorting cells stage by stage. When the empty space introduced by such alignment is taken into consideration, the total space in the layout would be very much parallel to the depth.

Section II below reviews the fast knockout technique introduced in [8], wherein the concentration algorithm is aimed at those cases with the number of inputs being a power of 2. In principle this algorithm is always applicable by always rounding up the number of necessary inputs to the next power of 2. However, such rounding up introduces inefficiency. In order to generate best known constructions, Section III presents a *conceptual* generalization of the fast knockout algorithm. The generalized algorithm involves a sub-algorithm for "pairing", which is discussed in Section IV. The fast knockout algorithm not only performs concentration but also linearly sort the winners. Section V adapts the generalized fast knockout algorithm into an algorithm *just* for concentration in order to strive for further efficiency. The performance of the algorithms proposed in this paper is evaluated in Section VI and compared with existing algorithms in the literature. Finally a brief epilogue concludes the results and suggests some directions of research on further improvement.

2 Fast Knockout

A knockout tournament, as in human athletic competitions, is a binary tree of one-to-one matches for selecting the champion. Knockout tournaments are sometimes cascaded for selecting a number of top n winners. Thus there are n tournament; the winner of the kth tournament gets the overall k^{th} place and all losers of the kth knockout tournament enter the $(k+1)^{st}$ tournament, where k<n. The n tournaments may be scheduled as partially overlapping in time. We shall refer to this type of comparison processes as *Conventional Knockout (CKO)* tournaments. Knockout tournaments have first been adopted in [10] for the construction of concentrators from 2×2 sorters in packet switching.

As two packets enter a 2×2 sorter, the numerical contents of their headers are compared against each other. When numbers, instead of human athletes, enter a network of comparators they obey laws in Boolean algebra. Fig.2.0. illustrates the Boolean functions executed by a network.

For a large network, the outputs can be very complex Boolean expressions in terms of variables representing the inputs. In general, it is not clear how to best translate such Boolean expressions into information that fits the purpose of concentration. CKO, as well as other concentration algorithms, has to somehow simplify the information carried by the Boolean expressions in order to relate the information to the purpose of concentration. Naturally the simplification process may incur loss of information, hence the resulting algorithm may lose optimality. One may regard the CKO as reducing the Boolean information obtained from comparisons to a *state vector* defined below.

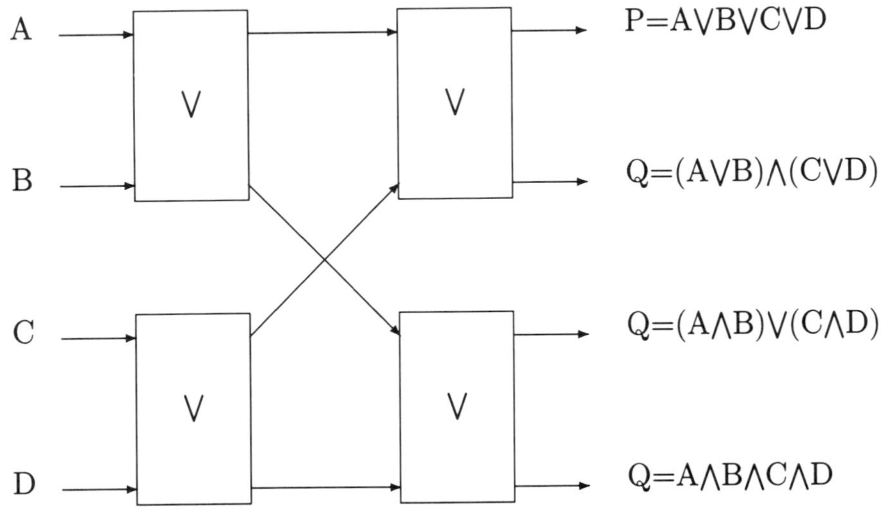

Figure 2.0. A 2-stage network of 2×2 sorter.

Definition 2.1 In a set of numbers where certain information about the ordering among them is known, define an *i-element* as any element that is known to be smaller than at least i other elements.

Definition 2.2 Fix a number n. If a set S can be partitioned into a disjoint union $S_0 \cup S_1 \cup \ldots \cup S_{n-1}$ such that S_i consists of x_i elements of S for all I, then S is said to be with the state vector $(x_0, x_1, x_2, \ldots, x_{n-1})$.

An m-by-n concentrator selects n among m inputs as its outputs, according to a certain criterion of comparison, and blocks the remaining m-n. A digital signal, or a packet, progressing through an m-to-n self-route concentrator is a binary number. Through each stage of routing cells, new information on the ordering among the m numbers is gathered. The CKO starts with the state vector (m, 0, 0, ... , 0) and terminates at the state vector (1, 1, 1, ... , 1). Thus the algorithm performs not only the concentration but also the sorting among the n winners.

Definition 2.3 The *disjoint* union of z sets in the state$(x_0,x_1,x_2,\ldots,x_{n-1})$ is said to be a set with the state vector z $\times(x_0,x_1,x_2,\ldots,x_{n-1})$

In the FKO algorithm for 2^k-to-n concentration/sorting described below, the Boolean information obtained from each stage of sorting cells will be translated into a state vector of the form $2^k\times$ $(x_0,x_1,x_2,\ldots,x_{n-1})$ with the initial state 2^k $\times$ (1, 0, 0, ... , 0). This slightly more complex form of state vector allows the advantage from the transitive law that governs the ordering on numbers. For example, let a number X be smaller than at least x numbers in a certain set and another number Y smaller than at least y numbers in a disjoint set, then the smaller between X and Y is smaller than at least x+y+1 numbers. When translated into knockout tournaments, this means that the loser in the match between X and Y can go directly into the $(x+y+1)^{st}$ tournament. For example, the last output of the network in Fig. 2.0 would be regarded as a 3-element, instead of a 2-element, by the FKO. This advantage speeds up the process of knockout.

The FKO algorithm for 2^k-to-n concentration/sorting [8]

Start with the state vector 2^k $\times$ (1, 0, 0, ... , 0). Apply an FKO *stage* for k times followed by CKO *stages* until the state vector becomes (1, 1, 1, ... , 1), where FKO and CKO stages are defined as follows.

CKO stage Consider a set in the state$(x_0,x_1,x_2,\ldots,x_{n-1})$ For 0≤i<n, form $\lfloor x_i/2 \rfloor$ pairs of i-elements for comparisons. The state vector of the set changes by

$$(x_0,x_1,x_2,\ \ldots,\ x_{n-1})$$
$$\rightarrow(\lceil\tfrac{x_0}{2}\rceil,\ \lfloor\tfrac{x_0}{2}\rfloor+\lceil\tfrac{x_1}{2}\rceil,\cdots,\ \lfloor\tfrac{x_{i-1}}{2}\rfloor+\lceil\tfrac{x_i}{2}\rceil,\cdots,\lfloor\tfrac{x_{n-2}}{2}\rfloor+\lceil\tfrac{x_{n-1}}{2}\rceil),$$

through this stage.

FKO stage Consider two disjoint sets S and T with the same state vector $(x_0,x_1,x_2,\ldots,x_{n-1})$. For i<n, pair every i-element of S with an i-elements of T for comparison. The loser in the comparison becomes a (2i+1)-element of S∪T. Through these comparisons, the state vector of S ∪ T changes from $2 \times (x_0,x_1,x_2,\ldots,x_{n-1})$ to $(x_0,\ x_0+x_1,\ x_2,\ x_1+x_3,x_4,\ x_2+x_5,\ \ldots,\ x_{2i},\ x_i+x_{2i+1},\ldots)$. Applying the same process to 2z disjoint sets with the same state vector, we can change the state vector of a set

$$2Z \times (x_0,x_1,x_2,\ldots,x_{n-1})$$
$$\rightarrow Z \times (x_0,\ x_0+x_1,\ x_2,\ x_1+x_3,x_4,\ x_2+x_5,\ \ldots,\ x_{2i},\ x_i+x_{2i+1},\ldots)$$

in a single stage.

The evolution processes of the state vector in the 8-to-4 concentration by the FKO algorithm and by CKO tournaments, respectively, are contrasted in Table 1.

	State vector	
state	FKO algorithm	CKO tournaments
0	8 × (1,0,0,0)	(8,0,0,0)
1	4 × (1,1,0,0)	(4,4,0,0)
2	2 × (1,2,0,1)	(2,4,2,0)
3	1 × (1,3,0,3)	(1,3,3,1)
4	1 × (1,2,1,2)	(1,2,3,2)
5	1 × (1,1,2,1)	(1,1,3,2)
6	1 × (1,1,1,2)	(1,1,2,2)
7	1 × (1,1,1,1)	(1,1,1,2)
8		(1,1,1,1)

Table 1. Evolution of the state vector in the 8-to-4 concentration/sorting by FKO & CKO.

Fig. 2.1 below lays out the four knockout tournaments in the 8-to-4 FKO concentrator/sorter. Meanwhile, Fig. 2.2 is a symbolic representation of the same network using similar symbols as in [6].

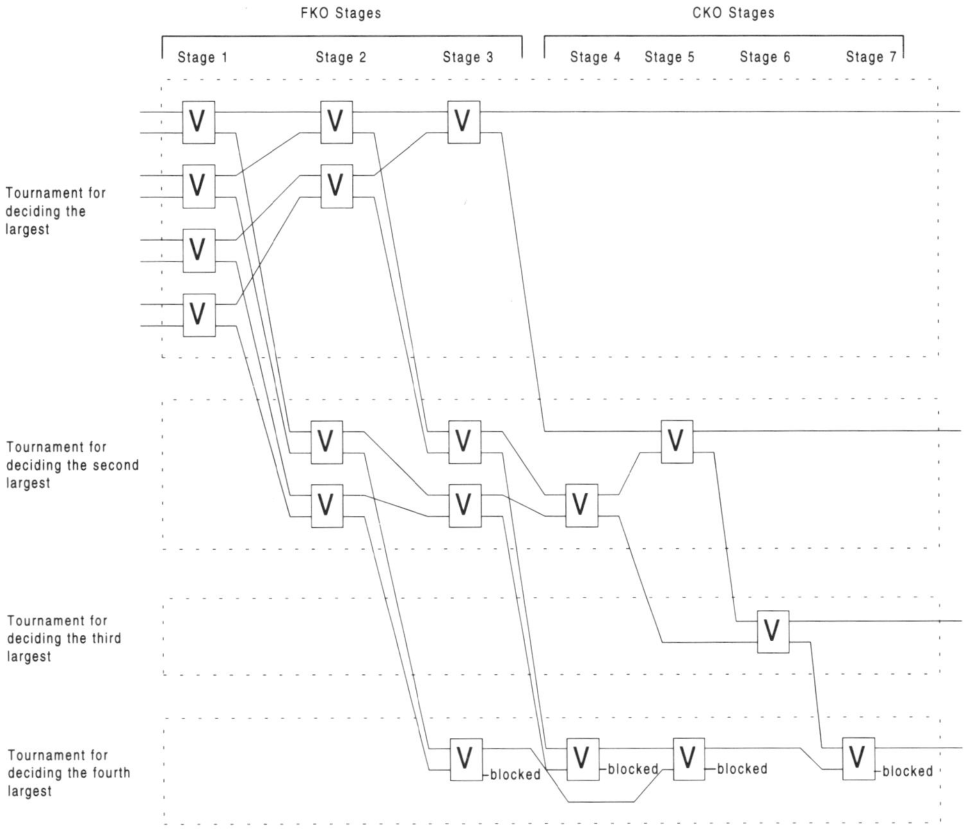

Figure 2.1. The four knock tournaments inside the 8-to-4 fast knockout concentrator/sorter.

Algorithms from the knockout approach are not very efficient when the concentration ratio is small. For example, a better 8-to-4 concentrator/sorter than both FKO and CKO is the classical Batcher sorter [1] on 8 elements, which requires only 6 stages. Despite this, all illustrative examples throughout this paper use small parameters m and n purely for the convenience in presentation.

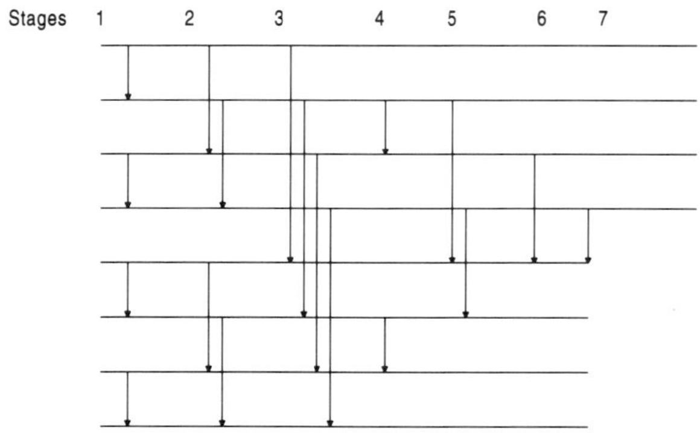

Figure 2.2. Symbolic representation of the 8-to-4 fast knockout concentrator/sorter.

3 Generalized Fast Knockout Algorithm

While taking the advantage of the transitive law, the FKO algorithm for 2^k-to-n concentration/sorting keeps the state vector in a simple form. Initially the state vector is in the form of $2^j \times (x_0, x_1, x_2, \ldots, x_{n-1})$, j>0, and an FKO stage applies. After k such stages, the state vector becomes simply in the form of $(x_0, x_1, x_2, \ldots, x_{n-1})$ and CKO stages take over. These are the only two types of the state vector throughout.

To generalize the FKO algorithm to m-to-n concentration/sorting, where m is not necessarily a power of 2, the goal again is to try to take the full advantage of the transitive law. However, the generalized algorithmic action must at all time maintain proper control on the complexity in the form of the state vector so that a polynomial-time software algorithm can generate the design of the hardware network.

To illustrate this point, consider the 12-to-5 example. The initial state vector is $12 \times (1,0,0,0,0)$. An FKO stage applies naturally and brings the state vector to $6 \times (1,1,0,0,0)$. A second FKO stage leads to $3 \times (1,2,0,1,0)$. Since 3 is an odd number, there is no natural way to apply a third FKO stage. On the other hand, if we transform the state vector from $3 \times (1,2,0,1,0)$ into $(3,6,0,3,0)$ to suit for CKO, that would mean surrendering the knowledge of the disjointness among three partially sorted sets. A better transformation of the state vector is from 3 × (1,2,0,1,0) into 2 × (1,2,0,1,0) + (1,2,0,1,0), where '+', as well as '×', indicates disjoint union of sets. When we allow this new compound form of the state vector, we can apply an FKO stage to the part represented by 2 × (1,2,0,1,0). That would take advantage of the disjointness between two of the three partially sorted sets and maintain their disjointness from the third set. As the algorithmic action gets more and more complicated, so does the form of the type of the state vector.

The Fast Knockout algorithm for m-to-n concentration/sorting

Start with the state vector m × (1,0,0,...,0). At every stage, perform the algorithmic action as prescribed in table 2 depending upon the type of the state vector.

Type	State Vector	Algorithm action
I	$(x_0,x_1,x_2,\ldots,x_{n-1})$	If$(x_0,x_1,x_2,\ldots,x_{n-1})$=(1,1,...,1) end;Else, apply a CKO stage.
II	2k × $(x_0,x_1,x_2,\ldots,x_{n-1})$	Apply an FKO stage
III	2k × $(x_0,x_1,x_2,\ldots,x_{n-1})$ $+(x'_0,x'_1,x'_2,\ldots,x'_{n-1})$	Apply FKO to the part represented by the first term and CKO to the remainder seperately.
IV	$(x_0,x_1,x_2,\ldots,x_{n-1})$ $+(x'_0,x'_1,x'_2,\ldots,x'_{n-1})$	Merge the two disjoint sets represented by the two terms with a "pairing" sub-algorithm
V	2k × $(x_0,x_1,x_2,\ldots,x_{n-1})$ $+(x'_0,x'_1,x'_2,\ldots,x'_{n-1})$ $+(x''_0,x''_1,x''_2,\ldots,x''_{n-1})$	Apply FKO to the part represented by the first term; merge the parts represented by the second and third terms with a "pairing" sub-algorithm

Table 2. Types of state vectors and the corresponding algorithmic actions.
Here '×' and '+' both indicate disjoint union of sets.

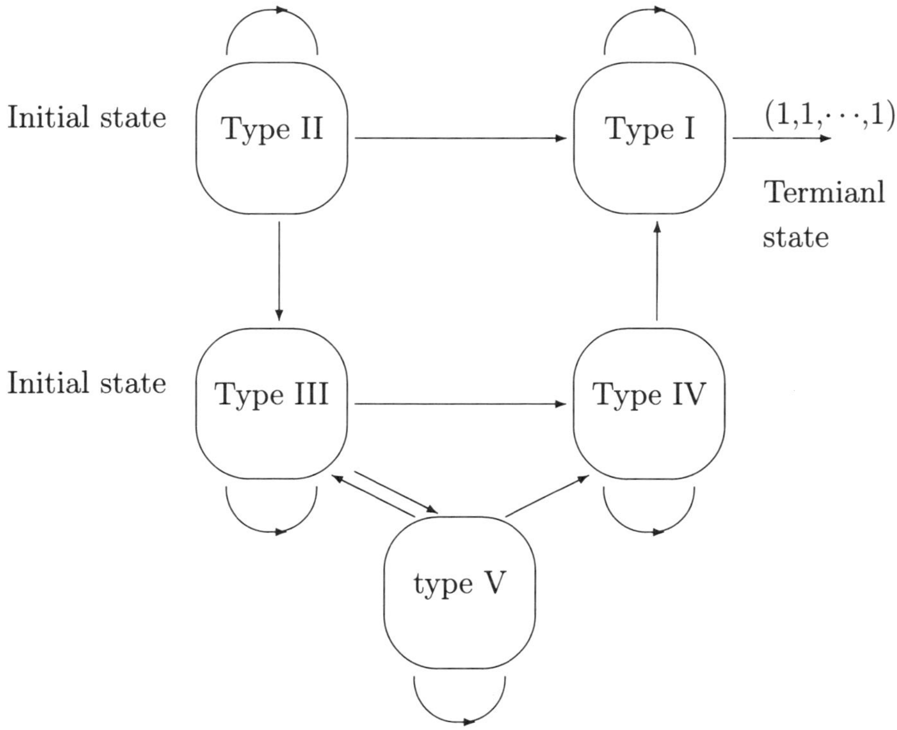

Figure 3.0. Transtion diagram of state vector.

The finite-state diagram in Fig. 3.0 describes the transition of the type of the state vector.

We now illustrate this generalized algorithm by the 12-to-5 example. After the second stage, the state vector becomes 2 x (1,2,0,1,0) + (1,2,0,1,0), which is of Type III. After another stage, the state vector becomes (1,3,0,3,0) + (1,1,1,1,0), which belongs to Type IV. The situation is illustrated by Fig. 3.1.

State	State Vector
0	$12 \times (1,0,0,0,0)$
1	$6 \times (1,1,0,0,0)$
2	$3 \times (1,2,0,1,0) = 2 \times (1,2,0,1,0) + (1,2,0,1,0)$
3	$(1,3,0,3,0) + (1,1,1,1,0)$

Table 3. The evolution of the state vector through the initial three stages of 12-to-5 FKO algorithm.

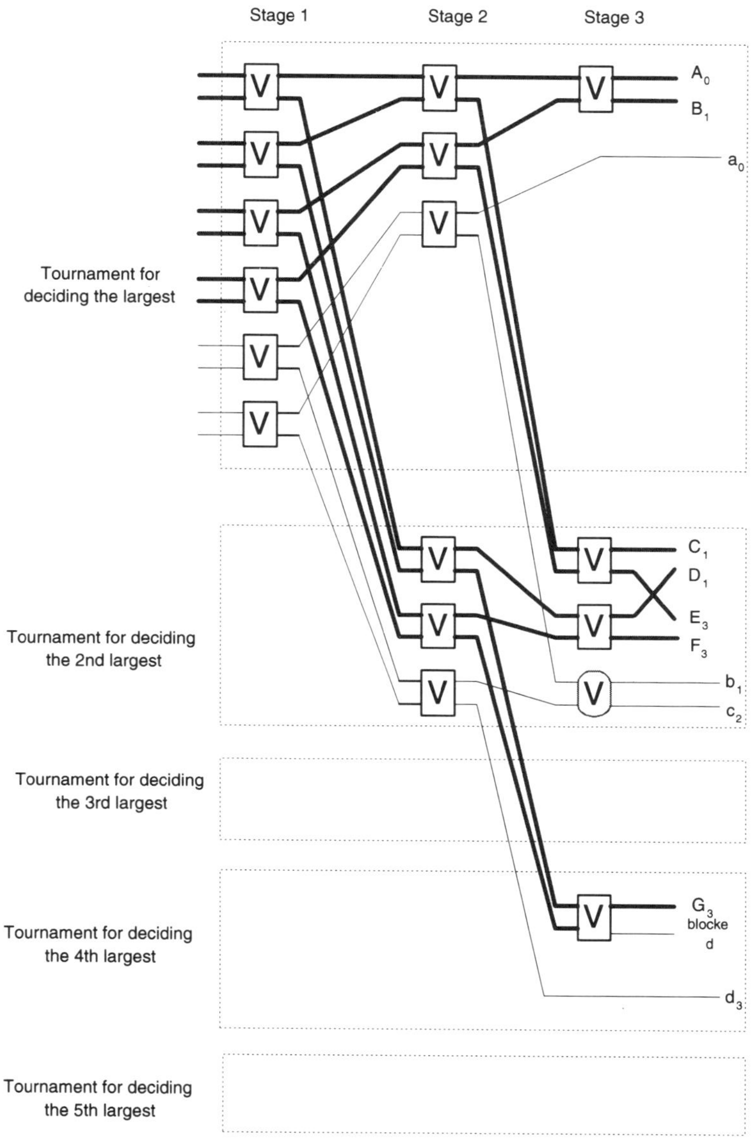

Figure 3.1. Initial three stages of the 12-to-5 FKO algorithm.

In the above 12-to-5 example, as depicted by Fig. 3.1, after the initial three stages the set of competing numbers that remain are divided into two disjoint subsets whose state vectors are (1,3,0,3,0) and (1,1,1,1,0), respectively. A pairing sub-algorithm is needed at stage 4 for merging the two disjoint subsets together in a way that takes the full advantage of the disjointness between the two subsets. The illustrative example will be continued in the next section, which deals with pairing sub-algorithms.

4 Paring Sub-algorithms

In Fig. 3.1, the initial 12 inputs can be divided into a subset of 8 elements and another of 4 so that there is no interaction between the two throughout the initial three stages. After three stages their elements are labeled as $\{A_0, B_1, C_1, D_1, E_3, F_3, G_3\}$ and $\{a_0, b_1, c_2, d_3\}$, respectively. A generic subscript j in a label indicates a j-element and the alphabetical order in the label follows the increasing order of the subscript. This labeling suggests the following simple but effective pairing algorithm.

The Left-first pairing sub-algorithm

Step 1. Label elements in one set by upper-case letters starting with 'A' and those in the other set by lower-case letters starting with 'a'. Compare elements labeled by the upper and lower cases of the same alphabetic letter. Such pairing would exhaust one of the two disjoint subsets, i.e., the one with the smaller cardinality.

Step 2. Apply CKO to the residual elements.

Since the two elements labeled by the upper and lower cases of the same alphabetical letter belong to two independent subsets respectively, the comparison takes advantage of the transitive law in the FKO style. We append to Fig. 3.1 a fourth stage according to the Left-first pairing sub-algorithm. The resulting network is illustrated in Fig. 4.1.

In general, a pairing sub-algorithm prescribes a whole stage of comparisons when the state vector is of Type IV. Figure 4.2 depicts the complete 12-to-5 FKO concentrator/sorter with Left-first pairing.

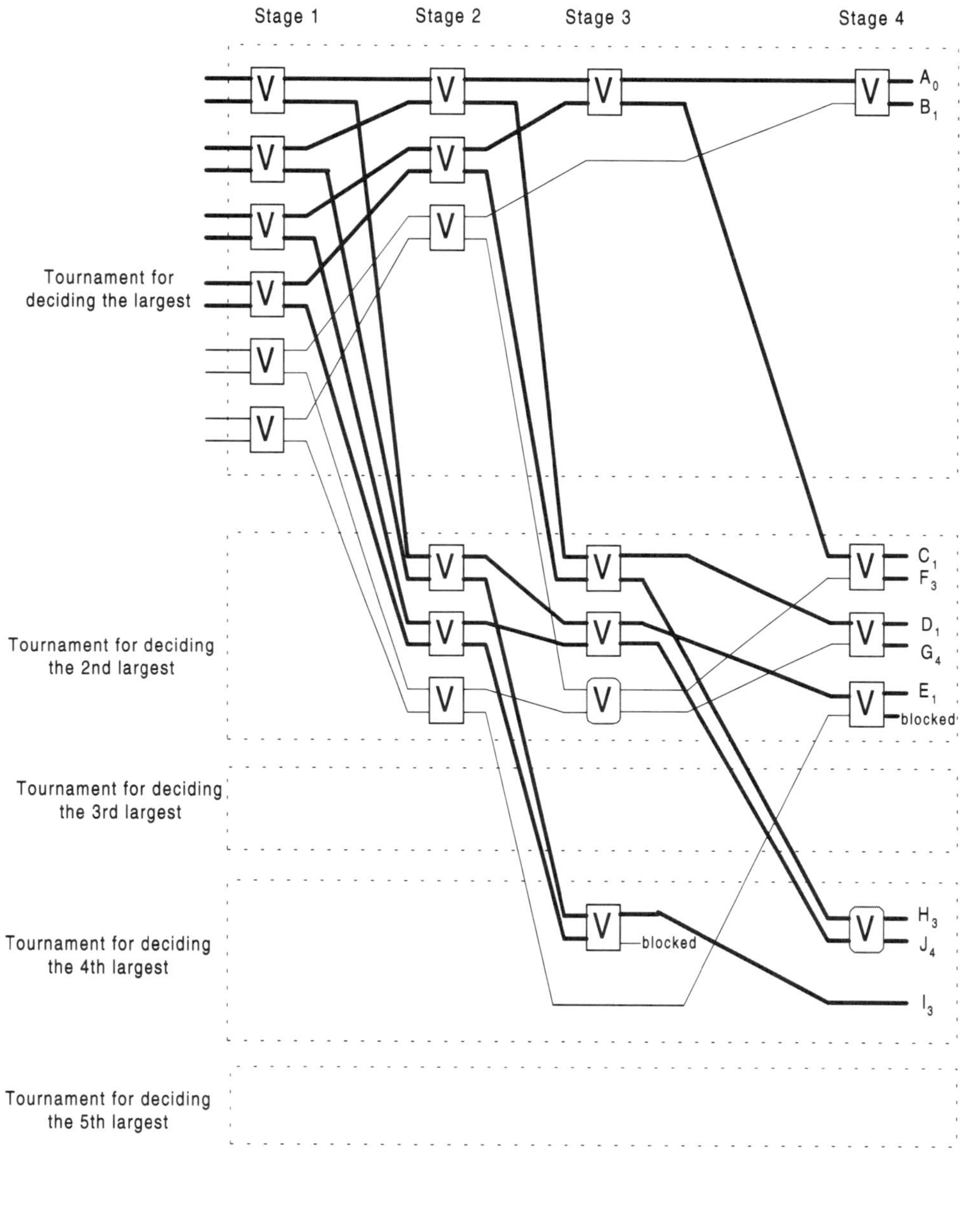

Fig. 4.1. Initial four stages of the 12-to-5 FKO concentrator/sorter with Left-first pairing.

We now introduce another sub-algorithm for pairing, called *Match-first*. Its performance will be compared against Left-first in Section VI.

The Match-first pairing sub-algorithm

Let A and B be two disjoint sets with state vectors $(a_0, a_1, a_2, \cdots, a_{n-1})$ and $(b_0, b_1, b_2, \cdots, b_{n-1})$, respectively. If there is no index j such that both a_j and b_j are nonzero, then perform the Left-first pairing sub-algorithm. On the other hand, if there is an index j such that both a_j and b_j are nonzero, then takes a j-element out of each set for comparison; then, recursively apply the match-first pairing sub-algorithm to the two residual sets.

The table below contrasts the two pairing sub-algorithms by the evolution of the state vector in the 12-to-5 FKO concentration/sorting. In general, the two pairing sub-algorithms may lead to different depths of the FKO concentrator/sorter. The performance of the two is discussed in Section VI.

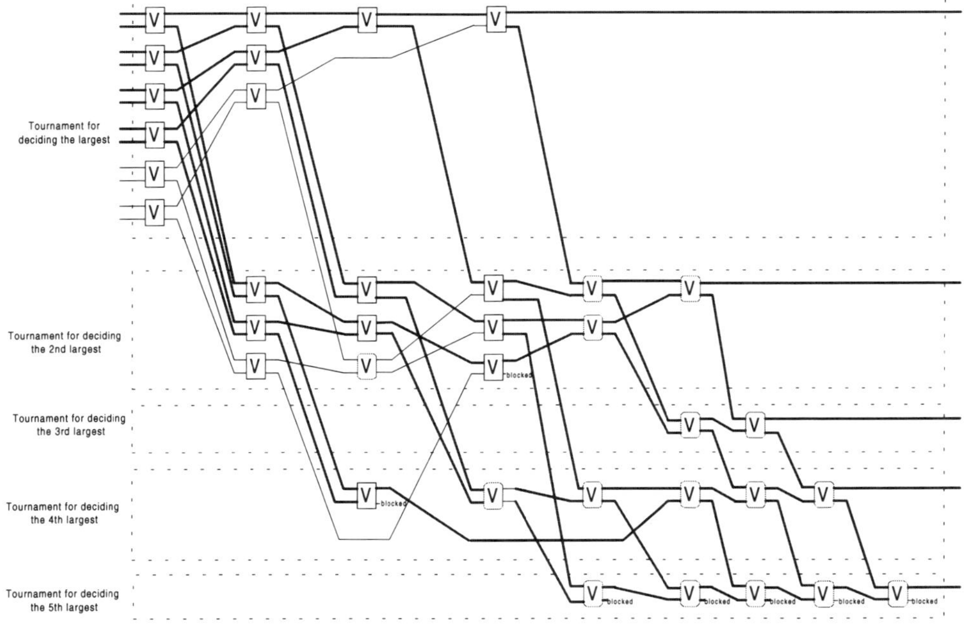

Fig. 4.2. Partition of the 12-to-5 concentration into 5 knockout tournaments.

Stage	Left-first pairing	Match-first paring
0	12 ×(1,0,0,0,0,0)	
1	6 ×(1,1,0,0,0,0)	
2	3 ×(1,2,0,1,0,0)	
	=2 ×(1,2,0,1,0,0)+(1,2,0,1,0,0)	
3	(1,3,0,3,0,0)+(1,1,1,1,0,0)	
4	(1,4,0,3,2,1)	(1,4,0,3,3,0)
5	(1,2,2,2,2,2)	(1,2,2,2,3,1)
6	(1,1,2,2,2,2)	
7	(1,1,1,2,2,2)	
8	(1,1,1,1,2,2)	
9	(1,1,1,1,1,2)	
10	(1,1,1,1,1,1)	

Table 4. Evolution of the state vector in the 20-to-6 concentration using two pairing sub-algorithms.

5 Algorithm for Concentration Without Necessarily Sorting the Winners

A sequence is said to be *circular unimodal* (or *circular bitonic* [1]) if, when the end is wrapped around to the beginning, it contains only one local maximum and one local minimum. A sequence of length jk is said to be *k-sorted* if it is the concatenation of j circular unimodal subsequences of length k such that every number in a subsequence is greater than every number in the next subsequence. In particular a 1- sorted sequence means a linearly sorted sequence. The technique of *k-sorting* is to k-sort a sequence with a smaller and smaller k until k=1. Clearly k-sorting can be applied to m-to-n concentration when n is any multiple of k. The technique of k-sorting has actually originated from topology, although only an algebraic version has appeared in the conference paper [7].

In [8], the multi-parametered *"General Sortout Algorithm "* for concentration was introduced, which combines FKO with the k-sorting together. One exemplifying representative of the General Sortout Algorithm was highlighted for its outstanding efficiency in many cases. This exemplifying algorithm may be described as follows.

Algorithm of m-to-n concentration for n $\leq$ m/2

Step 1. Apply any algorithm of concentration/sorting to perform $\lfloor m/2 \rfloor$-to-n and $\lceil m/2 \rceil$-to-n concentration/sorting on two separate sets. The results are two disjoint sorted lists of length n.

Step 2. For i = 0 to n-1, compare the i-element of the first list with the (n-1-i)-element of the second.

Fig. 5.1 below illustrates the concentrator by this algorithm. It is for concentration without necessarily sorting the n winners, because the n outputs are only guaranteed to be in the circular unimodal order but not necesssarily in the linear order.

The depth of the m/2-to-n FKO concentrator/sorter is typically much lower than that of the m-to-n counterpart when m is at least 16. When the FKO algorithm for concentration/sorting is adapted into an algorithm just for concentration as in Fig. 5.1, the efficiency is enhanced. The performance of the adapted concentration algorithm will be discussed in the next section.

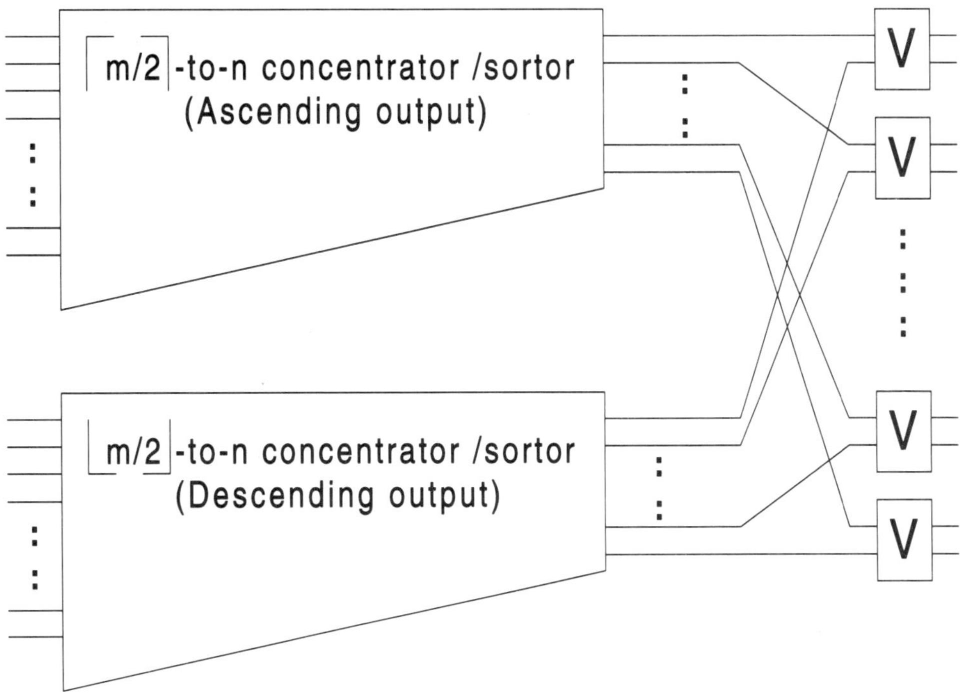

Figure 5.1. Adapting an algorithm for concentration/sorting into a pure concentration algorithm.

6 Performance comparisons

In order to make a comparison, *Left-first* and *Match-first pairing* sub-algorithms have been plugged into the generalized FKO algorithm for concentration/sorting. The depth of the m-to-n concentrator/sorter so constructed with m ranging from 8 to 2000 and n from 4 to min{16, m/2} has nonsurprisingly turned out far better than CKO. It is also better than using the 2^k-to-n FKO algorithm, where k = $\lceil \log_2 m \rceil$.

Between the two pairing sub-algorithms themselves, Match-first is superior in numerous cases, while Left-first is better in just six cases: 20-to-4, 20-to-5, 20-to-6, 21-to-4, 21-to-5, and 21-to-6 with the depth differential being just one stage in all six cases. Moreover, it is known that neither pairing sub-algorithm is optimal for the cases of 20-to-7, 20-to-8, 20-to-9, 20-to-10, 21-to-7, 21-to-8, 21-to-9, and 21-to-10.

When the FKO algorithm for concentration/sorting is adapted into an algorithm just for concentration as in Section V, the result has naturally turned out superior to using the FKO algorithm itself. The performance of this adapted algorithm is tabulated for m$\leq$512 in Table 5, where the sub-algorithm for pairing has been chosen to be the optimum between Left-first and Match-first. When n is large or when the concentration ratio is small, all algorithms from the knockout approach incur O(n) depth and are not efficient when compared with k-sorting [7] or with *"sortout"*[8]. Therefore the parameter n in Table 5 has been limited by n$<log_2$m. All values in Table 5 are the best known so far. In addition, there are other pairs of (m, n) such that the adapted FKO algorithm for concentration is optimal, e.g., when (m, n) = (8, 4).

Input	Number of Output n							
m	1	2	3	4	5	6	7	8
257-512	9	12	14	16	18	20	22	24
131-256	8	11	13	15	17	19	21	
129-130	8	10	12	14	16	18	20	
81-128	7	10						
69-080		9	12	14	16	18		
65-068	7		11	13	15			
49-064	6	9	10	12	14			
43-48		8						
35-42			10					
33–34	6	8	9	11	13			
27-32	5	7	8	10				
17-26	5	7	8	9				
11-16	4	6	7					
9-10	4	5	6					
7-8	3	4						
5-6	3	3						
3-4	2							

Table 5. Performance of adapting the FKO concentration/sorting into the algorithm for m-to-n concentration, $2^n <$ m $\leq$512, $n < log_2$m.

7 Epilogue

The "fast knockout" concept was introduced in [8] for the construction of concentrators from 2×2 sorters, and the algorithms therein have been particularly efficient when the number of inputs is a power of 2. We generalize the fast knockout algorithms to the case where the number of inputs is not necessarily a power of 2 in order to strive for efficiency in the general case. The efficiency is measured mainly by the depth in the construction of concentration network.

The most intriguing part is the sub-algorithm of "pairing", which roughly means mixing both fast knockout and conventional knockout in a single stage of 2×2 sorters. There has not been an optimal pairing sub-algorithm in general. It would be interesting to see further research in polynomial-time algorithms for pairing that lead to nearly optimal construction of concentrators of sizes in the practical range.

Another possible direction of further research is to improve the FKO algorithm itself for concentration/sorting. Conceivably, there could be a way to do m-to-n FKO with a large variety of types of the state vector instead of just the five types being used now. Yet another possible direction of further research is to look for better ways for adapting concentrator/sorter into concentrators without sorting in order for the efficiency.

All concentrators considered in this paper perform deterministic comparisons. However, concentration itself is statistical in nature. Loss is incurred concentration when too many inputs are active, and the parameters are engineered so that the loss is contained within tolerance. It would also be interesting to see applications of fast knockout to statistical concentration.

References

[1] Batcher, K. E., "Sorting networks and their applications", *Proc. AFIP 1968 Spring Joint Computer Conf.*, Vol. 32, (1968), pp 307-314.

[2] Chen, D. X. and Mark, J. M., "SCOQ: a fast packet switch with shared concentration and output queueing", *IEEE/ACM Trans. Networking*, Vol. 1, No.1 (February 1993), pp 142-151.

[3] Cormen, Leiserson, and Rivest, *Introduction to Algorithms*, The MIT Press, 1994.

[4] Giacopelli, J. N., Hickey, J. J., Marcus, W. S. , Sincoskie, W. D. and Littlewood, M., "Sunshine: a high-performance self-routing broadband packet switch architecture", *IEEE J. Select. Areas Commun.*, Vol 9, No 8 (October 1991), pp 1289-1298.

[5] Huang, A. and Knauer, S., "Starlite: a wideband digital switch", *Proc. GLOBECOM'84* (November 1984), pp 121-125.

[6] Knuth, D. E., *The Art of Computer Programming*, Volume 3, Addison-Wesley, 1973.

[7] Li, S.-Y. R., "Partial Sorting and Concentration by Parallel Networks", *Proc. Int. Workshop on Discrete Math. and Algorithms* (IWDMA'94) (December 1994), pp 27-43.

[8] Li, S.-Y. R. & Lau, C. M., "Concentrators in ATM Switching", *Computer Systems Science & Engineering*, Vol. 11, No.6 (November 1996), pp 335-342.

[9] Pippenger, N., "Self-Routing Superconcentrators", *J. of Computer and System Sciences*, Vol. 52 (1996) , pp 53-60.

[10] Wang, W. and Tobagi. F. A., "The Christmas-tree switch: an output queueing space-division fast packet switch base on interleaving distribution and concentration function", *Proc. INFOCOM'91*, Vol. 1 (April 1991), pp 163-170.

[11] Yeh, Y. S., Hluchyj, M. G. and Acampora, A. S., "The knockout switch: a simple, modular architecture for high performance packet switching", *IEEE J. Select. Areas Commun.*, Vol. 5, No 8 (October 1987), pp 1274-1283.

DIMACS Series in Discrete Mathematics
and Theoretical Computer Science
Volume **42**, 1998

Strictly Nonblocking Conferencing Meshes

Yi Du and Gerald M. Masson

ABSTRACT. This paper describes a conferencing server design based on an innovative configurable computing architecture to support the information transmission and distributed processing associated with creating and maintaining simultaneous disjoint conferences among sets of N conferees. The configurable conference servers we will consider consist of two components: a bidirectional $N \times M$ switching network used for routing information between the N conferees and a M node conferencing network comprised of processing elements which perform the actual signal processing required for a conference. For our configurable conferencing server to be strictly nonblocking, clearly the switching network must be strictly nonblocking; but so must the conference network. In this paper, we will consider r-dimensional conferencing meshes as the conference network component. An r-dimensional conferencing mesh is strictly nonblocking if, regardless of the existing conferences implemented, a new conference among any subset of the idle conferees can be implemented by means of a connected set of idle processing elements without any disturbance (rearrangement) to the existing conferences and without any information being passed-through processing elements allocated to other conferences. Using arguments employing the so-called isoperimetric ratios of the sizes of edge and node sets in a graph, we show that a necessary and sufficient condition such that an r-dimensional conferencing mesh of M processing elements provides strictly nonblocking conferencing to N conferees is that $M = O(N^{(r+1)/r})$, for fixed r. For general r-dimensional mesh, $M = O(r^{(r-1)/r}N^{(r+1)/r})$ nodes are sufficient to support strictly nonblocking conferencing for N conferees.

1. Introduction

In this paper, we study constructions of a novel architecture for configurable conferencing servers which supports the simultaneous transmission of processed messages among conference participants, referred to as conferees. Each conferee uses a generic send/receive device as a conferencing instrument which provides access to/from the server. Depending on the type of the application, the messages exchanged by conferees can involve audio, video, graphics, data, or essentially any such information medium, including combinations thereof.

The configurable conference server architecture of consideration consists of two components: a bidirectional $N \times M$ *switching network* used for routing information to/from a conference network, and an M node *conference network* comprised of

1991 *Mathematics Subject Classification.* Primary 68R10, 68M10; Secondary 05C35, 90B12.

The second author was partially supported by Compunetix Inc., Monroeville, PA [**Patent90**].

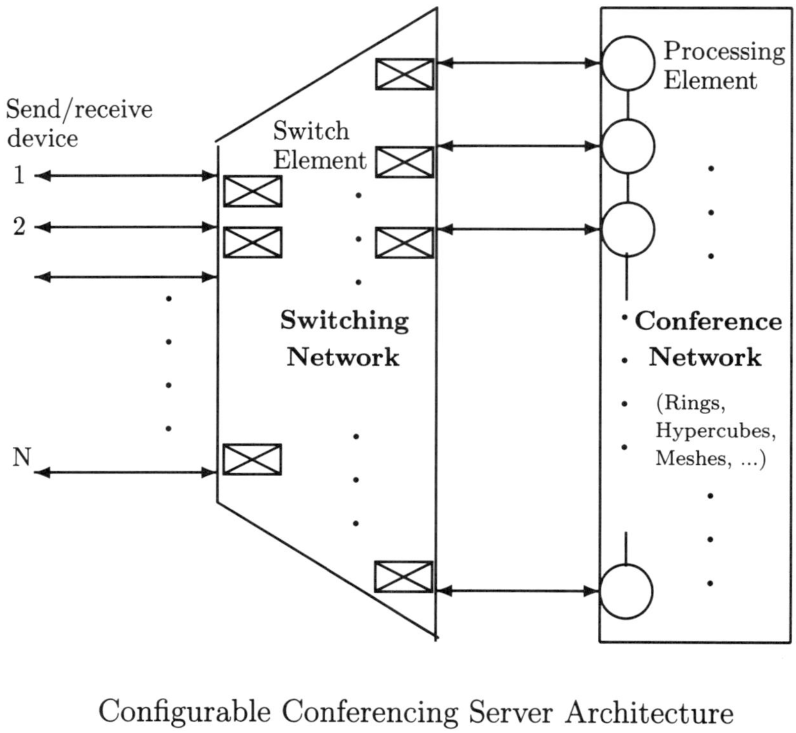

FIGURE 1. A general structure of a configurable conferencing servers. A conference among a specified set of conferees is implemented on the conference network by the processing of information exchanged among a set of directly connected processing elements, each of which is linked to exactly one of the send/receive devices associated with the conferees by means of the switching network.

processing elements (nodes) which perform the actual signal processing of the information associated with the conference. Figure 1 illustrates the general structure of the configurable conferencing server we are studying. As is shown, a conference among a specified set of conferees is realized on the conference network by processing the information exchanged among a set of directly connected processing elements, each of which is linked by means of the switching network to exactly one of the send/receive devices associated with the conferees.

We are interested in conferencing applications in which after a requested conference among a group of conferees has been set up, each conferee will remain in the conference until the entire conference is terminated. This is a reasonable assumption as very often the conference message streams being exchanged, such as video and audio streams for so-called video-conference applications, or graphics and data information for interactive distributed simulation environments, will demand a specified upper bound on the acceptable delays for message delivery, possibly with certain communication bandwidth requirements. In conferencing applications where high quality of service is demanded, these message streams can be quite sensitive to rearrangements and/or transmission disturbances resulting from reconfigurations of existing conferences. As a result, disruptions of these message streams due to additions or deletions of conferees within an established conference cannot always be tolerated.

Similarly, to support a conference of, say, k conferees, we assume that a connected set of k processing elements are allocated so that these processing elements are not involved in any way with other existing conferences which are supported by the server. Thus, these k processing elements associated with the conference form a connected component of nodes in the conference network of the server. This means that the bandwidth of the links connecting processing elements as well as the processing power of these allocated elements are available entirely for supporting a given conference.

In this paper, we consider r-dimensional conferencing meshes as the conference network component of our server architecture for providing strictly nonblocking conferencing capability. It is obvious that the $N \times M$ switching network component of our architecture must be strictly nonblocking. This can be provided in numerous ways, employing, for example, the well-known Cantor network construction [**Cantor72**]. Hence, we give this issue no particular attention in our discussions.

Our main focus will be the strictly nonblocking capability for the conference network component of our server. An r-dimensional conferencing mesh is *strictly nonblocking* if, regardless of the conferences already implemented on the r-dimensional mesh, a new conference among any subset of the idle conferees can always be implemented by means of a connected set of idle processing elements without any disturbance (rearrangement) to the existing conferences and without any information being passed through processing elements allocated to other conferences.

Using arguments employing the so-called isoperimetric ratios of the sizes of boundary edge and node sets in an r-dimensional mesh, we will show that a necessary and sufficient condition such that an r-dimensional conferencing mesh of M processing elements provides strictly nonblocking conferencing for a set of N conferees is that $M = O(N^{(r+1)/r})$ processing elements, for fixed dimension r. For general r-dimensional mesh, $M = O(r^{(r-1)/r}N^{(r+1)/r})$ nodes are sufficient to support strictly nonblocking conferencing for N conferees. Our analyses will be seen to give further insights into related results regarding the use of r-dimensional meshes in configurable conferencing server architectures.

2. Previous Research on Conferencing Server Architectures

Configurable conferencing server architectures similar to the one considered in this paper have been previously explored. Yang and Masson [**YM95**] have considered the use of a sequential ring of $M = O(N)$ processing elements for a multi-cast conference network in which rearrangements of existing conferencing configurations are permitted in order to satisfy a request of addition of a new conferee to a conference. Also utilizing a series of processing elements in the shape of a ring as the conferencing network component, Woodall [**Woodall74**] shows that for ring network $M = O(N^2)$ processing element nodes are necessary and sufficient to support strictly nonblocking conferencing for N conferees. Hypercube constructions for the conferencing network component have been studied [**HCM96a, HCM96b**], and it has been shown that $M = O(N\sqrt{\log N})$ nodes are sufficient for hypercubes to support strictly nonblocking conferencing for N conferees. The study in [**LM96**] considers the special case of a 2-dimensional mesh design and shows $M = O(N^{3/2})$ is both necessary and sufficient for nonblocking conferencing.

3. Conditions for a Strictly Nonblocking r-Dimensional Mesh

In this section we determine the necessary and sufficient conditions on the number of processing elements (nodes) required for strictly nonblocking conferencing meshes. In particular, we are interested in the number, M, of nodes in an r-dimensional mesh which can support conferences comprised of subsets of N conferees when a single link between two nodes has capacity to accommodate at most one conference. Our approach will be based on a general technique that can be deployed to obtain upper and lower bounds on M for general conferencing network structures with certain properties.

In the following, we first consider the necessary condition on M for r-dimensional meshes for strictly nonblocking conferencing capability for N conferees. Then the sufficient condition on M for strictly nonblocking r-dimensional meshes is considered.

3.1. Necessary condition on M for a strictly nonblocking r-dimensional mesh serving N conferees. Consider an r-dimensional (that is, an r-D) mesh with M nodes. Each node v in the r-D mesh can be specified by means of a vector listing its coordinates in each dimension: $v = \langle i_1^v, i_2^v, \ldots, i_j^v, \ldots, i_r^v \rangle$, where $1 \leq i_1^v, i_2^v, \ldots, i_r^v \leq M^{1/r}$. Value i_j^v indicates the coordinate of node v in the j dimension. Two nodes $v = \langle i_1^v, i_2^v, \ldots, i_r^v \rangle$ and $w = \langle i_1^w, i_2^w, \ldots, i_r^w \rangle$ are connected by a link in an r-D mesh if there exists a k $(1 \leq k \leq r)$, such that for all $j \neq k$ we have $i_j^v = i_j^w$ and, either $i_k^v = i_k^w + 1$ or $i_k^v = i_k^w - 1$.

Denote L as the *length* of any dimension of the r-D mesh. This length is measured in terms of the number of nodes in a specific dimension of the mesh; it follows that $L = M^{1/r}$. It should be noted that, depending on its location in an r-D mesh, every node has an edge (or link) degree between r and $2r$.

A *hyper-plane* in the r-D mesh consists of a set of nodes whose coordinate for one given dimension have the same value. For example, all the nodes with coordinates having the same value x $(1 \leq x \leq L)$ in dimension j (that is, all nodes with $i_j = x$) are contained in a hyper-plane in dimension j. We will denote as $H_j(x)$ the hyper-plane of nodes each with coordinate x in dimension j.

It is obvious that nodes in a hyper-plane comprise a connected graph. Thus, in satisfying a conference, assigning conferees in a conference to all the nodes in the same hyper-plane is an acceptable node allocation. However, it should be noted that doing so partitions the remaining nodes in the r-D mesh so that the size of future conferences is limited. This is a key observation to establishing a necessary condition on M for strictly nonblocking capability.

Thus, to obtain a lower bound on the number M of nodes needed for a strictly nonblocking r-D mesh, we consider a specific conference request sequence, and a method to allocate nodes to conferees to satisfy these requests. In particular, we refer to a *hyper-plane allocation* of nodes for satisfying conference requests wherein the conferees associated with a conference request are assigned to all the nodes in some hyper-planes in the r-D mesh. When the nodes in some hyper-planes are all assigned to a conference, the mesh becomes disconnected into several disjoint *connected components.* Obviously to guarantee a strictly nonblocking capability for possible future conferences requests by idle conferees, at least one of these connected components (namely, the one with most nodes) must be of sufficient size to support conference request among any remaining conferees.

To prove the main theorem for a lower bound on M, consider the following lemma which presents a regular scheme by which a hyper-plane separates an r-D mesh into several mutually disjoint sub-meshes, each has similar structure but with smaller size relative to the original mesh. Denote $B(L) = |\cup_{j=1}^{r} H_j(\lceil L/2 \rceil)|$ as the total number of nodes in r hyper-planes $H_j(L/2)$ $(1 \le j \le r)$ in an r-D mesh of length L in each dimension.

LEMMA 3.1. *An r-D mesh of length L in each dimension can be separated into exactly 2^r disjoint connected components by the r hyper-planes $H_j(\lceil L/2 \rceil) = \{\langle i_1, i_2, \dots, i_j, \dots, i_r \rangle : i_j = \lceil L/2 \rceil\}$, $1 \le j \le r$. Each of these disjoint connected components is a smaller r-D mesh with length at most $\lceil (L-1)/2 \rceil$ and has at most $(\lceil (L-1)/2 \rceil)^r$ nodes. The number of nodes in these hyper-planes is $B(L) = L^r - (L-1)^r$. Moreover, $B(L)$ is increasing on L.*

Proof: See [**DM97a**]. ∎

The following lemma is useful for proving a necessary condition on M.

LEMMA 3.2. *For integer $n \ge 0$, if $a \ge b \ge 0$, then $(a-b)^n \ge a^n - na^{n-1}b$.*

Proof: See [**DM97a**]. ∎

Now we prove the necessary condition on M to support strictly nonblocking conferencing among N conferees, by adopting a method of repeated hyper-plane allocation of nodes to satisfy a sequence of requests for conferences.

THEOREM 3.3. *$M = \Omega((N/r)^{(r+1)/r})$ nodes are necessary for an r-D mesh to support strictly nonblocking conferencing for N conferees. More precisely, M is lower bounded by $(N/(r+1))^{(r+1)/r}$.*

Proof: Assume an r-D mesh with M nodes is strictly nonblocking for N conferees. Suppose initially no conferences have been realized on the r-D mesh, so all M nodes are idle. Now suppose an initial conference request is made of size $B_1 = B(L) = L^r - (L-1)^r$. Since this is exactly the number of nodes in the hyper-planes $H_1(\lceil L/2 \rceil)$, $H_2(\lceil L/2 \rceil)$, ..., $H_r(\lceil L/2 \rceil)$, we can allocate all the nodes in these hyper-planes to the conferees for this requested conference. As discussed above, we refer to this as a hyper-plane allocation of nodes. By the lemma 3.1, such an allocation of nodes in these hyper-planes partitions the r-D mesh of M nodes with length $L = M^{1/r}$ into 2^r disjoint smaller r-D meshes, each with at most $(\lceil (L-1)/2 \rceil)^r$ nodes. Since we are assuming the r-D mesh is strictly nonblocking, the network must be able to satisfy all additional possible conference requests in a strictly nonblocking manner.

The number of nodes in each of the smaller sub-meshes resulting from the first hyper-plane allocation to satisfy the initial request is at most $S_1 = (\lceil (L-1)/2 \rceil)^r$. Clearly, additional conferencing requests among the remaining $F_1 = N - B_1$ idle conferees must now be satisfied within one of the 2^r disjoint r-D meshes, each with at most S_1 nodes. Thus,

$$S_1 \ge F_1 = N - B_1$$

where

$$S_1 \le M/2^r$$

and

$$B_1 = B(L) = L^r - (L-1)^r$$

Next suppose there are 2^r additional conference requests, each for a conference of appropriate size corresponding to a sub-mesh obtained from first step (of the same size to that of the r hyper-planes in the sub-mesh, which is no bigger than $B(\lceil \frac{L-1}{2} \rceil))$ conferees. These new conference requests can be satisfied with a second set of 2^r hyper-plane allocations of nodes, each within one of the 2^r disjoint r-D sub-meshes of at most $(\lceil \frac{L-1}{2} \rceil)^r$ nodes resulting from the first hyper-plane allocation. The total number of nodes used to satisfy these 2^r new conference requests is at most $B_2 = 2^r B(\lceil \frac{L-1}{2} \rceil)$, where

$$B_2 = 2^r B(\lceil \frac{L-1}{2} \rceil) \leq 2^r B(\frac{L}{2}) = 2^r((\frac{L}{2})^r - (\frac{L}{2} - 1)^r) = L^r - (L-2)^r$$

Also, note that using a second round of hyper-plane allocations of nodes within each of the 2^r disjoint r-D meshes results in another partitioning of each of these smaller meshes into 2^r still smaller disjoint r-D meshes, each now having at most $S_2 = (\lceil (\lceil \frac{L-1}{2} \rceil - 1)/2 \rceil)^r$ nodes. Because the overall r-D mesh is assumed to be strictly nonblocking, additional conferencing requests among the remaining $F_2 = N - B_1 - B_2$ idle conferees must now be satisfied within the still smaller 2^r disjoint r-D meshes, each with at most S_2 nodes. Thus,

$$S_2 \geq F_2 = N - B_1 - B_2$$

where

$$S_2 \leq \frac{S_1}{2^r} \leq \frac{M}{2^r}\frac{1}{2^r} \leq \frac{M}{(2^2)^r}$$

If we repeat the above conference request and hyper-plane allocation process k times for incoming conference requests of appropriate sizes, we would have $p^k = 2^{kr}$ disconnected sub-meshes each with S_k nodes, where B_k is the number of nodes assigned to conferees in the k-th step (which is the same number of nodes in the hyper-planes used in conference assignment in this step, as our conference request pattern indicates), and F_k idle nodes. Since the r-D mesh is strictly nonblocking,

$$S_k \geq F_k = N - B_1 - B_2 - \ldots - B_k = N - \sum_{i=1}^{i} B^i$$

where

$$S_k \leq \frac{M}{(2^k)^r}$$

$$B_k \leq p^{k-1}B(L/2^{k-1}) = (2^r)^{k-1}((L/2^{k-1})^r - (L/2^{k-1}-1)^r) = L^r - (L-2^{k-1})^r$$

$$B_i \leq L^r - (L-2^{i-1})^r,\ 1 \leq i \leq k$$

Thus for all k we need

$$\frac{M}{(2^k)^r} \geq N - \sum_{i=1}^{k} 2^{i-1} r M^{\frac{r-1}{r}} = N - \sum_{i=1}^{k} L^r - (L-2^{i-1})^r = N - (kL^r - \sum_{i=1}^{k}(L-2^{i-1})^r)$$

or,

$$\frac{M}{(2^k)^r} + (kL^r - \sum_{i=1}^{k}(L-2^{i-1})^r) \geq N$$

Consider $f(k) = \dfrac{M}{(2^k)^r} + (kL^r - \sum_{i=1}^{k}(L-2^{i-1})^r)$. By Lemma 3.2, we have

$$f(k) \leq \frac{M}{(2^k)^r} + kL^r - \sum_{i=1}^{k}(L^k - rL^{r-1}2^{i-1}) \leq \frac{M}{(2^k)^r} + rL^{r-1}(2^k-1) \leq \frac{M}{(2^k)^r} + rL^{r-1}2^k$$

$$= \frac{M}{(2^k)^r} + rM^{\frac{r-1}{r}}2^k = M^{\frac{r-1}{r}}\left(\frac{M^{\frac{1}{r}}}{(2^k)^r} + r2^k\right) = g(k)$$

$g(k)$ achieves its minimum when $M^{\frac{1}{r}} = (2^k)^{r+1}$, or $2^k = M^{\frac{1}{r(r+1)}}$. That is,

$$\min g(k) = M^{\frac{r-1}{r}}\left(\frac{M^{\frac{1}{r}}}{(M^{\frac{1}{r(r+1)}})^r} + rM^{\frac{1}{r(r+1)}}\right) = (r+1)M^{\frac{r}{r+1}}$$

However, since for each k we must have $g(k) \geq f(k) \geq N$, we have,

$$(r+1)M^{\frac{r}{r+1}} \geq N$$

or

$$M \geq (N/(r+1))^{(r+1)/r} = \Omega((N/r)^{(r+1)/r})$$

∎

Before considering sufficient conditions for r dimensional meshes, we consider the necessary condition for the case of an r-dimensional torus. An r-D torus is similar to an r-D mesh, except that every node now has an identical degree of $2r$, as the special situation of "border" has been eliminated in the r-D torus. Thus, the r-D torus has slightly more connectivity than that of a corresponding r-D mesh.

The arguments deployed in deriving necessary condition for r-D mesh can be used to obtain the following result for r-D torus.

THEOREM 3.4. *$M = \Omega((N/r)^{(r+1)/r})$ nodes are necessary for an r-D torus to support nonblocking conferencing for N conferees.*

3.2. Sufficient condition on M for a strictly nonblocking r-dimensional mesh serving N conferees. In this section we determine for an r-D mesh the number of nodes, M, sufficient to support strictly nonblocking N-node conferencing. To do this, we first determine the minimum number of boundary edges (or links) of a connected component of nodes with a specified size which can separate that component from all other components. In general, consider component C of graph $G(V, E)$. C is a subset of nodes in the graph; that is, $C \subset V$. The *edge boundary* $BE(C)$ of C is $BE(C) = \{(v,w) : (v,w) \in E(G), v \in C, w \notin C\}$, and each edge $e \in BE(C)$ is called a *boundary edge* for component C. The *node boundary* $BN(C)$ is $BN(C) = \{w : (v,w) \in E, v \in C, w \notin C, w \in V\}$, and each node $n \in BN(C)$ is called a *boundary node* for C.

For many graphs, a lower bound can be determined for the ratio of the number of boundary edges (nodes) for an component versus the number of nodes in the component. The *edge (node) isoperimetric ratio* [**BL91**] $\lambda_e(C)$ ($\lambda_v(C)$) of a component C of given size k is

$$\lambda_e(C) = \min_{|C|=k} \frac{|BE(C)|}{|C|}, \quad \lambda_v(C) = \min_{|C|=k} \frac{|BN(C)|}{|C|}$$

For some graphs, this ratio is known to be lower bounded by a constant which does not depend on the component size k. For example, expanders and superconcentrators are such classes of graphs [**Alon86, AM85**]. However, for other graphs, the isoperimetric ratio depends on the size k of the component. This property is particularly useful to us as given a component set of nodes of certain size, we can determine a lower bound for the number of edges necessary to separate the component from other nodes of the graph. In other words, if a fixed upper bound is known on the number of edges that can be used to separate a component, the size of the component cannot be arbitrarily small. The component size thus has a lower bound depending on the number of boundary edges used. This observation is the basis of our proof of the theorem to obtain an upper bound on M for the strictly nonblocking r-D mesh.

The following theorem by Bollobás and Leader [**BL91**] presents a key relationship between size of a component and the size of its edge boundary for r-D meshes.

THEOREM 3.5. **[Bollobás and Leader, 1991]**
Let A be a subset of connected nodes in an r-D mesh $[L]^r$ with $A \leq L^r/2$. Then

$$|BE(A)| \geq \min\{|A|^{1-1/k} k L^{(r/k)-1} : k = 1, 2, \ldots, r\}$$

We can derive the isoperimetric ratio for r-D meshes from this theorem. In the following, BE_i is used to denote $BE(C_i)$, the edge boundary for component C_i. Similarly, B_i denotes $B(C_i)$, the node boundary for C_i.

THEOREM 3.6. *Given connected components in an r-D mesh with size M, consider component C_i of size $|C_i|$ such that $M/|C_i| \geq e^r$. The edge boundary BE_i used to separate C_i from other components has size $|BE_i|$ satisfying the following inequality:*

$$\frac{|BE_i|}{|C_i|} \geq \lambda_e(C_i) = \frac{r}{|C_i|^{1/r}}$$

Proof: See [**DM97a**]. ∎

Now we consider the sufficient condition on M for an r-D mesh. The basic idea is to establish a lower bound on the maximum size idle node component (the one with the largest size) such that we can always find a component to allocate to the remaining conferees. Since at most N nodes are allocated to conferees, we only need to consider the case where all components have size less than N. This fact allows the use of the above theorem to determine the maximum component size, since, for any $|C_i|$, we have $M/|C_i| \geq M/N$. The following theorem deals with r-D mesh with *fixed* dimension; that is, r is a constant.

THEOREM 3.7. *An r-D mesh with $M = O(N^{(r+1)/r})$ nodes can support strictly nonblocking conferences for N conferees, when dimension r is fixed.*

Proof: Consider any configuration where a certain number of conferees (less than N, of course) are allocated arbitrarily to nodes. We want to establish a lower bound on the size of the maximum connected idle component among all those separated by the allocated conferee nodes. If the size of this maximum component is sufficiently large, as indicated by its lower bound for any number of allocated nodes, we can guarantee the existence of a component with enough size to allocate to any future conference request. The network thus provides a strictly nonblocking conferencing capability.

Consider an r-D mesh with M nodes which is to support strictly nonblocking conferencing applications. Denote X as the set of nodes already allocated to conferees, and C as the set of components of idle nodes, which will be referred to as *idle components.* Without loss of generality, suppose C consists of k ($k \geq 1$) connected components, separated by (boundary) nodes in X. So $C = \{C_1, C_2, \dots, C_k\}$. Since these components are disjoint, we have $|\cup C_i| = \sum |C_i|$. We also have $M = |X| + |\cup C_i| = |X| + \sum |C_i|$. Note that all nodes in $\cup C_i$ are idle, and each node v in X can be either a boundary node for one component, or not a boundary node for any component.

Each connected idle component is separated from other idle components by a set of boundary edges, which in turn are incident on boundary nodes in X. Denote such an edge boundary set as BE_i for component C_i if it is comprised of edges with one end in X and the other end in C_i. The edge boundary BE_i separates C_i from the other components. Since each boundary node must be allocated to a conferee, we have $\cup_{i=1}^k B_i \subseteq X$. Each $v \in \cup_{i=1}^k B_i$ can be incident to at most $2r$ boundary edges, while a boundary edge is always connected to a boundary node. Thus we have

$$|\cup_{i=1}^k B_i| \leq |X|$$

and

$$\sum_{i=1}^k |BE_i| \leq 2r |\cup_{i=1}^k B_i|$$

Thus,

$$|X| \geq \frac{1}{2r} \sum_{i=1}^k |BE_i| \tag{3.1}$$

and the number of idle nodes $M - |X|$ in the mesh is,

$$M - |X| = \sum_{i=1}^k |C_i| \tag{3.2}$$

Consider any component C_i. If there exists C_i such that $|C_i| \geq N$, then we have a component with sufficient size to realize any conference of no more than N conferees. Obviously, the mesh can support any feasible request in this case. Hence, we only need to consider the case when $|C_i| \leq N$. For such C_i, consider its edge boundary set BE_i. By Theorem 3.3, $M/|C_i| \geq (\frac{N}{r+1})^{(r+1)/r}/N = N^{1/r}/(r+1)^{(r+1)/r}$. When N is sufficiently large ($N \geq c_0$ where $c_0 = ((r+1)^{(r+1)/r} e^r)^r$ is a constant), $N^{1/r}/(r+1)^{(r+1)/r} \geq e^r$, so $M/|C_i| \geq e^r$. By Theorem 3.6, for such C_i, we have

$$\frac{|BE_i|}{|C_i|} \geq \frac{r}{|C_i|^{1/r}}$$

or

$$\frac{|BE_i|}{|C_i|} \geq \frac{r}{|C_{max}|^{1/r}} \tag{3.3}$$

here $C_{max} = C_j$ with $|C_j| = \max_{i=1}^{k} |C_i|$.

Note that in general for $a, b, c, d, e > 0$, if $\frac{a}{b} \geq e$ and $\frac{c}{d} \geq e$, then $\frac{a+c}{b+d} \geq e$. Thus by formula 3.1, 3.2 and 3.3 we have

$$\frac{|X|}{M-|X|} \geq \frac{\frac{1}{2r}\sum_{i=1}^{k}|B_i|}{\sum_{i=1}^{k}|C_i|} \geq \frac{1}{2r}\frac{\sum_{i=1}^{k}|B_i|}{\sum_{i=1}^{k}|C_i|} \geq \frac{1}{2r}\frac{r}{|C_{max}|^{1/r}} = \frac{1}{2|C_{max}|^{1/r}}$$

or,

$$|C_{max}| \geq \left(\frac{M-|X|}{2|X|}\right)^r \tag{3.4}$$

For every configuration, if the remaining largest idle component C_{max} has big enough size to support any conference among the remaining idle nodes, the r-D mesh is strictly nonblocking. That is, we need to guarantee that

$$|C_{max}| \geq N - |X|, \forall |X| \leq N \tag{3.5}$$

The above condition 3.5 is satisfied when the following holds:

$$\min_X \left(\frac{M-|X|}{2|X|}\right)^r \geq N - |X|$$

and from Inequality 3.4 it is sufficient that

$$\min_X \left(\frac{(M-|X|)}{2|X|}\right)^r + |X| \geq N \tag{3.6}$$

The above inequality 3.6 holds when the following is satisfied:

$$\min_X f(|X|) \geq N,\ f(|X|) = \left(\frac{M-N}{2|X|}\right)^r + |X|$$

The derivative $f'(x)$ of function $f(x)$ is $f'(x) = 1 - \left(\frac{M-N}{2}\right)^r \frac{r}{x^{r+1}}$, so $f(|X|)$ reaches its minimum when $|X|^{r+1} = r\left(\frac{M-N}{2}\right)^r$, or $|X| = \left(r\left(\frac{M-N}{2}\right)^r\right)^{\frac{1}{r+1}}$, and

$$\min_X f(|X|) = \left(\frac{M-N}{2\left(r\left(\frac{M-N}{2}\right)^r\right)^{\frac{1}{r+1}}}\right)^r + \left(r\left(\frac{M-N}{2}\right)^r\right)^{\frac{1}{r+1}} = \left(r^{\frac{1}{r+1}} + \frac{1}{r^{\frac{r}{r+1}}}\right)\left(\frac{M-N}{2}\right)^{\frac{r}{r+1}}$$

Thus we need,

$$\left(r^{\frac{1}{r+1}} + r^{-\frac{r}{r+1}}\right)\left(\frac{M-N}{2}\right)^{\frac{r}{r+1}} \geq N$$

or,

$$M \geq 2\left(\left(\frac{N}{r^{\frac{1}{r+1}} + r^{-\frac{r}{r+1}}}\right)^{\frac{r+1}{r}} \right) + N = 2\left(\left(\frac{1}{r^{\frac{r}{r+1}} + r^{-\frac{r}{r+1}}}\right)^{\frac{r+1}{r}} N^{\frac{r+1}{r}} \right) + N$$

Note that, for $r > 1$, we have $\left(\frac{1}{r^{\frac{1}{r+1}} + r^{-\frac{r}{r+1}}}\right)^{\frac{r+1}{r}} \leq \left(\frac{1}{r^{\frac{1}{r+1}}}\right)^{\frac{r+1}{r}} = \frac{1}{r^{\frac{1}{r}}}$,
Thus it suffices that

$$M \geq 2\left(\frac{1}{r^{\frac{1}{r}}} N^{\frac{r+1}{r}} + N\right) = O\left(\frac{1}{r^{\frac{1}{r}}} N^{\frac{r+1}{r}} + N\right)$$

When r is constant, $N = o(N^{(r+1)/r})$, so

$$M = O(N^{(r+1)/r})$$

∎

When the mesh dimension r is a fixed integer, from Theorem 3.3 we know the necessary condition is $\Omega(((N/r)^{(r+1)/r}) = \Omega(N^{(r+1)/r})$. Together with Theorem 3.7 we have the following result.

THEOREM 3.8. *For an r-D mesh with a fixed dimension r,* $M = \Theta(N^{(r+1)/r})$ *nodes are necessary and sufficient to support a strictly nonblocking conference applications for N conferees.*

3.3. Variable mesh dimensionality. The above discussions consider conditions for an r-D mesh under the assumption that the dimensionality r is a fixed constant. We now consider the implications of loosening this assumption such that r can be varied. That is, it is only required that $L^r = M$ is satisfied for r-D mesh with M nodes and length L.

For this case, we have the following theorem on the edge isoperimetric ratio for r-D mesh with arbitrary r value. Note that, the lower bound here is smaller compared to the case when r is a fixed constant.

THEOREM 3.9. *For any connected component* C_i *in an r-D mesh with size M, the edge boundary* BE_i *used to separate* C_i *from others has size* $|BE_i|$ *satisfying:*

$$\frac{|BE_i|}{|C_i|} \geq \frac{1}{|C_i|^{1/r}}$$

Proof: See [**DM97a**]. ∎

Using a similar argument as in Theorem 3.7, we can prove a more general result for an r-D mesh. Here there is no constraint on the dimensionality r.

THEOREM 3.10. *An r-D mesh with* $M = O(r^{(r-1)/r} N^{(r+1)/r})$ *nodes can support strictly nonblocking conference for N conferees.*

Proof: See [**DM97a**]. ∎

Note that this upper bound can be tighter, as for the fixed dimensionality case, for r values appropriately selected. Thus, Theorem 3.7 can be viewed as a special case of this more general theorem.

In the following, we discuss the sufficient conditions for a nonblocking r-D torus.

3.4. Results for the r-dimensional torus. The torus and the mesh have very similar structures, while the torus has slightly more edges. Bollobás and Leader [**BL91**] prove also result for the r-dimensional torus. Using similar reasoning to that in the proof of Theorem 3.7, we have the following theorems. Refer to [**DM97a**] for detailed proofs.

THEOREM 3.11. *An r-D torus with $M = O(N^{(r+1)/r})$ nodes can support strictly nonblocking conferences for N conferees, when r is fixed. For variable size r, $M = O(r^{(r-1)/r}N^{(r+1)/r})$ is sufficient for a strictly nonblocking r-D torus conference network for N conferees.*

when combined with Theorem 3.4, we have

THEOREM 3.12. *For an r-D torus with a fixed dimension r, $M = \Theta(N^{(r+1)/r})$ nodes is necessary and sufficient to support a strictly nonblocking conference network for N conferees.*

4. Conclusion

In this paper, we considered r-dimensional conferencing meshes as a configurable conference server component. Using an argument employing the isoperimetric ratio of the sizes of edge and node sets in a graph, we showed that a necessary and sufficient condition such that an r-dimensional conferencing mesh of M processing elements can provide strictly nonblocking conferencing capability among a set of N conferees is that $M = \Omega(N^{(r+1)/r})$ processing elements, when r is fixed. We also give the general bound of $M = O(r^{(r-1)/r}N^{(r+1)/r})$ for a strictly nonblocking r-dimensional mesh for N conferees for variable r values.

It is important to note that our approach to establishing the sufficient conditions on M for strictly nonblocking r-D meshes can be extended to general graph structures, such as, for example, expanders and lattices, or, indeed, to any graphs for which the isoperimetric ratios can be determined [**DM97b**]. Hence, we have established a fundamental relationship between the size requirements on M for strictly nonblocking conferencing among N conferees using certain graph structures and the isoperimetric ratios for those structures.

References

[Alon86] N. Alon. *Eigenvalues and Expanders*, Combinatorica: 85-98 **6**, 1986.

[AM85] N. Alon & V.D. Milman, λ_1, *Isoperemetric inequalities for graphs, and Superconcentrators*, J. Combin. Theory Ser. B: 73-88, **38** 1995.

[BL91] B. Bollobás & I. Leader. *Edge-Isoperimetric Inequalities in the Grid, Combinatorica*: 299-314, 11(4), 1991.

[Cantor72] D.G. Cantor. *On Non-Blocking Switching Networks*, Networks **1**, 367-377(1972).

[DM97a] Y. Du & G.M. Masson. *Configurable Conferencing Server Architectures Using High Dimensional Mesh Networks*, Technical Report TR-1997-12, Department of Computer Science, The Johns Hopkins University, 1997.

[DM97b] Y. Du & G.M. Masson. *Isoperimetric Inequalities and General Theorem on Sufficient Conditions for Strictly Nonblocking Conference Networks*, Technical Report TR-1997-14, Department of Computer Science, The Johns Hopkins University, 1997.

[HCM96a] J.F. Houlahan, L.M. Cowen, & G.M. Masson. *Hypercube Sandwich Approach to Conferencing*, Journal of Supercomputing, 10, 271-283(1996).

[HCM96b] J.F. Houlahan, L.M. Cowen, & G.M. Masson. *Wide-Sense Non-blocking Hypercube Sandwich Conference Networks*, Proceedings of 30th Annual Conference on Information Sciences and Systems, 222-227, 1996.

[LM96] Y. Li & G.M. Masson. *Strictly Non-blocking Mesh Conference Networks*, Proceedings of 30th Annual Conference on Information Sciences and Systems, 228-233, 1996.

[Woodall74] D.R. Woodall. *The Bay Restaurant – A Linear Storage Problem*, American Math. Monthly: 240-246, March 1974.

[YM95] Y. Yang & G.M. Masson. *Broadcast Ring Sandwich Networks*, IEEE Transactions on Computers, 44(10), 1169-1180(1995).

[Patent90] *Programmable Conference Modules for Switchable Ring Arrays*, U.S. Patent No. 4,094,948. Driscol, J., Law, E., Kancel, M., and Masson, G., assigned to Compunetix, Inc., Monroeville, PA, 1990.

Department of Computer Science, The Johns Hopkins University, Baltimore, Maryland 21218

E-mail address: yidu@cs.jhu.edu

Department of Computer Science, The Johns Hopkins University, Baltimore, Maryland 21218

E-mail address: masson@cs.jhu.edu

DIMACS Series in Discrete Mathematics
and Theoretical Computer Science
Volume **42**, 1998

An $O(N^{1.695})$ Permutation Routing Algorithm on Augmented Data Manipulators

Ehab S. Elmallah and Chin-Hung Lam

ABSTRACT. This paper shows an $O(N^{1.695})$ time algorithm for routing any arbitrary admissible permutation on Augmented Data Manipulator networks.

1. Introduction

The class of *Augmented Data Manipulators* (ADMs) is defined as follows: for $n \geq 1$, an ADM of size $N = 2^n$, has N inputs and outputs. The switch fabric has $n+1$ stages of switches, labeled from left to right, V_i for $i = n, n-1, \cdots, 0$ (e.g., see Figure 1.a). Each stage has N switches, and each switch has a unique address between 0 and $N-1$. Switches in the input stage V_n are all 1-to-3 multiplexers. Similarly, switches in the output stage are all 3-to-1 demultiplexers. Switches in all possible intermediate stages (i.e., stages with indices $i = n-1, n-2, \cdots, 1$) are 3×3 switches, in which at most one input-output connection can be active at any time. Links are organized into n stages, labeled from left to right L_i for $i = n-1, \ldots, 0$. Stage L_i connects switch $x \in V_{i+1}$ to switches $\{x, x+2^i, x-2^i\} \in V_i \pmod N$. A special case arises in the leftmost stage L_{n-1}, where $x + 2^{n-1} = x - 2^{n-1} \pmod N$, and hence each input switch has access to exactly two distinct switches in V_{n-1}.

1991 *Mathematics Subject Classification.* Primary 94Cxx.
This research is supported by NSERC Canada under grant number OGP 36899.

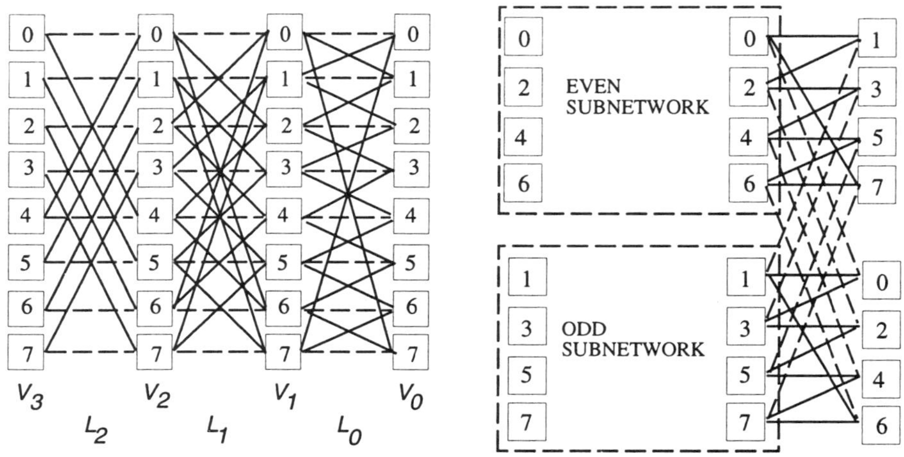

Figure 1.a An ADM with $N = 8$ inputs   1.b The even/odd recursive decomposition

The topology of the ADM networks has been originally introduced in [**Fen74**], and subsequently explored in [**SS78**]. Several combinatorial and enumeration results on routing permutations in ADM networks appear in the literature. In one direction, for any $N = 2^n$, it is known that the set of ADM admissible permutations is a proper superset of its counterpart in the *generalized cube network* [**SNKN89**] (or, equivalently, the *Omega* network [**Law75**] under a suitable relabeling of the terminals) (see, for example, [**Sie90**] and the references therein). The related problem of counting the number of ADM admissible permutations has been initiated in [**OS82**] and [**AS82**]. Later, Leland [**Lel83**] (see also [**Lel85**]) has shown a set of recurrence relations for the exact number $P(N)$ of ADM admissible permutations. The results shown in [**LL88**] give another distinct set of exact recurrence relations for $P(N)$. The following lower and upper bounds on $P(N)$ have also been shown in [**Lel83**]: $P(4) = 24$, $P(8) = 26496$, and for $N \geq 16$,

$$\phi^{N(\lg N-2)} 23.7^{\frac{N}{4}} \leq P(N) \leq \phi^{N(\lg N-3+\frac{1}{192})} P(8)^{\frac{N}{8}},$$

where $\phi = \frac{1+\sqrt{5}}{2}$ is the golden ratio.

In a second direction, several tag-based control algorithms have been devised for routing useful permutations in ADM networks (see, for example, [**MS82**], [**Lel85**], [**LL88**], and [**RFS92**]). Some algorithms in this class are designed to exploit the existing plurality of input-output paths to handle some possible fault patterns in the network. Viewed as universal routing algorithms, however, tag-based routing is typically very efficient but limited in power: any such algorithm realizes only a strict subset of the set of all ADM admissible permutations. To remedy the situation, several backtrack search algorithms that can route any arbitrary admissible permutation has been discussed in [**Lel83**], [**LL88**], and [**RFS92**]. None of the above formulations, however, provide a time complexity analysis that gives a worst case bound on the time required to decide whether a given permutation is an ADM admissible permutation. In fact, [**Lel83**] hints that such a universal backtrack search algorithm is expected to be extremely inefficient.

The main result of this paper (Theorem 4.1) fills the above apparent gap in the literature by showing an $O(N^{1.695})$ deterministic routing algorithm to solve the problem. The algorithm is similar to that given in [**LL88**] and [**RFS92**] in that all three algorithms exploit the basic combinatorial properties summarized in the next section.

2. Background Results

The correctness and timing of the routing algorithm hinges on the fundamental properties of ADM networks summarized in lemmas 2.1 and 2.2 below. The result in Lemma 2.1 has been first mentioned in [**Fen74**], and used subsequently in [**AS82, LL88, Lel83**]. Properties equivalent to Lemma 2.2 have also been used in [**AS82, LL88, Lel83**]. For our purpose, however, we use graph-theoretic concepts to present and prove the properties. Our objective is to expose the combinatorial properties of the network while minimizing the use of bit-level notation that is typically associated with tag-based control algorithms.

To start, we need the following definition: given a permutation π on a set of n-bit long integers, we say that π is *all-pairs parity switching* with respect to the ith bit $(0 \le i \le n-1)$, written $\pi \in \mathrm{PS}(i)$, if for any possible pair $(x, \pi(x))$ the ith bit of x is the complement of its counterpart in $\pi(x)$. Otherwise (i.e., if for some pair $(x, \pi(x))$ the above property fails), we write $x \notin \mathrm{PS}(i)$.

LEMMA 2.1. *(the even/odd decomposition) Let G be an ADM of size N then $G \setminus V_0$ (the subgraph obtained by deleting the right most stage V_0 from G) is a disjoint union of two ADM networks, denoted G_{even} and G_{odd}, each of size $N/2$, where G_{even} (G_{odd}) has all the even (respectively, odd) numbered switches of $G \setminus V_0$.*

PROOF. To show that $G \setminus V_0$ is a disjoint union of at least two components, note that only links in L_0 (more specifically, the $\pm 2^0$ connections) join two switches with different even/odd parities. To see that G_{even} (or G_{odd}) induces an ADM, note that stripping the least significant bit from each address in the subnetwork gives a properly labeled ADM of size $N/2$. □

Call the two subnetworks (G_{even} and G_{odd}) defined in Lemma 2.1 the *rank*-1 subnetworks of G (or, the (1)-subnetworks of G, for short). For $i \ge 2$, applying the recursive decomposition to each of the available $(i-1)$-subnetworks, gives rise to 2^i ADM subnetworks, each of size $\frac{N}{2^i}$; call such subnetworks the rank-i (or, the (i)-subnetworks) of G. It is also convenient, to view G itself as a rank-0 subnetwork. By inspection, each (i)-subnetwork is composed of switches in $V_n \bigcup \cdots \bigcup V_i$, whose addresses (written in binary) share a unique pattern in their least significant i bits.

LEMMA 2.2. *Let π be a permutation on $\{0, \cdots, N-1\}$ that is passable through an ADM network G of size N then*

1. *if $\pi \notin \mathrm{PS}(0)$ then there exists a unique setting of the switches in V_1 (equivalently, a perfect matching in L_0) to accept π.*
2. *Else, if $\pi \in \mathrm{PS}(0)$ then there exist at most 4 possible settings of the switches in V_1 to accept π.*

PROOF. To prove the lemma, it suffices to analyze the possible settings of the output switches in either G_{even} or G_{odd}; the lemma follows by applying the argument to both subnetworks. Without loss of generality, we focus on G_{even}. For our purpose, let $X \subset V_1$ be the subset of the even numbered switches of V_1, and let $Y \subset V_0$ be the subset of the odd numbered switches (V_0 is G's output stage). Note that the subnetwork induced by $X \bigcup Y$ is a cycle on N switches whose links are the union of two perfect matchings corresponding to connections with the $+2^0$ and the -2^0 offsets, respectively.

Routing π in G entails setting the switches in V_1 so as to induce a perfect matching in L_0. For any such collection of valid settings (as may be specified in the context below), denote by $X^=$, and $Y^=$ the possible subsets of X and Y, respectively, that use the identity connections. Note that the equality $|X^=| = |Y^=|$ holds true since π is of the same size as G. We are now ready to prove parts 1 and 2 of the lemma.

1. If $\pi \notin \mathrm{PS}(0)$ then, by definition, $|X^=| = |Y^=| \geq 1$ in any configuration that admits π. In fact, the sets $X^=$ and $Y^=$ are uniquely defined by the pairs $(x, \pi(x))$ that agrees on the least significant bit of their binary representations. Now, none of the switches in $X^=$ and $Y^=$ uses the links in the cycle induced by $X \bigcup Y$. Consequently, deleting $X^=$ and $Y^=$ from $X \bigcup Y$ splits the cycle into one (or more) path(s). By assumption π is admissible in G, and hence any such path has an even number of switches, and a unique possible perfect matching, as required.
2. If $\pi \in \mathrm{PS}(0)$ then, by definition, $|X^=| = |Y^=| = 0$ in any configuration that admits π. Consequently, there are two possible settings of X corresponding to the two possible perfect matchings in the cycle induced by $X \bigcup Y$. Applying a similar argument to G_{odd} gives a total of 4 possible settings of switches in V_1.

□

3. The Algorithm

In this section we show a polynomial time algorithm for solving the decision problem mentioned above. The routing algorithm (function `dfsADM` below) is a simple divide-and-conquer algorithm implemented as depth-first search. The algorithm uses a global matrix ADM in which element $ADM[i,j]$ stores a setting of switch j in the ith stage of switches. Function `dfsADM(`i, π`)` takes as input two parameters: an index i, $0 \leq i \leq n$, of some stage V_i of switches, and a permutation π on a subset of $\{0, \cdots, N-1\}$. For $i = 0$, π is the input permutation specified in an instance of the decision problem. For $i \geq 1$, an important invariant of the algorithm is that the parameter π corresponds to a proper subset of $\{0, \cdots, N-1\}$ of size $\frac{N}{2^i}$ in which all integers (written in binary) share a unique suffix of the i least significant bits. In any of the above cases, if $\pi(x)$ is defined for some integer x, then we interpret that $x \in V_{n-1}$ is an input switch and $\pi(x) \in V_i$ is some switch in the ith stage of switches. The recursive decomposition of ADM networks ensures that π corresponds to a permutation over the set of inputs of exactly one of the 2^i possible (i)-subnetworks, denoted G_i in function `dfsADM`.

If π is admissible then `dfsADM` returns zero and sets the appropriate elements in matrix ADM to correspond to a non-conflicting set of routes; otherwise an error value of -1 is returned. More specifically, for $i = n$, V_i is the leftmost stage, and step 1 of the function terminates the recursion successfully. For $i \leq n-1$, the algorithm tries to route π through G_i by first splitting π into two subpermutations: π^{even} and π^{odd} to be routed through

1. the last stage of links in G_i, and
2. the even and the odd subnetworks of G_i, respectively;

the splitting is done in step 2. Step 3 iterates two times, once for $\pi^{hlaf} = \pi^{even}$, and another for $\pi^{half} = \pi^{odd}$. In each iteration we denote by $G_{i+1} \subset G_i$ (read G_{i+1} is a proper subgraph of G_i) the unique $(i+1)$-subnetwork through which π^{half} should be routed. Based on Lemma 2.2, the algorithm subsequently tries to compute two objects: (1) a valid setting of the output switches of G_{i+1}, and (2) an induced permutation (denoted, $\pi^{=}_{i+1}$, π^{+}_{i+1}, or π^{-}_{i+1}), that together suffice to pass π^{half} in G_i. Recursion is then used to route each of the induced permutations.

```
function dfsADM (i, π) {
#     input:    an index i, 0 ≤ i ≤ n, of a stage of switches, and a permutation
#               π on a subset of {0,··· ,N − 1}, as defined above.  Throughout the
#               presentation, let G_i be the unique (i)-subnetwork for which
#               π is an input-output mapping (cf., the above discussion).
#     output:   an error return:  error = 0 only if G_i admits π.
      1.  if (i == n) return 0
      2.  for (x in π) {
              set π^even(x) = π(x) only if the ith bit in x's binary
              representation is zero; else, set π^odd(x) = π(x)
          }
      3.  for (π^half in {π^even, π^odd}) {
              # Let G_{i+1} ⊂ G_i be the unique (i + 1)-subnetwork for which
              # π^half is an input-output mapping;
              # denote by X ⊂ V_{i+1} the set of G_{i+1}'s output switches;
              # denote by Y ⊂ V_i the set of G_i's output switches;
              # denote by Y^idle ⊆ Y the subset of switches in Y that are
              # available to connections from X;
              3.1 if (π ∉ PS(i)) {
                      # By Lemma 2.2 (1), there is a unique subset X^= ⊆ X of
                      # switches that must be set to the identity connection:
                      3.1.1 find the unique subset X^=;
                      3.1.2 find whether there exists a conflict-free setting of
                            the switches in X\X^= to access switches in Y;
                      3.1.3 if no such setting exists then return (−1);
                      3.1.4 else, save the obtained setting in the global matrix
                            ADM;
                      3.1.5 mark the Y switches involved in the obtained settings
                            as unavailable for subsequent allocations;
                      3.1.6 given the settings in 3.1.2, let π^=_{i+1} be the permutation
                            that must be passed through G_{i+1} so as to admit π^half;
                      3.1.7 error = dfsADM(i + 1, π^=_{i+1})
              }
              3.2 else if (π^half ∈ PS(i)) {
                      # By Lemma 2.2 (2), if π^half is admissible in G_i then all
                      # switches in X are set to either the +2^i or the −2^i edges;
                      3.2.1 set X to use the +2^i connections; save the settings
                            in matrix ADM; let π^+_{i+1} be the permutation that should
                            be passed through G_{i+1} in order to admit π^half;
                      3.2.2 error = dfsADM(i + 1, π^+_{i+1})
                      3.2.3 if (error ≠ 0) {
```

```
                    • repeat step 3.2.1 for the -2^i connections;
                    • error = dfsADM(i+1, π⁻_{i+1})
                  }
            }
        }
    4.  return error
}
```

Correctness of `dfsADM` follows by Lemma 2.2 since it exhausts all possible ways of passing an admissible permutation π. Assuming a uniform cost model (see, for example, [**AHU74**]), the following timing result is then a direct consequence of the structure of the algorithm.

LEMMA 3.1. *Let $T(N)$ be the time required to test whether a given N-element permutation is admissible through an ADM of equal size, then for some constant c, $T(N)$ satisfies*

$$T(N) \le cN + 4T(\frac{N}{2}), \qquad T(1) \le c.$$

That is, $T(N) \in O(N^2)$.

The above algorithm has been implemented. For $N = 8$, all of the ADM admissible permutations (26,496) are verified. For $N = 16$ and 32, there are approximately 1.549×10^{12} and 1.169×10^{31} ADM admissible permutations to verify, respectively. In these two latter cases, an exhaustive verification is prohibitive; we choose to measure the percentage of the admissible permutations on a random sample of 10^6 permutations; the results are 7.689% for $N = 16$ (the recurrence relations in [**Lel83**] give $\approx$ 7.404%), and .004562% for $N = 32$ (versus $\approx$ .004443% from [**Lel83**]). For $N \ge 64$, the percentage of the admissible permutations in a random sample of size 10^6 is so small that no reliable figure can be given.

4. A Tighter Bound

An upper bound on T(n) is given by:

$$T(n) \le \sum_{i=1}^{n} 2^i \left(\frac{cN}{2^i}\right) F_i = cN \sum_{i=1}^{n} F_i$$

where the summation is of the computations done in a single invocation of function `dfsADM` (i.e., not involving the recursive calls) over all (i)-subnetworks, for $i = 1, 2, \cdots, n$. Specifically, 2^i is the number of (i)-subnetworks in the recursive decomposition, $(\frac{cN}{2^i})$ is the cost incurred by the nonrecursive part of `dfsADM` when the formal parameter π has $\frac{N}{2^i}$ integers, and F_i is an upper bound on the total number of induced permutations that the algorithm tries to route through any possible (i)-subnetwork.

For example, for stage $i = 1$, there are two (1)-subnetworks, each of size $\frac{N}{2}$. The maximum number of induced permutations generated and tested through each (1)-subnetwork results when $\pi \in$ PS(0); the corresponding upper bound is $F_1 = 2$. In general, for any $i \ge 1$, setting $F_i = 2^i$ gives the $T(N) \in O(N^2)$ bound mentioned above. The main result of this section is Theorem 4.1 below; the result is obtained by using the improved bound on F_i shown in part 3 of Lemma 4.3 (as can be readily verified).

THEOREM 4.1. *$T(N) \in O(N^{1+\lg\phi})$, where $\phi = \frac{1+\sqrt{5}}{2}$ is the golden ratio ($\lg\phi <$ .695).*

The proof of Lemma 4.3 has two basic ingredients: Lemma 4.2 and the recurrence relations in Lemma 4.3. To introduce the basic ideas we need the following notation. For any possible stage $i \geq 2$, let $G_i \subset G_{i-1} \subset G_{i-2}$ be three subnetworks of ranks i, $(i-1)$, and $(i-2)$, respectively. If $i = 1$ is implied by the context, then only G_i and G_{i-1} are defined.

At some point, the algorithm tries to route a given permutation, say π_{i-1} through G_{i-1}. If $\pi_{i-1} \notin \mathrm{PS}(i-1)$ then the algorithm sets at least one of G_i's output switches to the identity connection, and computes an induced permutation, denoted $\pi_i^=$ thereafter. Permutation $\pi_i^=$ has $\frac{N}{2^i}$ elements, and it is tested for admissibility in G_i. We henceforth refer to permutations obtained in this way as $(^=_i)$-induced permutations.

If, on the other hand, $\pi_{i-1} \in \mathrm{PS}(i-1)$ then the algorithm tries out two possible settings of the output switches of G_i (the $+2^{i-1}$ and the -2^{i-1} connections), and computes the two induced permutations π_i^+ and π_i^-, respectively. We call the resulting permutations (^+_i)- and (^-_i)-induced permutations. We are now ready to present the first ingredient:

LEMMA 4.2. *For any all-pairs parity-switching permutation $\pi_{i-1} \in \mathrm{PS}(i-1)$, let π_i^+ and π_i^- be as defined above, then at least one of π_i^+ and π_i^- is not in $\mathrm{PS}(i)$*

PROOF. For any input x in G_i, $\pi_i^+(x) = \pi_{i-1}(x) - 2^{i-1}$, and $\pi_i^-(x) = \pi_{i-1}(x) + 2^{i-1}$. Hence, the ith bit in the binary representation of $\pi_i^+(x)$ is the complement of its counterpart in $\pi_i^-(x)$. Consequently, the ith bit in the binary representation of x agrees with its counterpart in either $\pi_i^+(x)$ or $\pi_i^-(x)$. That is, either $\pi_i^+(x) \notin \mathrm{PS}(i)$, or $\pi_i^+(x) \notin \mathrm{PS}(i)$, as required. □

We next incorporate the above observation in a counting argument. First, write F_i as $F_i^= + F_i^+ + F_i^-$, where $F_i^=$, F_i^+, and F_i^- are upper bounds on the number of $(^=_i)$-, (^+_i)-, and (^-_i)-induced permutations generated for routing through G_i. To simplify the counting argument, we use the following charging rules:

1. If $\pi_{i-1} \notin \mathrm{PS}(i-1)$ then the induced permutation $\pi_i^=$ is always assumed to be in $\mathrm{PS}(i)$ (a worst case scenario).
2. Else, if $\pi_{i-1} \in \mathrm{PS}(i-1)$, then, by Lemma 4.2, at least one of the induced permutations π_i^+ and π_i^- is not in $\mathrm{PS}(i)$. Without loss of generality, we may assume that π_i^- is always not in $\mathrm{PS}(i)$. The other permutation, π_i^+ will always be assumed in $\mathrm{PS}(i)$ (the worst case).

LEMMA 4.3. *The following hold in any ADM network:*

1. *$F_0^+ = F_0^- = 0$, and $F_1^+ = F_1^- = 1$.*
2. *For any stage $i \geq 2$: $F_i^- = F_i^+$, $F_i^= = F_{i-1}^-$, and $F_i^+ = F_{i-1}^+ + F_{i-1}^=$.*
3. *For any stage $i \geq 1$: $F_i < 3(\frac{\phi^i}{\sqrt{5}} + 1)$, where $\phi = \frac{1+\sqrt{5}}{2}$ is the golden ratio.*

PROOF.

1. For stage $i = 0$ the initial conditions $F_0^+ = F_0^- = 0$ are trivially set. For stage $i = 1$, the worst case occurs if $\pi \in \mathrm{PS}(0)$, and hence $F_1^+ = F_1^- = 1$.
2. For any stage $i \geq 1$, the equality $F_i^- = F_i^+$ holds by the structure of the algorithm.

   For $i \geq 2$, to see that $F_i^= = F_{i-1}^-$, note that if $\pi_{i-2} \in \mathrm{PS}(i-2)$, then applying rule (2) to π_{i-2} shows that the induced permutation π_{i-1}^- (at level $i-1$) is not in $\mathrm{PS}(i-1)$. Consequently π_{i-1}^- induces a unique permutation $\pi_i^=$ that contributes to the count $F_i^=$. Moreover, given the charging rules (1) and (2) above, this is the only scenario under which an $(^=_i)$–induced permutation is generated.

   To see the third identity: $F_i^+ = F_{i-1}^+ + F_{i-1}^=$, for $i \geq 2$, note that every $(^+_{i-1})$-induced permutation π_{i-1}^+ that contributes to F_{i-1}^+ is assumed to be in $\mathrm{PS}(i-1)$ by rule (2); hence, π_{i-1}^+ gives rise to an (^+_i)-induced permutation. Likewise, each $(^=_{i-1})$-induced permutation contributing to $F_{i-1}^=$ is assumed to be in $\mathrm{PS}(i-1)$, and hence, it gives rise to an (^+_i)-induced permutation.
3. From (1) and (2) above, it follows that $F_i^+ = F_i^-$ = the ith Fibonacci number. In addition, $F_i^=$ is the $(i-1)$–Fibonacci number. Hence, $F_i^+ = \frac{\phi^i}{\sqrt{5}}$ (rounded to the nearest integer). The statement $F_i < 3(\frac{\phi^i}{\sqrt{5}} + 1)$ then follows since $F_i = F_i^= + F_i^+ + F_i^-$.

□

5. Concluding Remarks

A fundamental problem in switching networks is to devise an efficient routing algorithm that captures the full power of a given network. This paper contributes to the above direction be showing an $O(N^{1.695})$ (worst case) time algorithm for routing admissible permutations on ADM networks. Asymptotically, the running time grows at a rate that is comparable to the growth rate of the number of switches ($N \lg N$) that must be set to route any such permutation. Hence, the algorithm is asymptotically near optimal. Currently, the complexity of the above problem on the closely related class of Gamma networks [**PR84, VR86, LH88**], appears to be open.

References

[AHU74] A.V. Aho, J.E. Hopcroft, and J.D. Ullman, *The design and analysis of computer algorithms*, Addison-Wesley, 1974.

[AS82] G.B. Adams III and H.J. Siegel, *On the number of permutations performable by the augmented data manipulator network*, IEEE Transactions on Computers **C-31** (1982), 270–277.

[Fen74] T. Feng, *Data manipulating functions in parallel processors and their implementations*, IEEE Transactions on Computers **C-23** (1974), no. 3, 309–318.

[Law75] D.H. Lawrie, *Access and alignment of data in an array processor*, IEEE Transactions on Computers **C-24** (1975), 1145–1155.

[Lel83] Mary-Diane Palmer Leland, *Properties and comparison of multistage interconnection networks for SIMD machines*, Ph.D. thesis, Department of Computer Sciences, University of Wisconsin - Madison, 1983.

[Lel85] Mary-Diane Palmer Leland, *On the power of the augmented data manipulator network*, Proceedings of the International Conference on Parallel Processing, August 1985, pp. 74–78.

[LH88] K.Y. Lee and W. Hegazy, *The extra stage gamma network*, IEEE Transactions on Computers **C-37** (1988), no. 11, 1445–1450.

[LL88] K.Y. Lee and D. Lee, *On the augmented data manipulator network in SIMD environments*, IEEE Transactions on Computers **C-37** (1988), no. 5, 574–584.

[MS82] R.J. McMillen and H. J. Siegel, *Routing schemes for the augmented data manipulator network in an MIMD system*, IEEE Transactions on Computers **C-31** (1982), no. 12, 1202–1214.

[OS82] M.J. O'Donnell and C.H. Smith, *A combinatorial problem concerning processor networks*, IEEE Transactions on Computers **C-31** (1982), 163–164.

[PR84] D.S. Parker and C. S. Raghavendra, *The gamma network*, IEEE Transactions on Computers **C-33** (1984), no. 4, 367–373.

[RFS92] D. Rau, J. Fortes, and H. J. Siegel, *Destination tag routing techniques based on a state model for the IADM network*, IEEE Transactions on Computers **C-41** (1992), 274–285.

[Sie90] H. J. Siegel, *Interconnection networks for large-scale parallel processing*, McGraw-Hill Series in Computer Organization and Architecture, 1990.

[SNKN89] H.J. Siegel, W.G. Nation, C.P. Kruskal, and L.M. Napolitano, *Using the multistage cube network topology in parallel supercomputers*, Proceedings of the IEEE **77** (1989), no. 2, 1932–1953.

[SS78] H.J. Siegel and S.D. Smith, *Study of multistage SIMD interconnection networks*, Proceedings of the Fifth Annual Symposium on Computer Architecture, April 1978, pp. 223–229.

[VR86] A. Varma and C.S. Raghavendra, *On permutations passable by the gamma network*, Journal of Parallel and Distributed Computing **3** (1986), 72–91.

Department of Computing Science, University of Alberta, Edmonton, Alberta
E-mail address: `ehab@cs.ualberta.ca`

Department of Computing Science, University of Alberta, Edmonton, Alberta

DIMACS Series in Discrete Mathematics
and Theoretical Computer Science
Volume 42, 1998

Preventing Conflicts in Input Buffering Baseline-Based ATM Switches

Hasan Çam

ABSTRACT. This paper proposes a window-based contention resolution algorithm for a *blocking* ATM switch based on reverse baseline network with content addressable FIFO (CAFIFO) input buffers. The proposed algorithm prevents not only external conflicts but also internal conflicts, in addition to alleviating HOL blocking. This algorithm was obtained by adapting the ring reservation algorithm used on nonblocking ATM switches to a reverse baseline network. The fact that a nonblocking network is replaced by a $log_2 N$-stage reverse baseline network yields a significant economy in implementation. We have conducted extensive simulations to evaluate the performance of reverse baseline network using the proposed window-based contention resolution algorithm. Simulation results show that the throughput of reverse baseline network can be as good as the throughput of nonblocking switches if the window depth of input buffers is made sufficiently large.

1 Introduction

An ATM (Asynchronous Transfer Mode) switch routes cells arriving at its input ports to their desired output ports. A multistage ATM switch is usually constructed by interconnecting switching elements (SEs) of size (2×2) each in a specific topology using links. When more than

AMS subject classification: Primary: 68M10, 68M20; Secondary: 05A05, 05C70

one cell compete for the same internal link of ATM switch, an *internal conflict* (or blocking) occurs. When more than one cell compete for to the same output port, an *external* (or output) conflict occurs. An ATM switch with N inputs and N outputs is said to be *nonblocking* if no internal conflict occurs in any permutation of its inputs to outputs, where $N = 2^n$.

In a packet switch, two or more packets may arrive simultaneously on different inputs destined for the same output and, hence, external blocking is unavoidable in a packet switch. In order to smooth the statistical fluctuations in packet arrivals to the switch, three main approaches were proposed: *input buffering*, *output buffering*, and *completely shared buffering* [1]. To be able to implement output buffering, either the switch must operate as fast as the sum of the speeds of its input ports, or the buffers must have as many write ports as there are input ports to the switch, to be able to handle simultaneous packet arrivals [2]. Because implementing buffers with multiple write ports increases their size and performance, the Knockout Switch [3] implements output buffering by using an N to L concentrator at each output port to reduce the number of buffers needed to receive simultaneously arriving packets. Therefore, although output buffering provides the best performance in terms of switch throughput and cell delay under uniform traffic patterns, the implementation of input buffering is appealing because only one packet at a time arrives at the input port so that the buffer needs only a single write port.

Input buffering switches resolve contention for external conflict prior to switching. Some of the external conflict resolution algorithms are three-phase algorithm [4], scheduling CAM [5], and ring reservation [6]. When input buffers are FIFO (first in first out) queues, a serious problem called head-of-line (HOL) blocking occurs: a single packet at the head of the queue whose destination output port is busy can block all other packets in that queue from being transmitted even if their destination output ports are idle. This limits the maximum throughput of a nonblocking switch to 0.56 under uniform traffic, regardless of the selection policy used among the contending HOL cells [7]. Contention resolution algorithms including ring reservation algorithm suffer from HOL blocking. Three methods, namely, output expansion [8], channel grouping [9] and windowing [1] are the most known techniques to alleviate HOL blocking. In *output expansion*, extra internal switch links and output buffers are added in order to be able to forward more than one cell to each output port during the same cell interval. In *channel grouping*, the switch output ports are partitioned into a number of groups and cells are allowed to contend for a group, rather than a specific port. In *windowing* (also called bypass queue [10, 11]), cells behind the HOL cell are also allowed to contend for output port access.

Two types of windowing, namely, sequential windowing (also called serial bypass) and parallel windowing (also called parallel bypass) are possible. In *sequential windowing*, the first cells in a switch's input buffers first contend for outputs, then the losing input buffers that contain a second cell are allowed to contend a second time, those that lose in the second time and have a third cell are allowed to contend a third time, and so on until w times are contended, where w is the depth of window. In *parallel windowing*, all cells in the windows of all buffers contend in a single round with the winners proceeding to the outputs, with the condition that only one cell can be selected at most from a buffer. To preserve the cell sequence, among those cells destined to the same output port in the window of a buffer the cell closest to the head of the buffer is selected if one of them needs to be chosen. Although output expansion and channel grouping require fundamental changes in the switch, they were preferred over windowing because extra contention cycles were needed in windowing to query cells behind the HOL cell. But, Shultz and Gulak [12] have recently introduced a cost-efficient hardware for implementing parallel windowing such that w cells can be queried in every input buffer during a single contention (or arbitration) cycle, without any delay overhead. In this paper, we shall explain how to modify their hardware to implement parallel windowing on reverse baseline network.

Baseline-type networks are very attractive to ATM switch designers due to their desirable features. In particular, they have constant latency for any input-output pair, are self-routing and modular, support synchronous and asynchronous modes of operation, preserve cell-sequence due to path-uniqueness, are very suitable for VLSI implementation, and have a complexity of $O(N\ log_2N)$ compared to $O(N^2)$ of the crossbar switch. The major drawback of baseline-type networks is to have a very limited number of connection patterns, which causes them to suffer from internal conflicts, in addition to external conflicts. Our objective is to alleviate this drawback to the extent that the throughput of a reverse baseline network with content addressable input buffers can reach the throughput of a nonblocking network. Our algorithm uses parallel windowing for preventing both external blocking and internal blocking in a reverse baseline network. The fact that we replace all stages of a nonblocking switching network by log_2N stages of a baseline-type network yields a significant economy in implementation.

The rest of the paper is organized as follows. Section 2 describes how to identify internal conflicts of a reverse baseline network using balanced (incomplete) matrices, frames and bipartite matching graphs. Section 3 presents a parallel windowing algorithm for preventing both internal and external conflicts on reverse baseline network with content addressable FIFO (CAFIFO) input buffers. The hardware implementation of the proposed algorithm is also discussed. The simulation results under different input traffics are presented in Section 4. Concluding remarks are made in Section 5.

2 Identifying Internal Conflicts in Reverse Baseline Network

Figure 1 shows the model of an input buffering reverse baseline network that we consider in the paper. The SEs of the network do not have any buffer. Because reverse baseline network is a self-routing net-

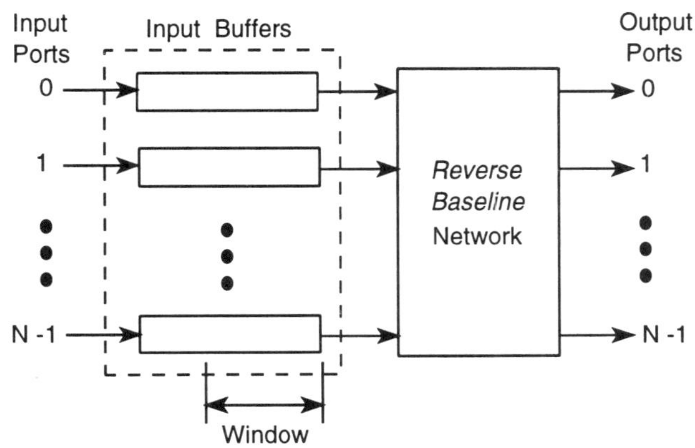

Figure 1: An input buffering reverse baseline network.

work, routing cells through the network is very simple: the destination address bits of cells are put into their headers and are used successively for routing as follows. To route a cell destined to an output port with binary representation $(d_1 d_2 \cdots d_n)$ in a self-routing ATM switch, an SE at stage k, $1 \leq k \leq n$ is set according to the bit d_k of the destination address of the cell arriving at an input of the SE: the cell is routed to the upper output of the SE if $d_k = 0$ and to the lower output if $d_k = 1$, regardless of whether the cell has arrived at lower or upper input of the SE. Internal conflict occurs when two cells arriving at a SE of stage k request the same output port of the SE because the d_k bits of their destination addresses are the same. Figure 2 illustrates the self-routing mechanism and internal conflict in a reverse baseline network.

In this section, we discuss how to grant admissions to the cells of input buffers of a reverse baseline network so that no internal (and external) conflict occurs. To avoid conflicts, the destination addresses of cells must meet the requirements identified by so-called frames [13]. These requirements can also be identified by bipartite matching graphs to be introduced later in this section. To introduce these requirements,

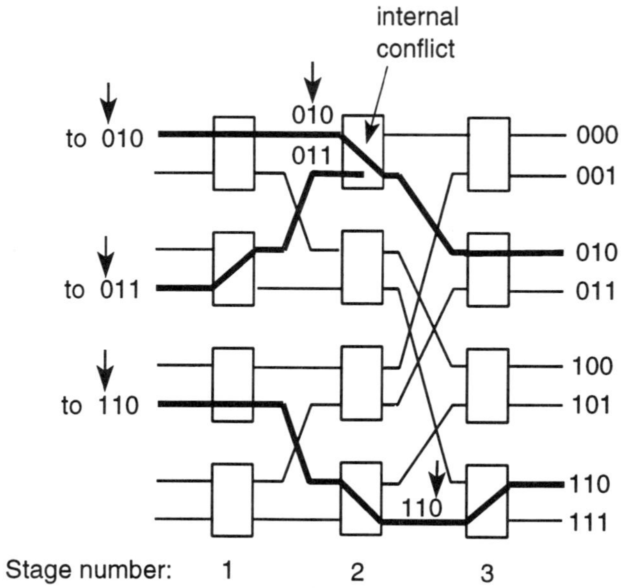

Figure 2: Internal blocking and self-routing in a reverse baseline network.

we first present some notation and definitions.

A matrix A with N rows and k columns is denoted by $A_{N\times k}$. The rows of any matrix are labeled in ascending order starting with 0. The ith row of a matrix $A_{N\times k}$ is denoted by $A_{N\times k}(i)$, for $0 \leq i \leq N-1$. A permutation h on the set $\{0, 1, \cdots, N-1\}$ is a bijective mapping and permutes the ordered list $0,\ 1, \cdots, N{-}1$ into $h(0),\ h(1), \cdots, h(N{-}1)$. A cyclic notation can be used to represent a permutation as the product of cycles, where a cycle $(c_0\, c_1\, c_2\ \cdots\ c_{k-1}\, c_k)$ means $h(c_0) = c_1$, $h(c_1) = c_2, \cdots,\ h(c_{k-1}) = c_k$, and $h(c_k) = c_0$. The concept of balanced matrices was introduced by Linial and Tarsi in [14]. We rephrase their definition below using permutations.

Definition 2.1 *(Balanced matrix)*: Let $N = 2^n$ and $n \geq 1$. A matrix $A_{N\times k}$ containing 0's and 1's only is *balanced* if either one of the following conditions is satisfied:

1. For $1 \leq k \leq n$, $A_{N\times k}$ consists of any k columns of the binary representation of a permutation on the set $\{0, 1, \cdots, N-1\}$, where $A_{N\times k}$ contains 2^k distinct rows, each being repeated 2^{n-k} times,

2. For $k > n$, every n consecutive columns form the binary representation of a permutation on the set $\{0, 1, \cdots, N-1\}$.

As an example, two balanced matrices D and E are shown below. But, notice that the matrix $[D\ E]$ is not balanced. When all entries of some rows of a balanced matrix are removed, to treat the remaining matrix balanced in this paper we introduce next a new term called balanced incomplete matrix.

$$D = [d_1\ d_2\ d_3] = \begin{bmatrix} 1 & 1 & 1 \\ 1 & 0 & 0 \\ 0 & 0 & 0 \\ 0 & 1 & 0 \\ 1 & 1 & 0 \\ 0 & 0 & 1 \\ 1 & 0 & 1 \\ 0 & 1 & 1 \end{bmatrix} \qquad E = [e_1\ e_2] = \begin{bmatrix} 1 & 0 \\ 1 & 0 \\ 1 & 1 \\ 1 & 1 \\ 0 & 1 \\ 0 & 1 \\ 0 & 0 \\ 0 & 0 \end{bmatrix}$$

Definition 2.2 *(balanced incomplete matrix)*: Let $M < N$ and $k \geq 1$. A matrix $A_{N\times k}$ is a *balanced incomplete matrix* if $A_{N\times k}$ consists of any M rows of a balanced matrix $A_{N\times k}$ plus $N - M$ blank rows that do not have any entry.

Figure 3.(d) shows a balanced incomplete matrix $B_{8\times 3}$ whose rows 3 and 6 are blank and the other six rows are identical to those $A_{8\times 3}$'s rows with the same row labels, shown in Figure 3.(b). Two more important concepts that we shall use are *frame* and *fit*, introduced by Çam and Fortes [13]. A *frame* looks like a grid with horizontal links of different lengths and is used to capture those permutations or partial permutations that can be realized on a network. Because we use only one type of frame corresponding to reverse baseline network, we simplify our frame definition given in [13]. A frame corresponding to the reverse baseline network with eight inputs is illustrated in Figure 3.(c). The definition of *fit* given in [13] is also modified slightly to incorporate balanced incomplete matrices, as well.

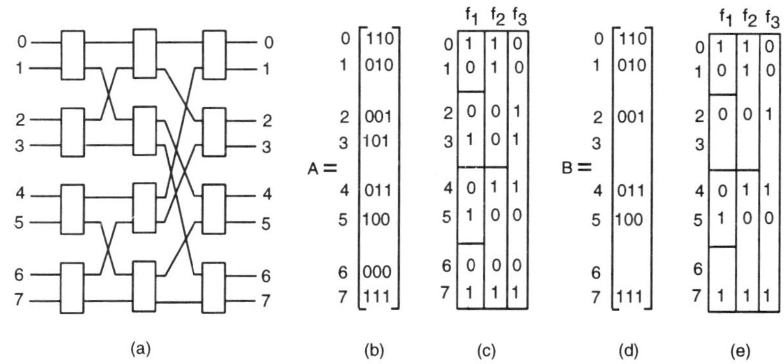

Figure 3: (a) A reverse baseline network with 8 inputs/outputs. (b) A balanced matrix $A_{8\times 3}$ which is the binary representation of permutation $\pi = (0\ 6)(1\ 2)(3\ 5\ 4)(7)$. (c) The frame corresponding to the reverse baseline network. Row i of the frame contains row i of $A_{8\times 3}$. $A_{8\times 3}$ fits the frame. (d) A balanced incomplete matrix, $B_{8\times 3}$, whose rows 3 and 6 do not have any entry. (e) Row i of the frame contains row i of $B_{8\times 3}$. $B_{8\times 3}$ fits the frame.

Definition 2.3 *(frame, rectangle of a frame)*: For $k \geq 1$ and $j = 1, 2, \cdots k$, a *frame* $F_{N\times k}$ consists of k columns and N rows labeled in ascending order, where columns are labeled f_j from left to right. For $1 \leq j \leq n$, column f_j consists of 2^{n-j} blocks of 2^j rows (entries) each. In the graphical representation of a frame, any polygon with four sides and four right angles is called a *rectangle* (or *block*) of a frame. (See Fig. 3.(c) for a frame).

Definition 2.4 *(Fit)*: Let $k \geq 1$, $j = 1, 2, ..., k$, and $H \leq N$. Consider a matrix $A_{H\times k}$ and a frame $F_{N\times k}$. When row i of $A_{H\times k}$ is placed into the row i of $F_{N\times k}$, the matrix $A_{H\times k}$ is said to fit the frame $F_{N\times k}$ if and only if for each j, each and every j column rectangles whose entries are located in columns $f_1, f_2, \cdots, f_j$ of $F_{N\times k}$ contains a balanced matrix or a balanced incomplete matrix.

Both matrices A and B shown in Figure 3.(b) and (d) fit the frames shown in (c) and (e). It follows from Theorem IV.1 of [13] that, if a balanced matrix $A_{N\times n}$ or balanced incomplete matrix $B_{N\times n}$ fits the frame $F_{N\times n}$, then the corresponding complete or partial permutations are realized by the reverse baseline network with N inputs/outputs. As an example, consider the reverse baseline network with 8 inputs/outputs shown in Figure 3.(a), the balanced matrix $A_{8\times 3}$ corresponding to the binary representation of the permutation $\pi = (0\ 6)(1\ 2)(3\ 5\ 4)(7)$, and the frame $F_{8\times 3}$ of Figure 3.(c) corresponding to the reverse baseline network. When row i, $0 \leq i \leq 7$, of $A_{8\times 3}$ is used as the routing tag for the ith input of the reverse baseline network, no conflict occurs in the SEs of the network and a path is established between input i and output $A_{8\times 3}(i)$ (i.e., the output whose label equals the contents of row i of $A_{8\times 3}$). Therefore, the reverse baseline network realizes the permutation π. Now, let us place the ith row of $A_{8\times 3}$ into the ith row of the frame as shown in Fig. 3.(c). The first k columns, $1 \leq k \leq 3$, of the frame consists of 2^{3-k} rectangles, each having 2^k rows and k columns. Because $A_{8\times 3}$ fits the frame and the balanced incomplete matrix $B_{8\times 3}$ is obtained from $A_{8\times 3}$ by removing all entries of rows 3 and 6, $B_{6\times 3}$ also fits the frame. Note that the matrix enclosed by any rectangle of the frames is balanced and represents a complete or partial permutation on $\{0, 1, \cdots, 2^k\}$.

We relate balanced matrix, balanced incomplete matrix, and frame to the problem of cell scheduling in input buffers of a reverse baseline network as follows. The objective in cell scheduling is to form a balanced matrix from the selected cells' destination addresses such that the balanced matrix fits the frame. If it is not possible to form a balanced matrix with N rows, then those cells whose destination addresses violate the conditions of fit are ignored in the selection process so that a balanced incomplete matrix is formed.

2.1 Matching and Cell Scheduling

A *bipartite graph* is one whose vertex set can be partitioned into two subsets X and Y, so that each edge has one end in X and one end in Y. A subset M of edges in a graph G is called a *matching* if no two edges of M have a vertex in common. The problem of scheduling optimally cells, located in input buffers with window depth w, on a nonblocking switch can be translated into the problem of finding a *maximum matching* [15]: when the two subsets X and Y of a bipartite graph correspond to the number of inputs and outputs, respectively, of a switch such that vertex i (corresponding to input i) of set X is connected by a link to a vertex j (corresponding to output j) if one of the w cells in input buffer i is destined to output j. Fig. 4.(b) shows a bipartite graph corresponding to the cells of input buffer windows shown in 4.(a). Note that a matching guarantees only that the selected cells' destination addresses are distinct. If the switch is nonblocking, then finding a maximum matching of this bipartite graph suffices to determine the cells that should be granted admission to the switch. A matching link between vertex i of set$X$ and vertex j of set Y indicates that the cell with the destination address j in input buffer i should be granted admission to the switch. When the switch fabric is, however, a reverse baseline network, a matching must satisfy those additional constraints defined next.

Definition 2.5 *(Partition k, Matching Constraint k)*: For $1 \leq k \leq n-1$, *partition k*, denoted *partition$_k$*, corresponds to two partitioning

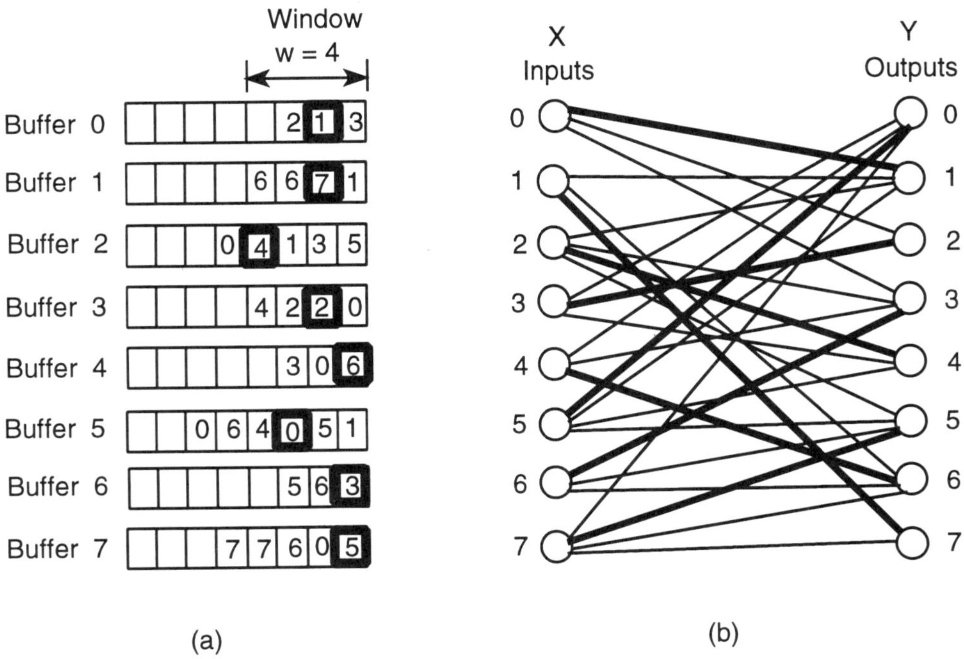

Figure 4: (a) Input buffers of a nonblocking switch. (b) Bipartite graph corresponding to the cells in the windows of buffers. The bold edges represent a maximum matching graph corresponding to the selected cells encircled by thick solid squares.

operations: 1) the input set X of a bipartite graph is partitioned into 2^{n-k} sets of 2^k vertices each, starting with the topmost vertex down to the bottom vertex in order, and 2) the output set Y of a bipartite graph is partitioned into 2^k sets of 2^{n-k} vertices each, starting with the topmost vertex down to the bottom vertex in order. (Fig. 5 shows $partition_1$ and $partition_2$ for $n = 3$). A *matching constraint* k, denoted $cons_k$, requires that a matching of bipartite graph corresponding to the cells in the windows of buffers can have at most one edge between any input set and any output set in the $partition_k$ of the graph.

Matching constraint 1 guarantees that the two entries of any block of the first column of the frame form neither the set {0,0} nor {1,1} when a balanced (incomplete) matrix formed by the destination addresses of the selected cells are placed into the frame. Matching constraint 2 assures that no rectangle of two columns located in the first two columns of the frame contains an unbalanced matrix. In general, matching constraint k guarantees that no rectangle of k columns located in the first k columns of the frame contains an unbalanced matrix. Therefore, if all $n-1$ matching constraints are satisfied in a bipartite graph matching, then the cells corresponding to the matching can be routed through the reverse baseline network without having any internal/external conflict. Thus, selecting cells optimally from input buffers of a reverse baseline network is equivalent to finding a maximum matching that satisfies all $n-1$ matching constraints.

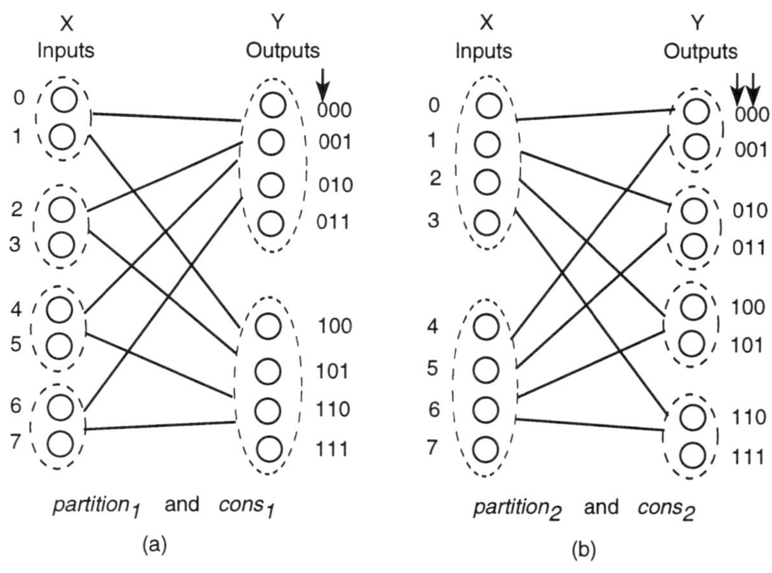

Figure 5: Partitions and matching constraints of a graph for $n = 3$. (a) $partition_1$ and $cons_1$. (b) $partition_2$ and $cons_2$.

3 Parallel Windowing for Reverse Baseline Network

Shultz and Gulak [12] introduced hardware that allows parallel windowing in a single arbitration cycle for nonblocking switches with CAFIFO input buffers. Their hardware basically implements ring reservation algorithm [6] which resolves external conflicts on a nonblocking switch with input buffers. Ring reservation mechanism has a distributed architecture as shown in Figure 6 and, therefore, does not require any centralized arbitration processor. A string of 1-bit tokens rotates around

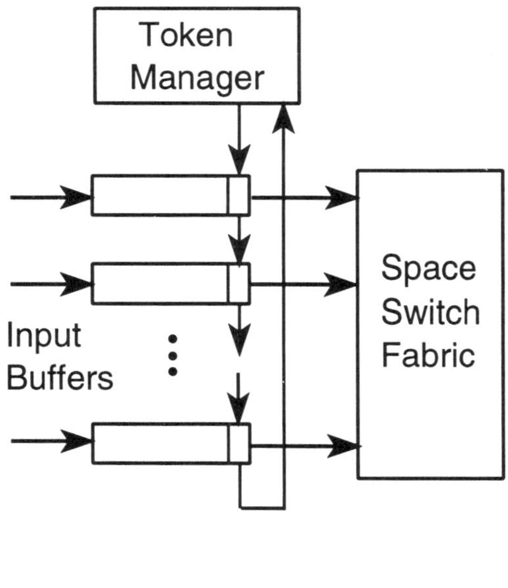

Figure 6: Ring reservation mechanism.

a ring of input buffers in pipelined fashion (one bit per clock tick), with each bit representing the status of an output port (0= FREE, 1= TAKEN). Each token is associated with an output port. At each buffer, a counter keeps track of the output port associated with the token bit. That is, if 1-bit tokens are, for instance, rotating in the direction of buffer j to buffer $(j+1)$ *mod* N, then the counter of buffer j is initially set to j and is reduced cyclically by 1 at every bit clock cycle. If the output port corresponding to the present 1-bit token is the destination of the first cell of buffer cell and the 1-bit token equals 1, then the cell is blocked. If the output port is the destination of the first cell of buffer and the corresponding 1-bit token is 0, then the cell is granted admission and the 1-bit token is set to 1. This technique can be pipelined with the switching function.

In case of CAFIFO-based input buffers, all cells in the windows are searched in an arbitration cycle to determine if their destination addresses match the output port associated with the present token. If a cell's destination address in a buffer matches the output port represented by the current 1-bit token, the cell is marked and the 1-bit token is set to 1 so that no other cell from any other buffer can be selected for the same destination in an arbitration cycle. All token bits pass through all input CAFIFOs during one *arbitration cycle.* The cells selected in an arbitration cycle enter the switch fabric during the next arbitration cycle. The length of an arbitration cycle is the same as that of a *time slot.* An arbitration cycle takes N bit clock cycles for N input buffers simply because each 1-bit token visits every buffer. The maximum value of N is determined by the ratio of the frequency of the bit clock to the frequency of cell interval clock [12]. Figure 7.(a) shows the cells selected in an arbitration cycle from CAFIFO buffers of a nonblocking switch.

We propose algorithm RB-CAFIFO, shown in Figure 8, for selecting cells in an arbitration cycle from the CAFIFO input buffers of a baseline network. In RB-CAFIFO, an arbitration cycle is implemented in n passes, and pass k requires 2^k bit clock cycles, for $1 \leq k \leq n$. (See three passes in Figure 7.(b) for $N = 8$). Thus, the total number of bit clock cycles required in an arbitration cycle is $2N-2$. Pass k guarantees that when the k leftmost bit(s) of the destination address of the selected cell from buffer i are placed into row i of the frame corresponding to reverse baseline network, the resultant matrix of order $N \times k$ fits the first k columns of the frame.

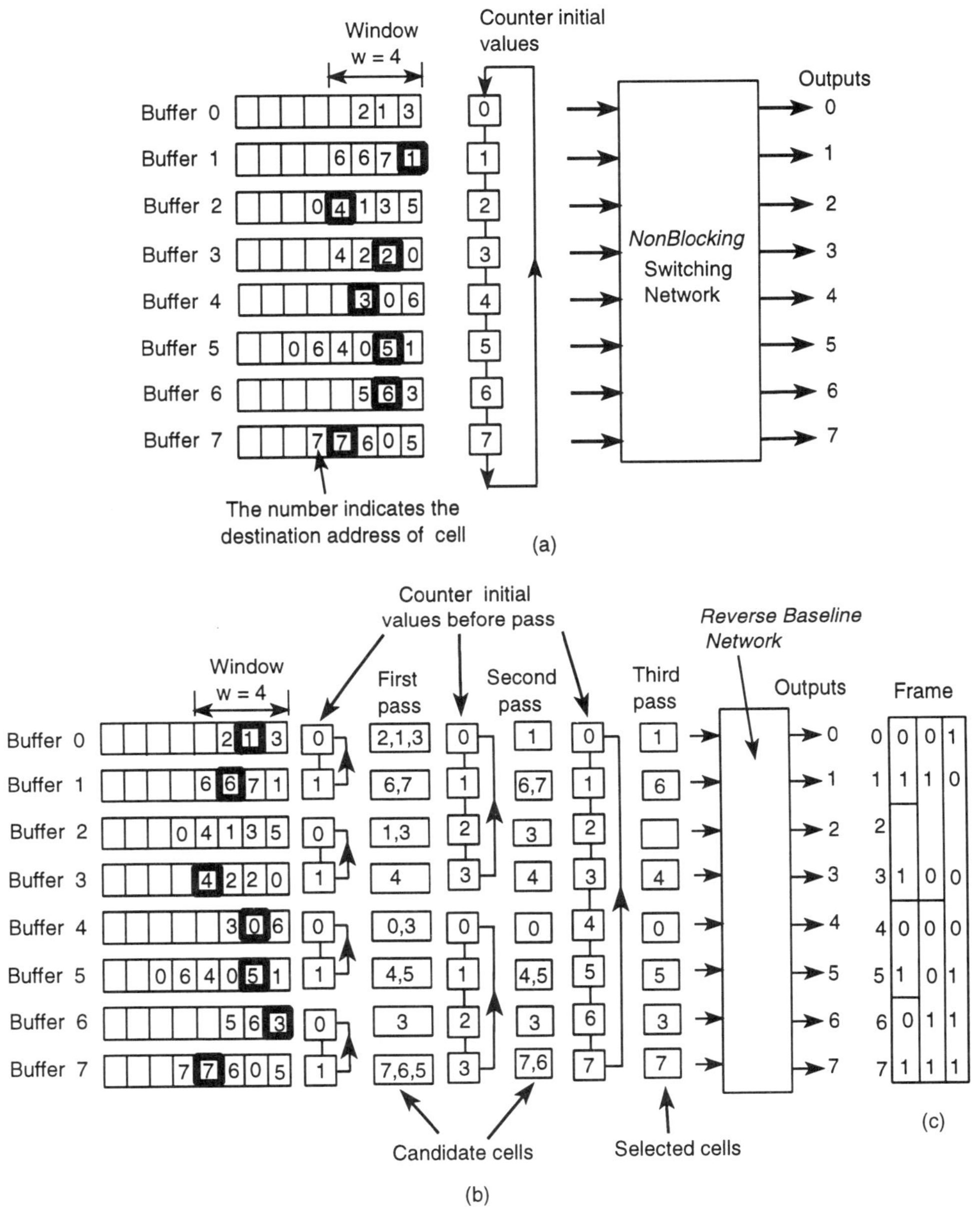

Figure 7: (a) Thick solid squares denote the cells selected in arbitration cycle of the CAFIFO algorithm of Shultz and Gulak for *nonblocking* switch with eight inputs. (b) Thick solid squares denote the cells selected in arbitration cycle; no cell is selected from buffer 2. In pass k, counter value is compared with the k leftmost bits of cells' destination addresses. (c) The balanced incomplete matrix formed by the destination addresses of selected cells fits the frame.

```
Algorithm RB-CAFIFO
begin
1. Initially, all the valid bit entry fields of all cells in each buffer are zero;
   all 1-bit tokens are zero; output ports are lebeled from 0 to N − 1;
2. w ← window_size;
3. for each buffer i, 0 ≤ i ≤ N − 1, in each arbitration cycle do in parallel with other buffers
      comment: k denotes the label of a pass.
4.    for k ← 1 to n do
5.       buffer[i] ← 0;
6.       buffer_counter ← i mod 2^k;
7.       set the current 1-bit token to 0;
8.       for j ← 1 to 2^k do
9.          if (buffer[i] = 0) and (the current 1-bit token is 0) then
10.            consider those cells that are located within the window_size w and their
              all valid bit entry fields are set to 1 in each one of the previous passes;
11.            if (the k leftmost bits of these cells' destination addresses are equal to
              the binary representation of buffer_counter) then
12.               buffer[i] ← 1;
13.               set the current 1-bit token to 1;
14.               for each matching cell of the buffer, set its kth pass valid bit entry field to 1;
15.         endif
16.         send the current 1-bit token to the next buffer in the cycle of 2^k elements;
17.         if (buffer_counter > 0) then
18.            buffer_counter ← (buffer_counter − 1) mod 2^k;
19.         else buffer_counter ← 2^k − 1;
20.      endfor
21.   endfor
22.   if (buffer[i] = 1) then
23.      select the cell whose all valid bit entry fields
           are 1 and it is the closest to the front of buffer;
24. endfor
end
```

Figure 8: The proposed algorithm RB-CAFIFO for reverse baseline network with CAFIFO input buffers.

Lines 3 to 24 of algorithm RB-CAFIFO are executed by each input buffer in each arbitration cycle. The variable k denotes the label of a *pass*. In the beginning of pass k, $buffer[i]$ is set to 0 in line 5, which means that the k leftmost bits of the destination addresses of cells in buffer i can be compared with the number that the current 1-bit token is representing to check if they match. Note that the destination address represented by the current 1-bit token is equal to the value of the $buffer\_counter$. If a match occurs in a pass (line 11), then $buffer[i]$ is set to 1 so that its cells are not searched until the end of the pass (line 12). Also, the current 1-bit token is set to 1 to indicate that it is taken in a pass (line 13). The kth pass valid bit entry field of every matching cell is set to 1 in line 14. (It is assumed that all pass valid bit entry fields of all cells are 0 in the beginning of arbitration cycle). Whether a match occurs or not in a bit clock cycle, the 1-bit token is sent to the next buffer (line 16), and the value of $buffer\_counter$ is decremented cyclically by 1 (lines 17 to 19). A comparison between cells' destination addresses and the buffer counter value takes place if both the 1-bit token is 0 (i.e., free) and no cell of the buffer matches earlier in a bit clock cycle of the same pass (line 9). At the end of pass n, if $buffer[i]$ is set to 1 (line 22), then there exists at least one cell in the buffer that matched successfully in all n passes. The cell closest to the front of the buffer is selected and will be launched into the switch fabric in the next arbitration cycle. The balanced (incomplete) matrix formed by the destination addresses of the selected cells fit the frame of reverse baseline network. As an example, see Figure 7.(b) and (c). Also, Figure 9 shows activity in a CAFIFO buffer and the related fields. The 1's in the first field indicate which cells are included in the window. Note that, if a cell is not available in a buffer location, its window inclusion bit is always 0. The bit 1 in the field of selected entry bits indicates that the corresponding cell is selected from the buffer to enter the switch fabric in the next arbitration cycle.

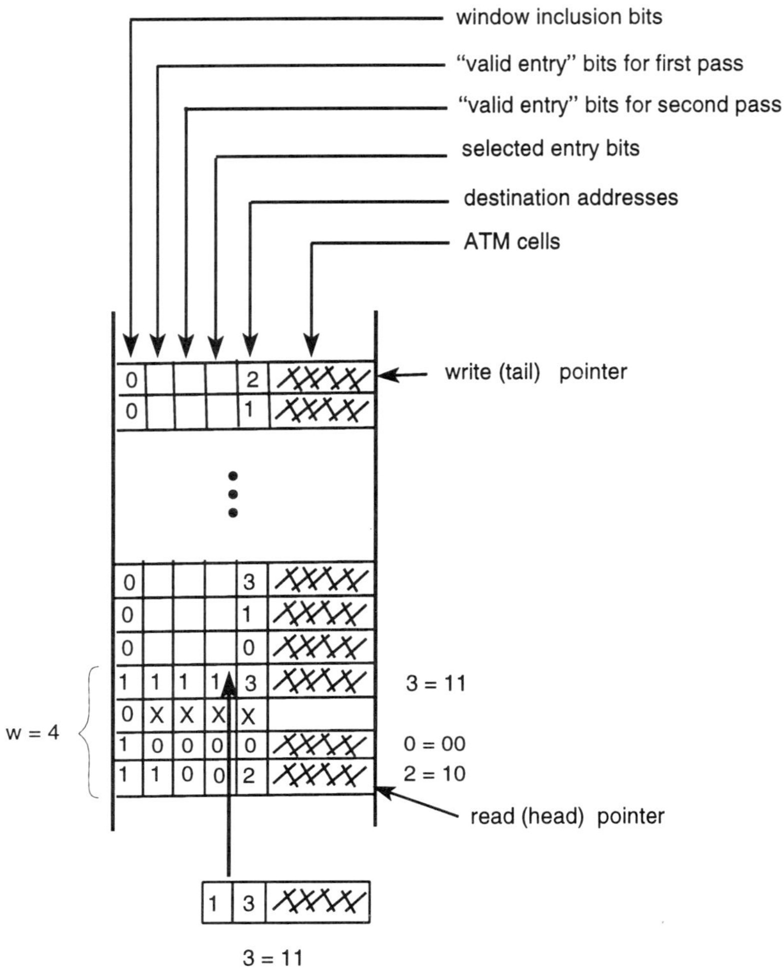

Figure 9: CAFIFO operation example of algorithm RB-CAFIFO for a single input buffer of window size four and $N = 4$. If the first field entry (i.e., window inclusion bit) is 1, the corresponding cell is included in the window search. All 1-bit tokens are 0 in the beginning of each pass. The tokens that are set to 1 at pass n represent those output ports for which a cell is selected from a buffer. This figure shows that the cell with destination address 3 is selected, and the corresponding 1-bit token is also set to 1.

4 Simulation Experiments

The performance of an interconnection network is measured by its throughput and delay. In evaluating the performance of a self-routing ATM switch, cell loss rate, being the fraction of cells lost within the switch, is usually more important than delay. In the proposed input buffering switch, no cell is lost within the switch fabric because only those cells that will not have any internal and external conflict enter the switch fabric. A cell is lost only if it is sent to a full input buffer. In the simulations, we assumed that an input buffer can store up to 20 cells. Throughput (also called unnormalized throughput) is defined as the average number of cells which are successfully delivered to the output ports by the switch per time slot per input line [16]. The normalized throughput is the fraction of incoming cells that are successfully delivered to the output ports per time slot per input line. Simulation results are presented in this section by plotting unnormalized throughput against cell arrival rate. Because those cells that are not selected in an arbitration cycle are kept in the input buffers, almost all the inputs of an ATM switch have at least one cell in their input buffers after some transient period, regardless of the value of cell arrival rate. Cell arrival rates somewhat affect the number of cells in buffers. By controlling cell arrival rates, cell lost probability in input buffers can be improved [17].

A traffic model is determined by: 1) the destination request distribution of arriving cells, and 2) the process describing the arrival of cells at the input ports of the network. The simplest traffic model is the uniform traffic in which the process describing the arrival of cells at the input ports is an independent and identically distributed Bernoulli process with the load parameter p $(0 < p \leq 1)$. This implies that a cell arrives at a given input port in a given time slot with probability p. As for the destination request distribution of arriving cells under uniform traffic, each cell chooses its destination port uniformly among all output ports, independently from all previous and current arriving cells. For a network with N output ports, the probability that an arriving cell chooses a particular output port is $1/N$ under uniformly distributed destination addresses. Although uniform traffic is not encountered in real ATM switches, it provides a common platform to compare different switches. In ATM switches, message generation patterns usually follow Poisson or bursty traffic patterns rather than uniform traffic. In Poisson traffic, message interarrival times are exponentially distributed.

We run simulations under uniform and Poisson traffic patterns. If the input traffic is Poisson distribution, messages arrive at each input according to Poisson distribution in the sense that the message length is exponentially distributed. The message length is expressed in terms of the number of cells. Thus, cell arrival rate in Poisson distribution is equal to the product of message arrival rate and the mean length of a message in units of cells. In our simulations under Poisson traffic, the destination address of a message is determined uniformly, but *all the cells of a message* are assigned *the same destination address.* We assumed that the average length of a message is two due to that multiple sources are assumed to multiplex their cells into a single buffer, as shown in Figure 10. The random number generator used in our simulation programs written in C language is not ideal and sometimes assigns the same destination address to the cells of more than one message in a buffer. This results in an adverse effect on the throughputs of both nonblocking and reverse baseline switches. In the simulations, nonblocking switch uses the ring reservation scheme with CAFIFO implementation proposed by Shultz and Gulak [12], while reverse baseline network employs algorithm RB-CAFIFO for cell selection. Each simulation result is an average of 8,000 time slots after a transient period of 2,000 time slots.

Figure 11 shows the throughput of a reverse baseline network against cell arrival rate under uniform input traffic for $N = 16$. Note that the throughput of reverse baseline network increases significantly as the window size, w, gets larger. Figure 12 illustrates the throughputs of nonblocking and reverse baseline networks against cell arrival rate under uniform traffic for $N = 16$. In contrast with the reverse baseline

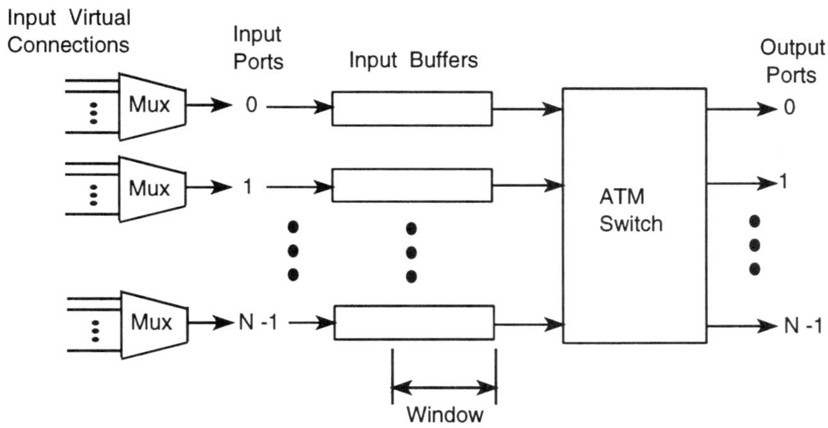

Figure 10: Switch model used in simulations.

network, the throughput of a nonblocking switch shows little improvement at the high cell arrival rates as the window size increases.

Figure 13 shows the throughput of a reverse baseline network against cell arrival rate under *Poisson* distribution for $N = 16$. Note again that the throughput of reverse baseline network increases significantly as the window size, w, gets larger. In contrast with the reverse baseline network, the throughput of a nonblocking switch tends to saturate when the window size is increased beyond six, as shown in Figure 14. This figure shows that at high cell arrival rates, the throughput of reverse baseline with $w = 20$ gets very close to that of nonblocking switch with $w = 2$.

The total delay of a cell consists of two components: the *input buffer delay* that a cell spends from the time of its arrival at an input buffer until it enters the first stage of ATM switch fabric, and the *routing delay* of a switch fabric from its inputs to outputs. The routing delay through reverse baseline network is much smaller than the routing delay through a multistage nonblocking switch like batcher-banyan network. But, in determining the total delay, the input buffer delay plays

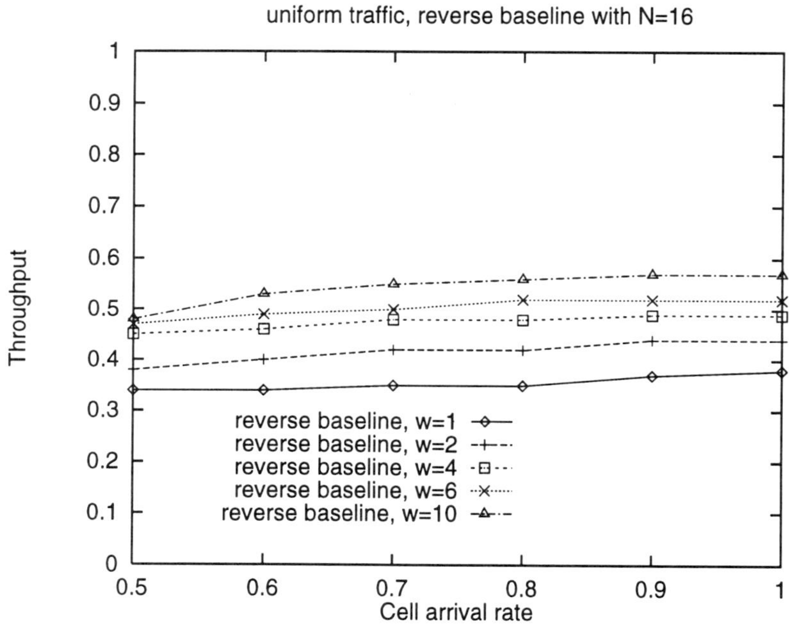

Figure 11: Throughput versus cell arrival rate characteristics for the baseline and nonblocking switches with input buffers under the uniform traffic pattern, $N = 16$, and $w = 4$.

much more important role than the routing delay as the window size increases. When the window size is very large in the input buffers of reverse baseline network, the cells close to the front of buffers may be assigned higher priorities in cell scheduling to prevent them from having excessive latencies. If the throughputs of nonblocking switch and reverse baseline network with input buffers are the same, their average total cell delays are not expected to differ significantly. Therefore, in the simulations, we focused on throughput measurements only for comparing reverse baseline network with nonblocking switches.

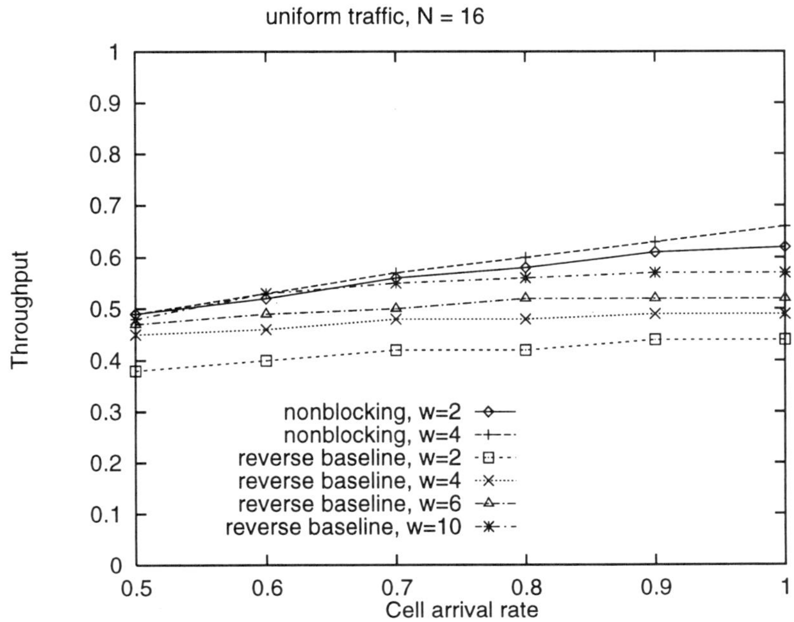

Figure 12: Throughput versus cell arrival rate characteristics for the baseline and nonblocking switches with input buffers under the uniform traffic pattern and $N = 16$.

5 Conclusion

If the limited number of connection patterns of baseline-type networks are not taken into consideration, they become very attractive to ATM switch designers due to their desirable features. In particular, they have constant latency for any input-output pair, are self-routing and modular, preserve cell-sequence due to path-uniqueness, are very suitable for VLSI implementation, and have an area complexity of $O(N\ log_2 N)$ compared to $O(N^2)$ of the crossbar switch. Therefore, we have been motivated to utilize window-based contention resolution algorithms for not only preventing external blocking and alleviating HOL blocking but also preventing internal blocking. This paper has proposed a parallel windowing algorithm, along with its hardware implementation, for

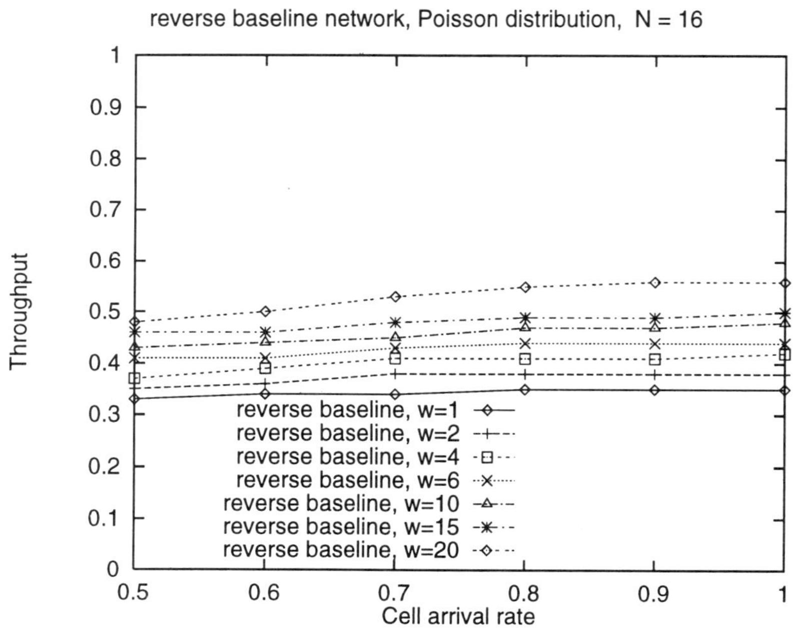

Figure 13: Throughput versus cell arrival rate characteristics with respect to different window sizes for the nonblocking and blocking switches under the poisson distribution and $N = 16$.

reverse baseline network with CAFIFO input buffers. Balanced incomplete matrices and matching with some constraints on bipartite graphs have been introduced to identify those cells that will not have any conflict on reverse baseline network when their destination addresses are used as routing tags.

We have conducted extensive simulations to compare the performance of a nonblocking switch with reverse baseline network, both of which having CAFIFO input buffers. Although throughput continually improves in a reverse baseline network as the window size increases, this is not true for nonblocking switches. In fact, it is stated for sequential windowing in [1] that throughput of nonblocking switches with input buffers starts diminishing as window size is increased beyond four. Our simulation results indicate that, when parallel windowing is used,

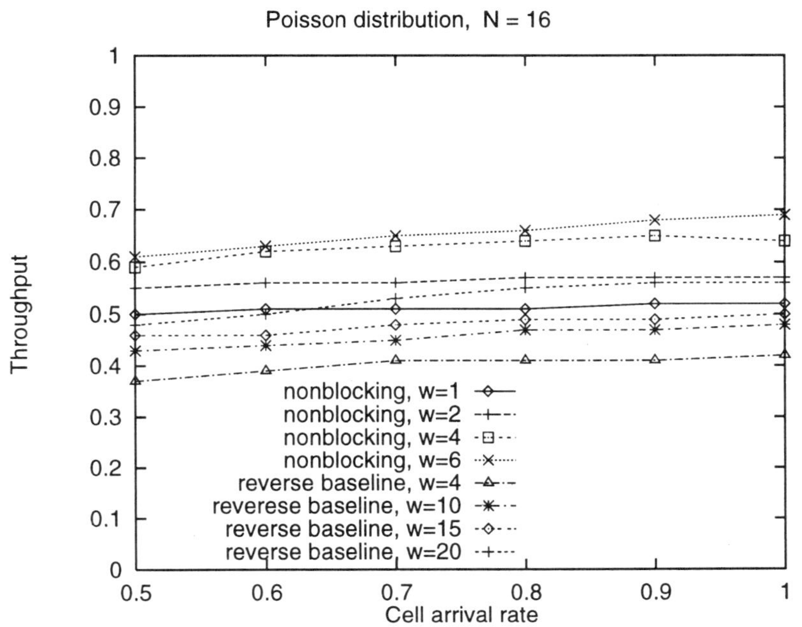

Figure 14: Throughput versus cell arrival rate characteristics with respect to different window sizes for the nonblocking and blocking switches under the poisson distribution and $N = 16$.

throughput of nonblocking switches tends to saturate as window size is around six. Also, our simulation results indicate that throughput of reverse baseline network can be as good as the throughput of a nonblocking switch if the window size is made sufficiently large. Therefore, reverse baseline network with CAFIFO input buffers can also be considered as a desirable ATM switch fabric. Due to the fact that input buffers are content addressable FIFOs, increasing the window size does not increase the hardware. But, when the window size is increased very much, some cells should be prevented from having excessive latencies by assigning higher priorities to the cells close to the head of buffers.

References

[1] M.G. Hluchyj and M.J. Karol, "Queueing in high-performance packet switching," *IEEE J. Select. Areas Commun.*, Vol. 6, No. 9, pp. 1587-1597, Dec. 1988.

[2] Y. Tamir and G.L. Frazier, "High-performance multi-queue buffers for VLSI communication switches," in *Proc. of the 15th Annual Int. Symp. on Computer Architecture,* pp. 343-354, 1988.

[3] Y.S. Yeh, M.G. Hluchyj and A.S. Acampora, "The knockout switch: a simple, modular architecture for high-performance packet switching," *IEEE J. Select. Areas Commun.*, Vol. SAC-5, No. 8, pp. 1274-1283, Oct. 1987.

[4] J. Hui and E. Arthurs, "A broadband packet switch for integrated transport," *IEEE J. Select. Areas Commun.*, Vol. 5, No. 8, pp. 1264-1273, Oct. 1987.

[5] M. Akata, S. Karube, T. Sakamoto, T. Saito, S. Wakasugi, S. Yoshida, H. Matsuno and H. Shibata, "A scheduling content-addressable memory for ATM space-division switch control," in *ISSCC Dig. Tech. Papers,* pp. 244-245, 1991.

[6] B. Bingham and H. Bussey, "Reservation-based contention resolution mechanism for batcher-banyan packet switches," *Electron. Lett.*, Vol. 24, No. 13, pp. 772-773, June 1988.

[7] M.J. Karol, M.G. Hluchyj and S.P. Morgan, "Input versus output queueing on a space-division packet switch," *IEEE Trans. Commun.*, Vol. com-35, No. 12, pp. 1347-1356, Dec. 1987.

[8] T.T. Lee, "A modular architecture for very large packet switches," in *Proc. GLOBECOM,* pp. 1801-1809, 1989.

[9] A. Pattavina, "Multichannel bandwidth allocation in a broadband packet switch," *IEEE J. Select. Areas Commun.*, Vol. 6, No. 9, pp. 1489-1499, Dec. 1988.

[10] K.W. Sarkies , "The bypass queue in fast packet switching," *IEEE Trans. Commun.*, Vol. 39, No. 5, pp. 766-774, May 1991.

[11] J.S. Turner, "Queueing analysis of buffered switching networks," *IEEE Trans. Commun.*, Vol. 41, No. 2, pp. 412-420, Feb. 1993.

[12] K.J. Schultz and P.G. Gulak, "Multicast contention resolution with single-cycle windowing using content addressable FIFO's," *IEEE/ACM Trans. Networking* , Vol. 4, No. 5, pp. 731-742, Oct. 1996.

[13] H. Cam and J.A.B. Fortes, "Frames: a simple characterization of permutations realized by frequently used networks," *IEEE Trans. on Comput.*, Vol. 44, No. 5, pp. 695-697, May 1995.

[14] N. Linial and M. Tarsi, "Interpolation between bases and the Shuffle-Exchange network," *Europ. J. Combinatorics,* Vol. 10, pp. 29-39, 1989.

[15] Y.K. Park and G. Lee, "NN based ATM cell scheduling with queue length-based priority scheme," *IEEE J. Select. Areas Commun.*, Vol. 15, No. 2, pp. 261-270, Feb. 1997.

[16] R.Y. Awdeh and H.T. Mouftah, "Survey of ATM switch architectures," *Computer Networks and ISDN Systems* , Vol. 27, pp. 1567-1613, 1995.

[17] A. Pattavina, "Nonblocking architectures for ATM switching," *IEEE Commun. Mag.*, No. 2, pp. 38-48, Feb. 1993.

King Fahd University of Petroleum and Minerals
Computer Engineering Department
Dhahran 31261, Saudi Arabia
e-mail: cam@ccse.kfupm.edu.sa

DIMACS Series in Discrete Mathematics
and Theoretical Computer Science
Volume **42**, 1998

Routing Strategy and Performance Evaluation of Multiple-Ring ShuffleNet Topology for High Speed Wavelength-Division Multiplexed Optical Communications

Dongjie Huang and Kamran Kiasaleh, Senior Member, IEEE

ABSTRACT. In this paper, the performance of a recently proposed topology, hereafter referred to as multiple ring ShuffleNet (MRSN), for wavelength-division multiplexed (WDM) optical communications is investigated. In the proposed topology, a multiple-ring ShuffleNet is realized via providing ring connectivity for the nodes arranged in every column of a ShuffleNet. A distributed routing algorithm is proposed for the MRSN. Since a finite link capacity is considered here, Erlang B formula is used to obtain a performance measure in terms of the percentage of blocked traffic for fault-free as well as for networks with single link failure. The reliability of the proposed MRSN network in terms of the above performance measure with a link capacity constraint is then obtained and is compared to those of ShuffleNet, Manhatan Street Network (MSN), de Bruijn graph-based, and Kautz graph-based networks. Two scenarios are considered. In one case, it is assumed that, upon a single link failure, no attempt is made to redistribute traffic. In an alternate scenario, upon a single failure, the optimization algorithm is invoked and the resulting traffic blocking probability is computed for the optimized traffic distribution. Under a uniform input traffic assumption, the numerical results indicate that the MRSN outperforms (yields smaller blocking probability for a given link capacity) its counterparts for networks of size greater than 8 for both scenarios.

1. Introduction

Optical wavelength-division multiple-access (WDM) technique takes advantage of the broad Lightwave bandwidth of an optical fiber to offer multiple-access capability to a large number of network subscribers via non-overlapping assignments of wavelength channels [**1**]. This multiple-access scheme has been the subject of intense study for broadband applications in recent years. The means by which WDM networks are realized may be via broadcast-and-select scheme [**2, 3**] or wavelength routing [**1, 4**]. In the broadcast-and-select scenario, transmitted signals from all users are summed by a passive coupler and the combined signal is broadcast, perhaps via a star coupler, to all the receivers in the network. Because of the use of a shared medium, this technique does not entertain the possibility of wavelength reuse. Unfortunately, at the present time, the number of independent wavelength channels of an optical fiber that can be utilized does not permit the realization

1991 *Mathematics Subject Classification.* Primary 94C10, 94A99.

of a WDM that can accommodate a significant number of independent users. In order to realize a scalable network which does not require a large number of wavelength channels, wavelength reuse becomes necessary [**5**]. It is therefore of interest to exploit network topologies that afford wavelength reuse in a WDM environment.

To that end and with the availability of acousto-optic tunable filters, it is possible to exploit both spatial and wavelength domains for WDM networks, and thereby enhance the multiple-access capability of the WDM scheme. Such filters can independently and simultaneously select an arbitrary subset of the available wavelengths on one fiber and route them to another. In this scheme, information from one node must travel through a number of intermediate nodes in order to reach its final destination. Due to this very attribute, the resulting WDM network is considered to be a multi-hop network, which allows for wavelength reuse to take place. In such networks, each node not only functions as an active user in the network, but also provides switching and routing services for other network subscribers. This distributed switching characteristic is desirable since it provides a higher degree of survivability when compared with its broadcast-and-select counterpart. That is, a failure of an optical component in one node would only disable communication among a subset of users in the network, whereas the failure of the broadcast element in the broadcast-and-select WDM network results in a total network failure. Obviously, transmission delay due to routing in a multi-hop environment is greater than that of the broadcast-and-select scheme, and must be taken into account in the networking planning of a WDM network.

To simplify routing and flow control, regular virtual topologies, built on top of an actual network's physical topology, are adapted to construct logical connection patterns to support both broadcast-and-select and wavelength-routing approaches. In recent years, regular topologies have been proposed for both virtual and physical connectivity for multi-hop, wavelength-routing applications. Of those, the most noteworthy structures are ShuffleNet [**6**], generalized shuffle net (GSN) [**7**], Manhattan street networks (MSN) [**8**] and de Bruijn graph-based networks [**9**]. Two important attributes that any proposed topology for multi-hop WDM communications must possess are ease of routing and flow control. Other issues, such as the network's ability to handle link failures in an efficient manner, cost, etc., are also of importance. The selection of a specific (virtual) topology for a multi-hop WDM network, however, is ultimately linked to the characteristics of the network, such as diameter, average number of hops, capacity and delay . From a user's perspective, in the face of a link capacity constraint[1] and for a given traffic load, the network must exhibit a high degree of connectivity or low end-to-end delay and insure a small probability of traffic loss.

Because of the limited number of wavelengths, when compared with the number of users, the number of required wavelengths must be computed in the design of any multi-hop WDM topology. It is also important to define connectivity for a multi-hop topology. In [**10**], connectivity between two nodes is realized via an optical path (via wavelength-routing). We have adopted a similar measure of connectivity in this paper. Traffic blocking probability is also an important performance measure for a multi-hop WDM network which is used in this paper to establish performance. In the past, the traffic blocking probability has been obtained for

[1]In WDM scenario, link capacity may be regarded as the number of available wavelengths used on an optical fiber.

static and dynamic traffic demands where the number of available wavelengths is a fixed number. In [**11**], a wavelength-convertible optical network was introduced to improve wavelength efficiency. In another study [**12**], a bound on the number of required wavelength channels is obtained for a WDM network with wavelength routing, wavelength changing, and circuiting switching.

In a recent study [**13**], we proposed the multiple-ring ShuffleNet topology for the realization of multi-hop WDM optical networks. In this paper, we present the performance of this topology and compare the result to those of other well-known networks. Another important feature of any multiple-access scheme is its ability to deal with failure(s) in an effective manner. The reliability and availability of Manhatan street networks were obtained in a recent work [**14**], where the impact of a single or multiple ring failures was to cause the creation of two or more disconnected subsets. In this paper, the reliability of the proposed WDM network topology[**15**], defined in terms of the percentage of the blocked traffic, is obtained in the face of single link failures (due to faulty optical components or fiber cuts). With the aid of this performance measure, the proposed MRSN is compared to ShuffleNet, GSN, MSN, Kautz graph-based [**16, 17**], and de Bruijn graph-based networks.

The remainder of this paper is organized as follows. In Section II, the structure of an MRSN is described. Subsequently, a routing algorithm is suggested in Section III for the proposed MRSN networks.

In Section IV, a performance measure based upon the percentage of the blocked traffic is defined. Moreover, the simplex algorithm is used to arrive at an algorithm for routing traffic that minimizes the network's blocking probability. Section V is dedicate to the performance analysis of the proposed and aforementioned regular topologies using the performance measures obtained in Section III. Two scenarios are considered. In one case, it is assumed that, upon a single link failure, no attempt is made to redistribute traffic. In an alternate scenario, upon a single failure, the optimization algorithm is invoked and the resulting traffic blocking probability is computed for the optimized traffic distribution.

Finally, in Section VI, conclusions are drawn based upon the results of Sections V.

2. Multiple-Ring ShuffleNet

Multiple-ring ShuffleNet is a modification of the ShuffleNet topology which has been studied as a regular topology for multi-hop WDM communication. Several variations of the ShuffletNet topology have been proposed [**18**]. The majority of these topologies are concerned with providing interconnections between different columns of a SuffleNet, without providing connectivity among the nodes in the same column, and yet maintain the "shuffle" connectivity among the columns. Since an MRSN is constructed by a ShuffleNet with directional rings on every column, the nodes may be reached via both column and row links.

In an MRSN connectivity graph, $N = kp^k$ $(k = 2, 4, \cdots; p = 1, 2, \cdots)$ nodes are arranged in k columns of rings with p^k users each. Hereafter, p is referred to as the node degree of the ShuffleNet. This parameter denotes the number of transmitter/receiver pairs per node of the ShuffleNet (i.e., the number of input or output links associated with each node in the topology). To define the routing structure, an addressing scheme similar to that of ShuffleNet [**6**] is adopted here. In an MRSN connectivity pattern with $N = kp^k$ nodes, each node can be identified by column and row coordinates (C, R), where $C \in \{0, 1, \cdots, k-1\}$ and $R \in \{0, 1, \cdots, p^k - 1\}$. Columns are labeled 0 through $k-1$ from left to right and rows are labeled 0 through $p^k - 1$ from top to bottom. For routing purpose, row addresses are also expressed as base p digits. That is,

$$R = (r_{k-1}r_{k-2}\cdots r_0)_p \qquad 0 \le r_i \le p-1 \ (i = 0, 1, \cdots, k-1). \tag{2.1}$$

This may also be expressed as

$$R = r_{k-1}p^{k-1} + r_{k-2}p^{k-2} + \cdots + r_1 p + r_0. \tag{2.2}$$

According to the above addressing scheme, a node with an address (C, R), where $R = (r_{k-1}r_{k-2}\cdots r_1 r_0)_p$, has links to p nodes with address $((C+1) \mod k, R')$, where $R' = (r_{k-2}r_{k-3}\cdots r_1 r_0 r')_p$ ($0 \le r' \le p-1$) and another node with an address (C, R''), where

$$\begin{aligned} R'' &= \left(R + p^k - (-1)^C\right) \mod p^k \\ &= \left(p^k + r_{k-1}p^{k-1} + r_{k-2}p^{k-2} + \cdots + r_1 p + r_0 - (-1)^C\right) \mod p^k. \end{aligned} \tag{2.3}$$

In the above, p^k is added to insure a nonnegative expression. Note that an MRSN has a degree of $p+1$, since there exists an additional path connecting each node of any column of the ShuffleNet to its adjacent node. Fig. 1 shows the connectivity for an 64-node ($p = 2, k = 4$) MRSN. In the figure, the first column is repeated on the right to show the cylindrical connectivity.

3. Routing

From the connectivity pattern of an MRSN, a distributed routing algorithm is developed. The knowledge of the ShuffleNet routing algorithm is beneficial to the understanding of the MRSN routing algorithm. The algorithm [**6**] selects a single path for every source and destination node pair.

For an intermediate node (C_a, R_a) and destination node (C_d, R_d), we define $X_{a,d}$ as

$$X_{a,d} = \begin{cases} (k + C_d - C_a) \mod k; & C_d \neq C_a, \\ k; & C_d = C_a. \end{cases} \tag{3.1}$$

The parameter $X_{a,d}$ will be used extensively in what follows.

ShuffleNet Routing Algorithm. In a ShuffleNet, given an intermediate node $(C_a, R_a) = (C_a, r_{a,k-1}r_{a,k-1}\cdots r_{a,0})$ [2] and a destination node $(C_d, R_d) = (C_d, r_{d,k-1}r_{d,k-2}\cdots r_{d,X_{a,d}-1}\cdots r_{d,0})$, the algorithm routes a message along a shuffle link to the next node with the address:

$$\left((C_a + 1) \mod k, (r_{a,k-2}r_{a,k-3}\cdots r_{a,1}r_{a,0}r_{d,X_{a,d}-1})_p\right).$$

This process is repeated until the destination node is reached.

Let $L(R_a, X_{a,d})$ and $F(R_d, X_{a,d})$ denote the operations needed to obtain the last and the first $(k - X_{a,d})$ digits of the current and destination row addresses, respectively. Then

$$\begin{cases} L(R_a, X_{a,d}) &= \left(r_{a,k-1-X_{a,d}}r_{a,k-2-X_{a,d}}\cdots r_{a,0}\right)_p = \left(R_a \mod p^{k-X_{a,d}}\right) \\ F(R_d, X_{a,d}) &= \left(r_{d,k-1}r_{d,k-2}\cdots r_{d,X_{a,d}}\right)_p = \left\lfloor R_d \,/\, p^{X_{a,d}} \right\rfloor \end{cases}$$

---

[2] We use $r_{Q,k-1}r_{Q,k-2}\cdots r_{Q,0}$ to denote the row address of node Q

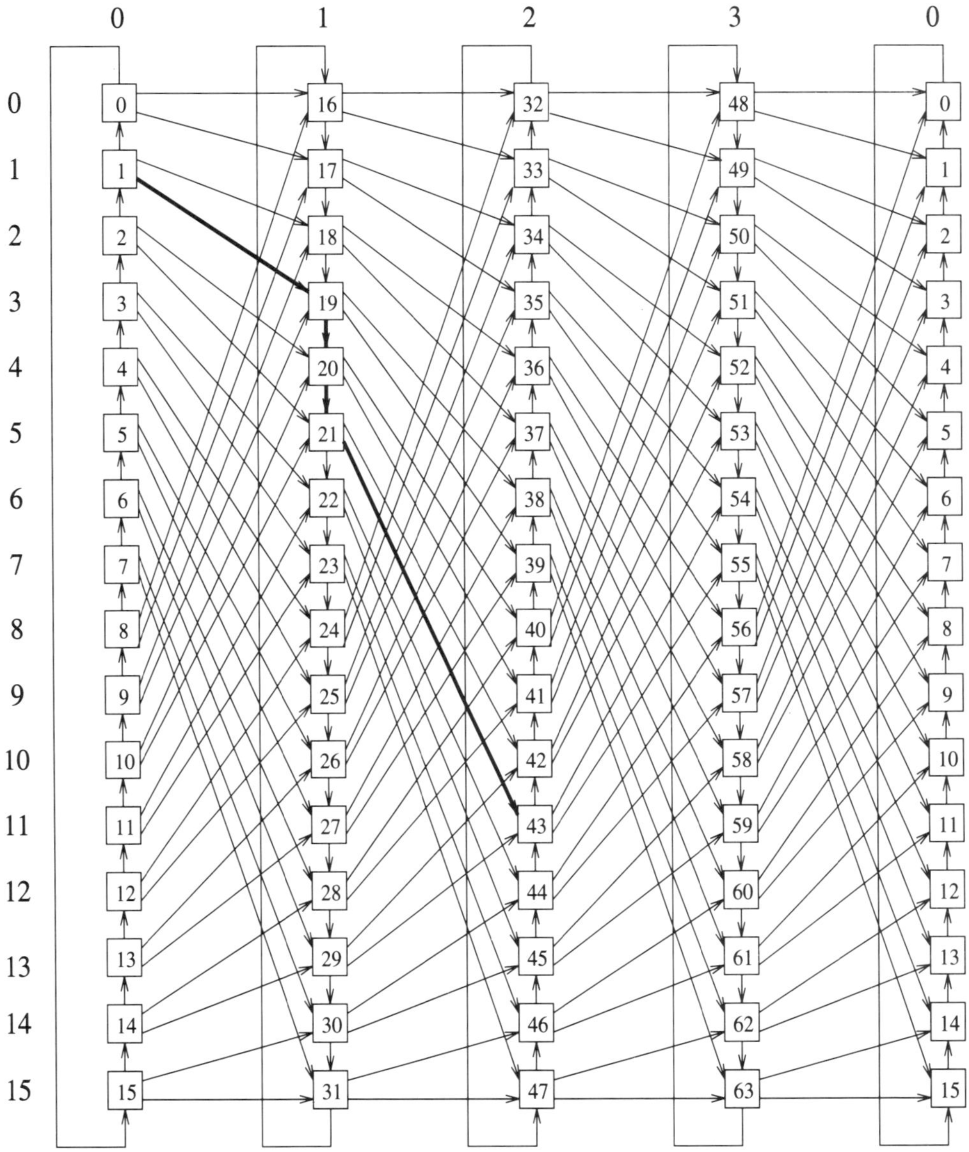

FIGURE 1. A (2, 4) Multiple Ring ShuffleNet (MRSN)

where $\lfloor x \rfloor$ denotes largest integer equal to or less than x. With the ShuffleNet routing rule and addressing scheme, the node after $X_{a,d}$ hops from a source node will have the address $(C_d, r_{a,k-1-X_{a,d}} r_{a,k-2-X_{a,d}} \cdots r_{a,0} r_{d,X_{a,d}-1} \cdots r_{d,0})$. Therefore, if $L(R_a, X_{a,d}) = F(R_d, X_{a,d})$, the message can reach its destination by $X_{a,d}$ hops. Otherwise, $X_{a,d}+k$ hops are required to reach the destination node using ShuffleNet routing.

MRSN Routing Algorithm. A routing algorithm is proposed for MRSN by taking advantage of additional rings. Two parameters, $t_{a,d}$ and $Y_{a,d}$, are introduced for the routing. If the source and destination nodes are on the same column, $t_{a,d}$

Paths for $X_{a,d} = k$
1. The first path uses shuffle links by ShuffleNet routing algorithm with k hops.
2. The second path only using ring links on the current column with $t_{a,d}$ hops:
(3.2) $$t_{a,d} = \left[(R_a - R_d)(-1)^{C_a} + p^k\right] \bmod p^k$$

TABLE 1. Paths proposed for $X_{a,d} = k$ in MRSN

denotes the number of hops from the source node (C_a, R_a) to the destination node (C_d, R_d) using ring links. If the two nodes are on different columns, $Y_{a,d}$ denotes the number of hops from the source node (C_a, R_a) to its first ShuffleNet matching node [3] (C_a, R_a^m). We also assume that $Y_{a,d} = t_{a,d}$ if the two nodes are on the same columns.

We first consider the routing where the source node and destination node are on the same column ($X_{a,d} = k$). Two paths from source node to destination node are proposed in table 3. If $t_{a,d} > k$, the first path is the shortest one, otherwise, the other path is selected.

If source node and destination node are on different columns, we first consider routing for a ShuffleNet matching node. The number of ShuffleNet matching nodes on column C_s equals $p^{X_{s,d}}$, since $p^{X_{s,d}}$ nodes can be reached by shuffle link in $X_{s,d}$ hops.

Three paths are given at table 2 when the source node (C_a, R_a) and its destination node are on different columns. The parameter $Y_{a,d}$ in the second path described in table 2 can be obtained as:

(3.4)
$$Y_{a,d} = \begin{cases} \left[\Big(L(R_a, X_{a,d}) - F(R_d, X_{a,d})\Big)(-1)^{C_a} + p^{k-X_{a,d}}\right] \bmod p^{k-X_{a,d}}; & X_{a,d} \neq k, \\ t_{a,d}; & X_{a,d} = k. \end{cases}$$

The node (C_b, R_b) in the third path is selected for the following reason. In the column next to the column C_a, p nodes with addresses $\Big((C_a+1) \bmod k, p(R_a \bmod p^{k-1}) + r\Big)$ $(r = 0, 1, \cdots p-1)$ have direct shuffle links from the node (C_a, R_a). If $Y_{a,d} \neq 0$, the node (C_b, R_b) requires the smallest number of hops to its first ShuffleNet matching node (C_b, R_b^m), since the other $p-1$ nodes reached by a shuffle link from (C_a, R_a) need to use (C_b, R_b) as an intermediate node to reach (C_b, R_b^m). (Note that rings are directional and a hop is made to the outer most node allowed by the hop.)

Based on paths provide by the above two tables, a routing algorithm is proposed in table 3. The algorithm examines the parameters $t_{a,d}$, $Y_{a,d}$ and $Y_{b,d}$ at every intermediate node, and finds the next immediate node $(C_{\text{next}}, R_{\text{next}})$ along the path for the given current node (C_a, R_a) and destination node (C_d, R_d). The storage

[3] If the node (C_s, R_s^m) has the property that its row address satisfies $L(R_s^m, X_{s,d}) = F(R_d, X_{s,d})$, then (C_s, R_s^m) is called a ShuffleNet matching node for the node (C_s, R_s) on column C_s when the destination node address is (C_d, R_d).

Paths for $X_{a,d} \neq k$

1. The first path for (C_a, R_a) to (C_d, R_d) is set up by ShuffleNet routing algorithm. It requires either $X_{a,d}$ or $X_{a,d} + k$ hops.
2. The second path (Fig. 2) is set up first by ring links from (C_a, R_a) to the first ShuffleNet matching node (C_a, R_a^m) with $Y_{a,d}$ hops, and then by ShuffleNet routing algorithm from (C_a, R_a^m) to (C_d, R_d). The total number of hops needed is $Y_{a,d} + X_{a,d}$.
3. The third path (Fig. 2) is set up first by shuffle link from (C_a, R_a) to a node with the address:

$$(C_b, R_b) = \Big((C_a + 1) \bmod k, p(R_a \bmod p^{k-1}) + p - p^{\frac{1-(-1)^{C_a}}{2}}\Big). \tag{3.3}$$

   - if $X_{b,d} \neq k$, the rest of the path uses ring links from (C_b, R_b) to the first ShuffleNet matching node (C_b, R_b^m). Then, the ShuffleNet routing algorithm is used for the path from (C_b, R_b^m) to (C_d, R_d). The total number of hops needed is $Y_{b,d} + X_{b,d} + 1$.
   - if $X_{b,d} = k$, the rest of the path follows the scenarios in table 3. The total number of hops needed is either $t_{b,d} + 1$ or $k + 1$.

TABLE 2. Paths proposed for $X_{a,d} \neq k$ in MRSN

Routing Algorithm

$Path = \{(C_a, R_a)\}$
WHILE $(C_a, R_a) \neq (C_d, R_d)$
Obtain $t_{a,d}$, $Y_{a,d}$ and $Y_{b,d}$ from equations (3.2),(3.4)

1. if ($X_{a,d} = k$ and $t_{a,d} > k$) or if ($X_{a,d} \neq k$ and $Y_{a,d} = 0$), then
   $(C_{\text{next}}, R_{\text{next}}) = \big((C_a + 1) \bmod k, (r_{a,k-2} r_{a,k-3} \cdots r_{a,1} r_{a,0} r_{d,X_{a,d}-1})_p\big)$
2. if ($X_{a,d} = k$ and $t_{a,d} \leq k$) or if ($X_{a,d} \neq k$ and $0 < Y_{a,d} < k$ and $Y_{a,d} < Y_{b,d}$), then
   $(C_{\text{next}}, R_{\text{next}}) = \big(C_a, [R_a - (-1)^C + p^k] \bmod p^k\big)$
3. otherwise, $(C_{\text{next}}, R_{\text{next}}) = (C_b, R_b)$

$(C_a, R_a) = (C_{\text{next}}, R_{\text{next}})$
$Path = \{Path, (C_a, R_a)\}$
END WHILE

TABLE 3. Routing algorithm for MRSN

variable *Path* records the sequential intermediate nodes from the source node to the destination node.

As an example, we consider the MRSN given in the Fig.1 with $p = 2$ and $k = 4$. There are $N = k \times p^k = 64$ nodes in the network. To illustrate the routing algorithm, we assume that the source node is $(0, 1)$ and the destination node is $(2, 11)$. As marked in Fig.1, the routing algorithm will traverse the path between

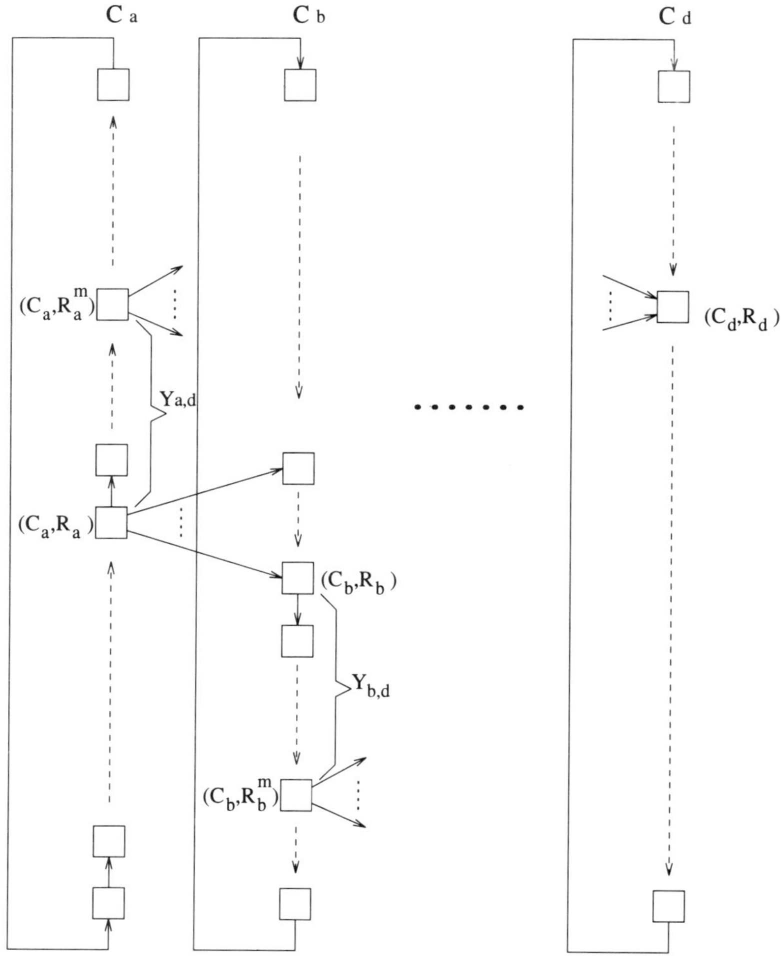

FIGURE 2. Paths selection for $X_{a,d} \neq k$ in MRSN

the two nodes, namely:

$$(0,1) \xrightarrow{(1)} (1,3) \xrightarrow{(2)} (1,4) \xrightarrow{(3)} (1,5) \xrightarrow{(4)} (2,11) \tag{3.5}$$

In this example, it takes 4 hops to send a message to the destination node. At iteration 1, $X_{a,d} = 2$, we can obtain the value of $Y_{a,d} = 3$, $Y_{b,d} = 2$. According to step 3 of Table 3, the next node on the path will be:

$$(C_{\text{next}}, R_{\text{next}}) = \Big((0+1) \bmod 4, (0011)_2\Big) = (1,3).$$

Now, at iteration 2, we have $X_{a,d} = 1$, and can also obtain $Y_{a,d} = 2$ and $Y_{b,d} = 11$. From step 2 of table 3, the data should be routed through ring link to the next node:

$$(C_{\text{next}}, R_{\text{next}}) = \Big(1, (3 - (-1)^1 + 16) \bmod 16\Big) = (1,4).$$

Similarly, the data again follows the ring link in column 2 to the node $(1,5)$ at the iteration 3. Finally at iteration 4, we have $(C_a, R_a) = (1,5)$, so $X_{a,d} = 1$ and

Node	Hop-1	Hop-2
1	2	2
2	2	2
3	1	1
4	1	1
5	1	1
6	2	2
7	2	2

TABLE 4. The number of hops from node 0 to all nodes in (2,2) MRSN. Hop-1 is the number of hops for the path generated by MRSN routing algorithm, Hop-2 is the number of hops for the path generated by Dijkstra's shortest path algorithm.

$Y_{a,d} = 0$. The next node is then:

$$(C_{\text{next}}, R_{\text{next}}) = \Big((1+1) \bmod 4, (1011)_2\Big) = (2, 11).$$

After iteration 4, we have $(C_a, R_a) = (C_d, R_d)$. The destination node is reached and the algorithm terminates. Thus we have successfully found a path between two nodes $(0,1)$ and $(2,11)$, and recorded all the nodes for it. That is, $Path = \Big\{(0,1),(1,3),(1,4),(1,5),(2,11)\Big\}$.

Tables 4 and 5 give the number of hops for the paths generated by MRSN routing algorithm and Dijkstra's algorithm [**19**]. For all given examples, MRSN routing algorithm produces the shortest path. The characteristics of MRSN routing is the topic of further study.

4. Network Reliability

In this Section, the ratio of the blocked traffic is used as a performance measure to compare the performance of the aforementioned topologies for a given traffic load in the presence of single link failure. It is important to note that a WDM network, regardless of its actual physical topology, may be represented by a graph $G(V, E)$ where V denotes a set of N nodes with E directional links connecting the nodes in V. Here, The traffic blocking ratio is used as an objective function for an optimization algorithm that balances traffic for a given network condition [**20**]. Similar to the analysis in [**20**], the expected traffic blocking ratio in the presence of a single link failure is obtained here.

In the study, we assume that the link (i,j), which connects node i to node j, possesses a capacity of C_{ij} channels, and that the traffic on this link is a random variable with an intensity of a_{ij} channels. Using the Erlang loss formula, $B(C_{ij}, a_{ij})$, the probability of finding all channels busy on (i,j), may be expressed as

$$B(C_{ij}, a_{ij}) = \frac{a_{ij}^{C_{ij}}/C_{ij}!}{\sum\limits_{k=0}^{C_{ij}} a_{ij}^k/k!}. \tag{4.1}$$

Node	Hop-1	Hop-2	Node	Hop-1	Hop-2
0			32	2	2
1	4	4	33	2	2
2	4	4	34	2	2
3	4	4	35	2	2
4	4	4	36	3	3
5	4	4	37	3	3
6	4	4	38	4	4
7	4	4	39	4	4
8	4	4	40	4	4
9	4	4	41	4	4
10	4	4	42	4	4
11	4	4	43	4	4
12	4	4	44	3	3
13	3	3	45	3	3
14	2	2	46	3	3
15	1	1	47	3	3
16	1	1	48	3	3
17	1	1	49	3	3
18	2	2	50	3	3
19	3	3	51	3	3
20	4	4	52	3	3
21	5	5	53	3	3
22	5	5	54	3	3
23	5	5	55	3	3
24	5	5	56	3	3
25	5	5	57	3	3
26	4	4	58	3	3
27	4	4	59	3	3
28	3	3	60	4	4
29	3	3	61	4	4
30	2	2	62	4	4
31	2	2	63	4	4

TABLE 5. The number of hops from node 0 to nodes 32-63 in (2,4) MRSN. Hop-1 is the number of hops for the path generated by MRSN routing algorithm, Hop-2 is the number of hops for the path generated by Dijkstra's shortest path algorithm.

As was noted above, in a multi-hop setting, traffic must travel through a number of nodes, and thus it is of interest to obtain the blocking probability of a message when a number of predefined links are used for routing. Let P_{BP} denote the blocking probability of a message travelling over a link that connects a source, say s, to a destination, say t, via links (i_1, j_1), (i_2, j_2), $\cdots$, (i_n, j_n). Using (4.1), we have

$$
\begin{aligned}
P_{BP} &= 1 - \prod_{k=1}^{n} (1 - B(C_{i_k j_k}, a_{i_k j_k})) \\
&\approx \sum_{k=1}^{n} B(C_{i_k j_k}, a_{i_k j_k})
\end{aligned}
\tag{4.2}
$$

where it is assumed that the blocking probability of a multi-link path is identical to the probability of blocking in any number of the links in the path and that the link blocking events are mutually independent. Obviously, the former assumption is only valid for a relatively small probability of blocking.

The expected portion of the traffic blocked on (i, j) may now be expressed as $a_{ij} B(C_{ij}, a_{ij})$. Let X_{ij}^{st} and d^{st} denote the traffic flow (in Erlang) due to source-destination pair $s-t$ travelling through link (i, j) and the source-destination offered load, respectively. We may now define an objective function T that will be minimized as follows:

$$T = \sum_{i,j} a_{ij} B(C_{ij}, a_{ij}), \tag{4.3}$$

subject to the flow conservation

$$\sum_j X_{ij}^{st} - \sum_j X_{ji}^{st} = \begin{cases} d^{st} & \text{if } i = s; \\ -d^{st} & \text{if } i = t; \\ 0 & \text{otherwise} \end{cases} \quad \forall\ i, s, t, \tag{4.4}$$

and nonnegativity assumption

$$X_{ij}^{st} \geq 0 \quad \forall\ i, s, t.$$

Furthermore, the traffic flow on link (i, j) is given by

$$a_{ij} = \sum_{s,t} X_{ij}^{st}. \tag{4.5}$$

Note that there are no capacity constraints in this formulation. In the analysis here, however, the routing algorithm uses (4.3) as the objective function and obtains a set of possible paths in the event that the optimization does not yield a unique solution. The shortest path among the possible solutions is then selected for routing.

The routing model defined above is a nonlinear multi-commodity problem where traffic on a given link may consists of payloads for many source-destination node pairs. By assuming that all commodities have the same priority, the objective function is known to be convex [**20**]. We resort to the decomposition method for the separation of commodities. For an N-node network, the idea of decomposition approach is to distribute link load for commodities in such a manner that the solution of the N subproblems yield a solution to the overall optimization problem. At each iteration, flow distribution is updated and N minimal cost network flow problems are solved. Each subproblem corresponds to the optimization of flow for the commodity from the same node basis. Hence, the combined flow obtained from solving the individual subproblems provides a feasible solution for the original problem. It is well-known that the possible source-destination paths for the commodity from a source node forms a spanning tree.

The convex simplex method [**21**] is adapted here to optimize each tree topology according to the objective function. In the convex simplex method, the flows due to each commodity are partitioned into two sets, known as basic and non-basic sets. All the commodity flows associated with the spanning tree are in the basic set, and thus the commodity flows associated with the non-basic set are those not in the spanning tree. Let $\mathcal{T}_b$ be the corresponding spanning tree of a source node b. Also, let $(i_1, i_m) \notin \mathcal{T}_b$, path $P = \{i_1, i_2, \cdots, i_m\}$ denote the intermediate nodes linking i_1 and i_m in the tree, and i_c $(1 \le c < m)$ be the ancestor of i_1 and i_m in P. Assuming that $a_{ij}^b = \sum_t X_{ij}^{bt}$ (a_{ij}^b is the traffic intensity carried by (i,j) due to the source node b, and thus $a_{ij} = \sum_b a_{ij}^b$), for the minimal cost flow problem, the reduced cost for any non-basic link (i_1, i_m) can be computed using

$$t_{i_1 i_m}^b = -\sum_{k=1}^{c-1} \frac{\partial T}{\partial a_{i_{k+1} i_k}^b} + \sum_{k=c}^{m-1} \frac{\partial T}{\partial a_{i_k i_{k+1}}^b} - \frac{\partial T}{\partial a_{i_1 i_m}^b}. \tag{4.6}$$

Because there are no capacity constraints in (4.4), the size of the non-basic set takes on its lower bound, which is zero. If $t_{i_1 i_m}^b > \varepsilon$ $(\varepsilon \ge 0)$, then increasing flow on (i_1, i_m) may force this link to join a basic set. The magnitude of the flow change in this algorithm is equal to the magnitude of the flow of a commodity in a link in the basic set. This in turn forces the flow on that link to zero. For optimization, the traffic flows are adjusted only if the resulting change has a smaller reduced cost than that of the previous step. The spanning tree is modified by replacing the zero flow link in the basic set with (i_1, i_m) in the non-basic set. The algorithm does not search for another commodity until no improvement can be gained for the present tree. This is a nonlinear problem, and thus the globally optimum solution is obtained by continuously optimizing each commodity. The search procedure is repeated by gradually reducing ε. The search terminates when ε satisfies a criterion or no further improvement can be made. The gradual reduction of ε will prevent the algorithm from spending a considerable amount of time optimizing only one commodity.

5. Results and Discussion

The network topologies considered here include those of ShuffleNet, Manhatan Street network (MSN), de Bruijn graph-based network, Kautz graph-based network, and the recently proposed GSN[4] and the proposed MRSN with node degrees of 2 and 3. The objective is not to be as exhaustive as possible. Rather, the objective is to compare MRSN to some well-known as well as a number of recently proposed architectures. Two distinct approaches are taken in the sequel for the performance characterization of these networks. In one approach, the blocking probability of each network is computed as a result of a random single failure. In this approach, no attempt is made to redirect traffic, and hence the traffic on the failed link is considered to be lost. In an alternate approach, in the event of a failure, the optimization algorithm discussed above is invoked to redistribute traffic. A blocking probability is then computed for the newly optimized network. Note that, in the second approach, it is assumed that the network is capable of redirecting traffic within a predetermined time frame to avoid the onset of a catastrophic failure.

In addition, for comparison purposes, two different scenarios are considered. In one case the performance measures are compared for networks with equal number of nodes. This leads to networks with various node degrees. Although this may

[4]A (p, k, n) GSN has np^k nodes where p is the node degree associated with each node, and n is the number of columns. Nodes in adjacent columns of GSN are connected as those in ShuffleNet

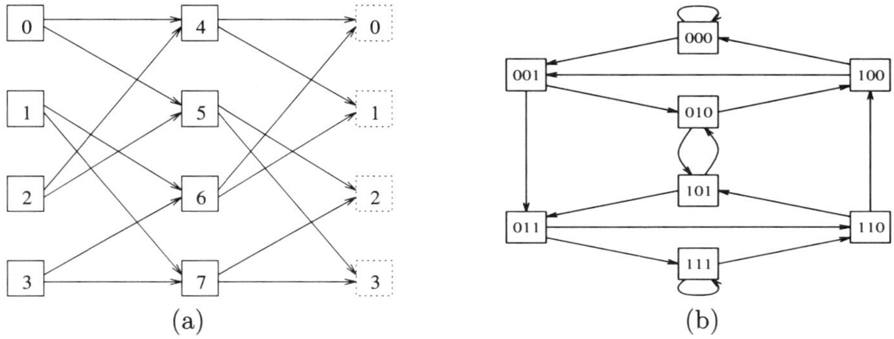

FIGURE 3. (a) A (2, 2) ShuffleNet. (b) A (2, 3) de Bruijn graph.

appear as an unfair comparison (MRSN and ShuffleNet with equal number of nodes, for instance, possess different node degrees), in some instances where an existing network is considered, the cost of increasing the node degree may be justified by the enhanced protection against failure offered by such a change. In the other scenario, it is assumed that the networks possess the same node degree, leading to a comparison between networks of various sizes.

The traffic distribution is assumed to be uniform. We consider an $N = 8$ node network with $d^{st} = 1$ for all s, t pairs ($s \neq t$). That is, the traffic intensity between any node and other seven nodes is assumed to be 1. Hence, the traffic intensity originated from any node is 7. This condition is kept intact as the size of the network is increased. That is, the total traffic offered to the network by a user is kept at 7.

The above assumption is motivated by the fact that a limited number of transmitters and receivers are assigned to each node and the traffic intensity originated at any node must remain stable as the network size grows. Hence, for a network with $N > 8$ nodes, the source-destination traffic intensity is determined as follows: $d^{st} = (\frac{7}{N-1})$. [5]

For an 8-node ShuffleNet (Fig. 3), the simplex optimization algorithm discussed in the previous section is invoked. The initial traffic is assumed to be uniform. The routing decision is made when the link capacity is fixed. The lost traffic is obtained as a function of the link capacity. Note that the Erlang B formula depends on the traffic load and the link capacity. In general, the optimized flow distribution to minimize lost traffic may vary as the link capacity is varied. In our case, each link has the same capacity, and thus the minimization of the objective function tends to distribute traffic flow evenly among all links. The even distribution of the traffic flow will minimize the lost calls in the event of a link failure.

The result of the optimization leads to a rather simple traffic distribution. That is, half of the links are assigned to carry 8 units of traffic while the other half are to carry only 6 units. The asymmetry in the assigned traffic distribution is attributable to the structure of the ShuffleNet. To elaborate, considering links $(0, 4)$ and $(0, 5)$, the traffic carried by $(0, 4)$ can only reach one new node by one more hop while traffic on $(0, 5)$ can reach two new nodes. It is conceivable that failure of $(0, 5)$ will cause more traffic loss and blockage than that caused by the failure of $(0, 4)$. Using

[5]This restriction, however, is imposed for the ensuing numerical analysis, and hence should not be regarded as a general guideline in determining the traffic intensity per node.

simulation, it can easily be shown that when $(0,4)$ is not available, the maximum traffic carried by a link is 12, whereas if $(0,5)$ is not available, the maximum traffic carried by a link increases to 16. This implies that the failure of $(0,5)$ results in more traffic loss than that caused by the failure of $(0,4)$ link. For this reason, the optimization algorithm yields uneven traffic distribution.

The simplex algorithm is also used to obtain the optimized flow distribution for a de Bruijn graph based network. In an 8-node de Bruijn graph (Fig. 3), the optimization algorithm leads to a scenario where traffic carried by any link ranges from 7 to 10. The algorithm is not able to find a better traffic distribution. When $(4,1)$ fails, the maximum traffic carried by any link (after optimization) increases to 18. The failure of $(0,1)$ will reduced the network into two disjoint entities. In this network, note that four links may cause this problem. From these observations, the above de Bruijn graph based network will experience a catastrophic failure in the event of some single link failures.

The above analysis gives some indication that the optimized traffic distribution takes on a non-uniform shape in an 8-node ShuffleNet or de Bruijn network with two transceivers per node. Now, we examine the proposed MRSN network. For an 8-node MRSN, the optimized traffic distribution ranges from 3 to 4 for every links in the network. Note that this is a reduction of 50 percent, as compared to the 8-node ShuffleNet, in the traffic carried by some links in the optimized network. In the event of a single failure, the maximum traffic on a link increases to 6 after optimization, whereas under a similar circumstance, an 8-node ShuffleNet must be able to carry a traffic of size 16 in some of its links. As will be seen in the ensuing numerical results, this drastic reduction in the size of traffic carried by a link due to failure is to be expected of MRSN. We now proceed to compute the blocking probability of the aforementioned networks in the presence and absence of a single failure. First, in Fig. 4, the blocking probabilities for MRSN, ShuffleNet, de Bruijn graph-based, and Kautz graph-based networks are computed for a node degree of 2. Note that MRSN cannot be included here since a node degree of 2 implies $p = 1$. In this figure, it is assumed that no attempt is made to optimize the traffic distribution after a link failure. It is quite clear that a blocking probability floor is developed for all networks in the event of a single link failure. In the absence of failure, MSN seems to be the superior network. In the presence of a single failure, however, the Kautz graph-based network yields the smallest blocking probability. In Fig. 5, networks whose node degrees are fixed at 3 are compared. Note that since the node degree is kept constant, the network size does not remain constant. In the absence of a link failure, MRSN proves to be quite superior to its counterparts. In the presence of a link failure, however, the ShuffleNet network provides the smallest blocking probability. One reason for this inconsistency is due to the fact that a fixed traffic intensity is considered here, and hence a large network is less strained by a single link failure than its counterparts. Note that, in Fig. 4, the Kautz network is the largest network, and thus it yields the best performance under failure. In Fig. 5, the ShuffleNet is the largest network, resulting in the best performance in the event of a link failure. Moreover, we are not optimizing traffic after a failure, and thus these results are somewhat distorted. Let us now consider the equal-size network scenario. In Fig. 6, the networks of equal size are compared. This naturally leads to nonidentical node degrees. In this case, the traffic intensity remains the same. Quite clearly, the MRSN outperforms its counterparts across the board. This observation is also valid for networks of size 24 (Fig. 7) and 64

(Fig. 8). Even in the event of failure the blocking probability floor of the MRSN is half that of its counterparts.

Next, we proceed to consider the second, and more interesting approach. That is, we assume that networks considered here, in the event of a single link failure, can reroute traffic using the optimization algorithm discussed above. Once again, we first consider that the node degrees of all the networks are identical. This is followed by an analysis of equal-size networks. In Figs. 9 and 10, the networks of node degrees 2 and 3, respectively, are compared. Obviously, for $p = 2$, the MRSN reduces to its MSN counterpart. One striking feature of these plots is the disappearance of the blocking probability floor, which is expected in view of the fact that traffic is redistributed upon failure. As in the previous case, the MSN network proves to be superior to others compared in Fig. 9. However, when all the aforementioned networks are compared for node $p = 3$ in Fig. 10, the proposed MRSN emerges as a superior network. Networks with identical sizes are then compared in Figs. 11, 12, and 13. Rather surprisingly, the MRSN seems to emerge as the superior network in all three cases. The degree by which this network outperforms the others is rather astounding. For example, in the case of a 64-node network and for a blocking probability of 0.01, the MRSN requires a link capacity that is almost half that of its counterparts. This clearly demonstrates the ability of the proposed MRSN to take advantage of the ring architecture to redistribute traffic and to maintain a uniform traffic distribution even in the event of a single link failure. Although it appears initially that one must consider networks whose node degrees are identical, as noted earlier, it is more practical to consider a situation where an existing network is to be re-connected or rearranged to reduce its susceptibility to a single link failure. In the case of ShuffleNet, this implies adding an additional link (or adding an additional pair of transmitter/receiver) to upgrade the existing ShuffleNet. This seems to be a modest increase in cost, especially in the case of large networks where MRSN seems to be most effective.

6. Conclusion

In this paper, the recently proposed multiple-ring ShuffleNet (MRSN) network was investigated. The performance of an MRSN network was established in terms of the ratio of blocked traffic due to single link failures or, in the absence of failure, due to finite channel capacity. An algorithm for distributing traffic that minimizes the blocking probability was also suggested. Under two scenarios the performance of the MRSN networks were compared to those of its counterparts. For networks with identical sizes, it was shown that the proposed MRSN can outperform its counterparts across the board. More importantly, in the event of a single link failure, the blocking probability of the MRSN remains far smaller than those of its counterpart.

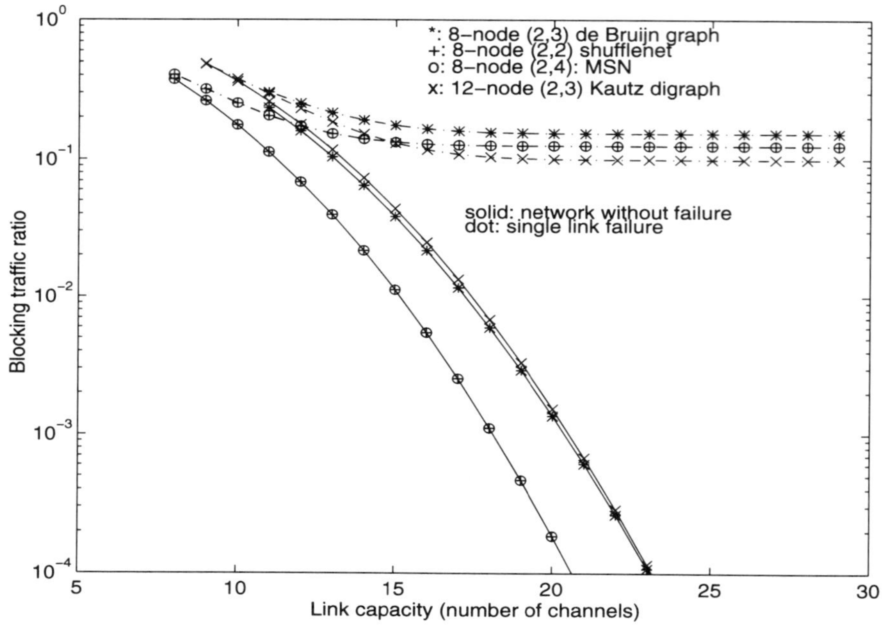

FIGURE 4. Traffic blocking probability as a function of link capacity for networks of degree 2. No attempt is made to optimize traffic distribution upon failure.

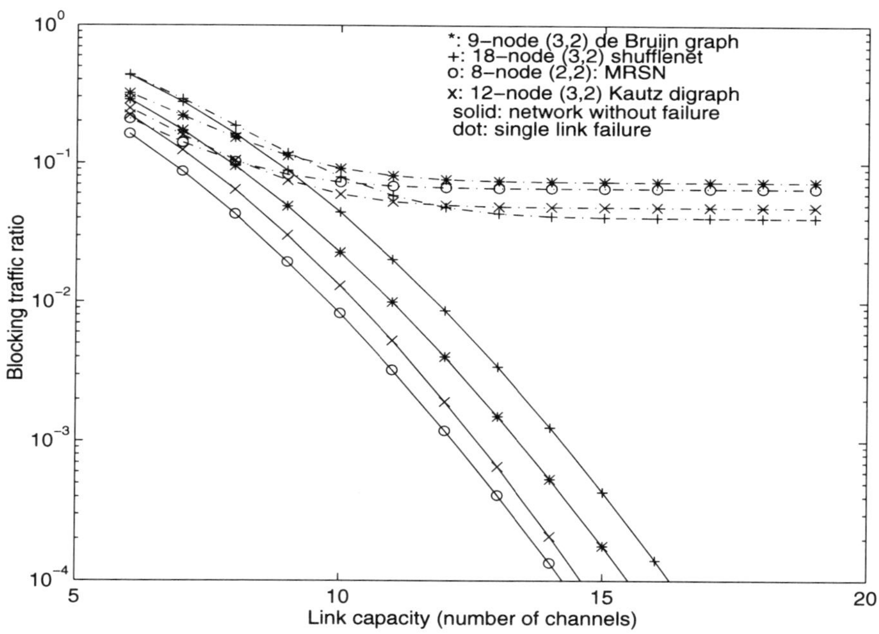

FIGURE 5. Traffic blocking probability as a function of link capacity for networks of degree 3. No attempt is made to optimize traffic distribution upon failure.

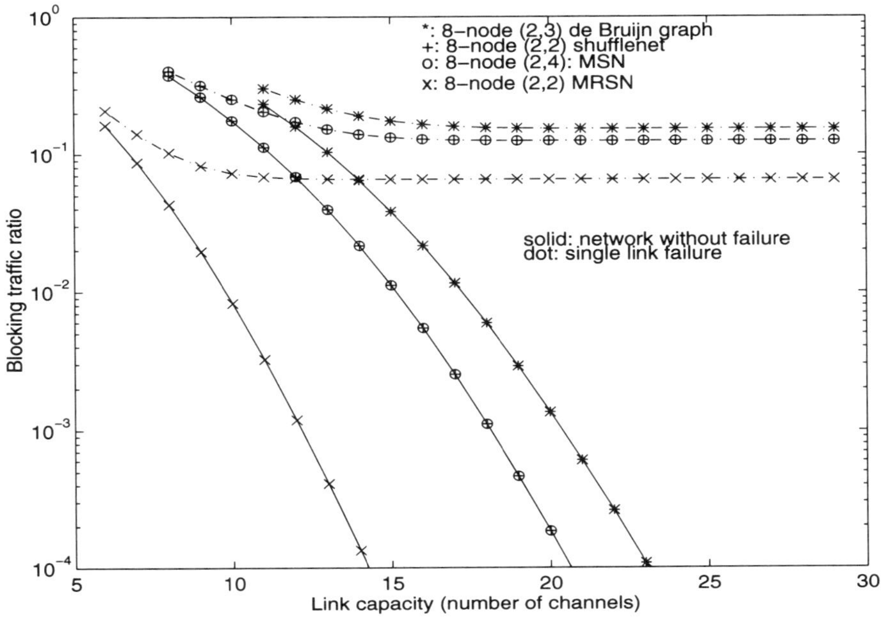

FIGURE 6. Traffic blocking probability as a function of link capacity for networks of size 8. No attempt is made to optimize traffic distribution upon failure.

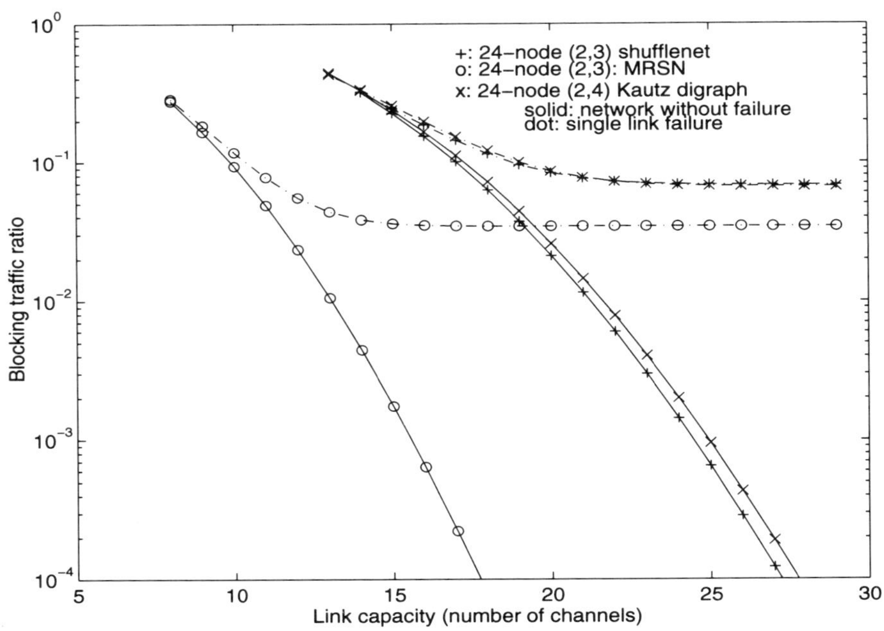

FIGURE 7. Traffic blocking probability as a function of link capacity for networks of size 24. No attempt is made to optimize traffic distribution upon failure.

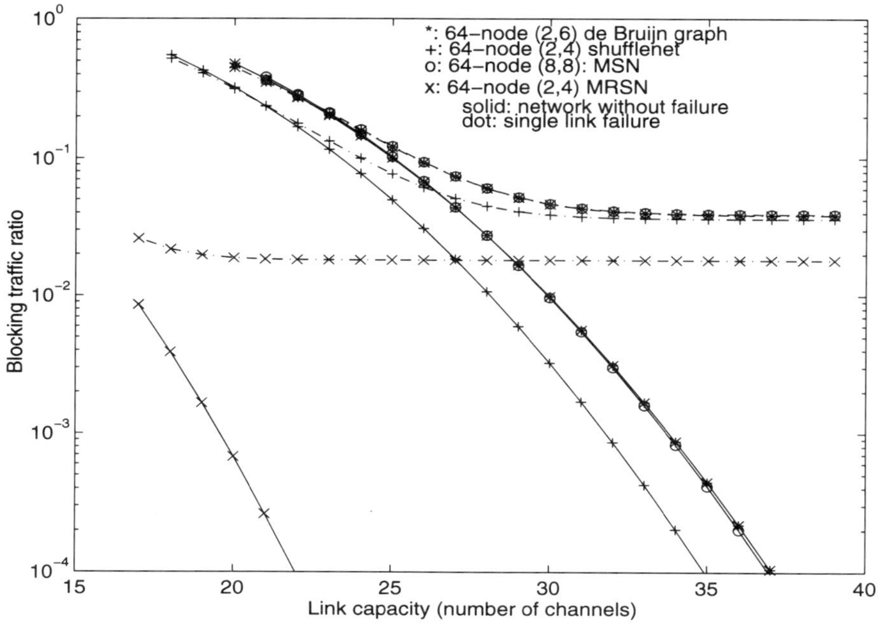

FIGURE 8. Traffic blocking probability as a function of link capacity for networks of size 64. No attempt is made to optimize traffic distribution upon failure.

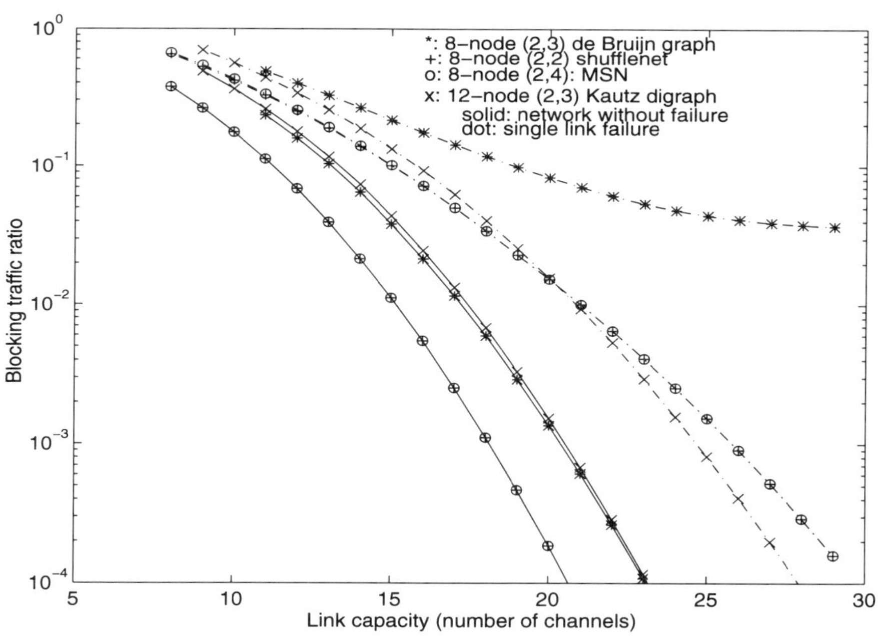

FIGURE 9. Traffic blocking probability as a function of link capacity for networks of degree 2. The traffic distribution is optimized upon failure.

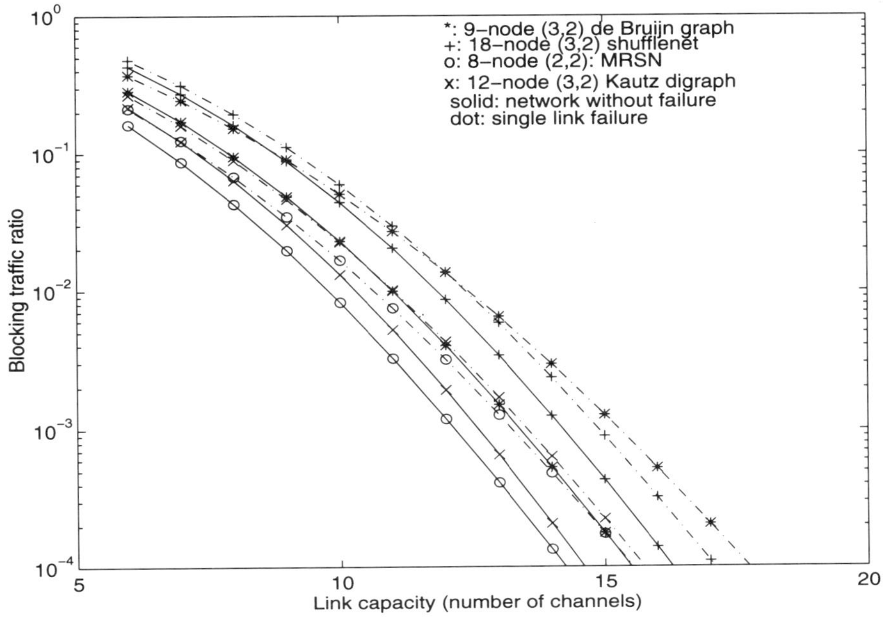

FIGURE 10. Traffic blocking probability as a function of link capacity for networks of degree 3. The traffic distribution is optimized upon failure.

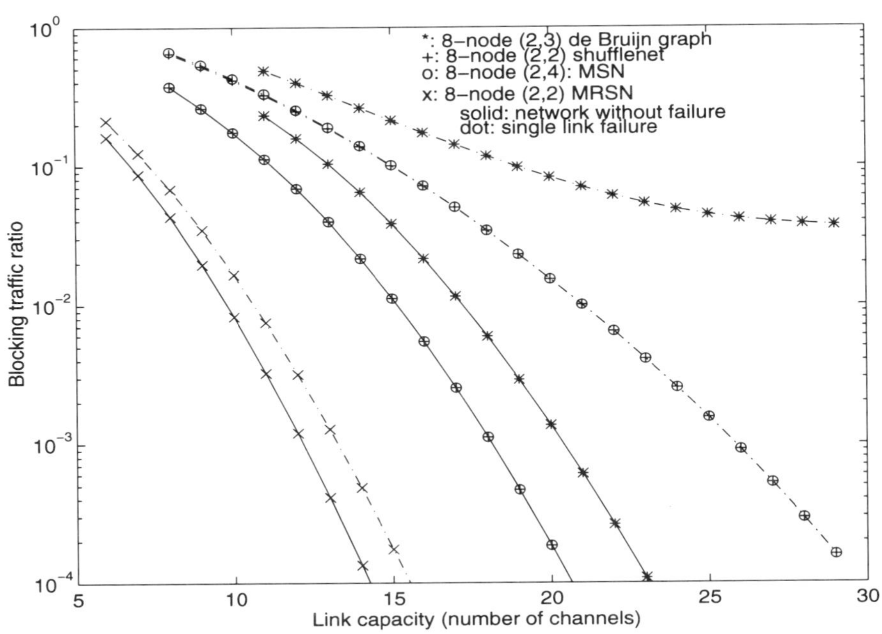

FIGURE 11. Traffic blocking probability as a function of link capacity for networks of size 8. The traffic distribution is optimized upon failure.

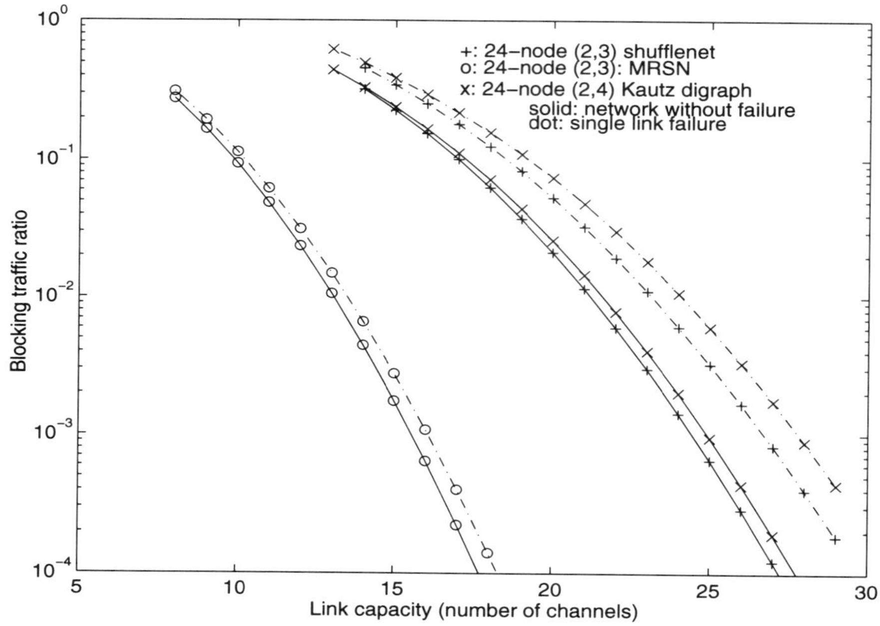

FIGURE 12. Traffic blocking probability as a function of link capacity for networks of size 24. The traffic distribution is optimized upon failure.

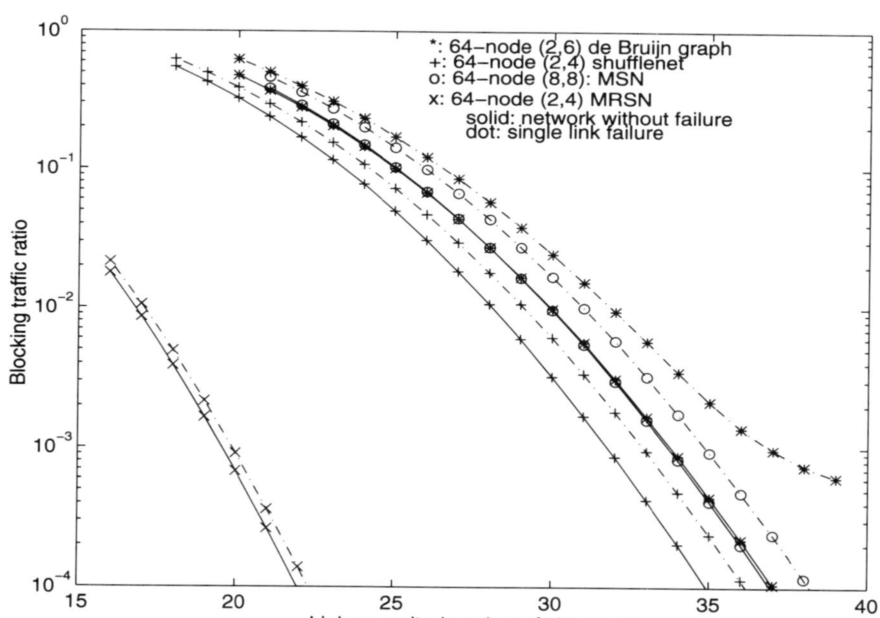

FIGURE 13. Traffic blocking probability as a function of link capacity for networks of size 64. The traffic distribution is optimized upon failure.

References

[1] C. A. Brackett, "Dense wavelength division multiplexing networks: Principles and applications," *IEEE J. Select. Areas Commun.*, vol. SAC-8, pp. 948–964, Aug. 1990.
[2] A. S. Acampora and M. J. Karol, "An overview of lightwave packet networks," *IEEE Network*, pp. 29–41, Jan. 1989.
[3] B. Mukherjee, "WDM-based local lightwave networks part I : Single-hop systems," *IEEE Network*, pp. 12–27, May 1992.
[4] B. Mukherjee, "WDM-based local lightwave networks part II: Multi-hop systems," *IEEE Network*, pp. 20–32, July 1992.
[5] C. A. Brackett, A. S. Acampora, J. Sweitzer, and et. al., "A scalable multiwavelength multihop optical network: A proposal for research on all-optical networks," *J. Lightwave Technol.*, vol. JLT-11, pp. 736–753, May/June 1993.
[6] M. G. Hluchyj and M. J. Karol, "ShuffleNet: An application of generalized perfect shuffles to multihop lightwave networks," *J. Lightwave Technol.*, vol. JLT-9, pp. 1386–1397, Oct. 1991.
[7] S.-W. Seo, P. Prucnal, and H. Kobayashi, "Generalized Multihop Shuffle Networks," *IEEE trans. Commun.*, vol. COM-44, pp. 1205–1211, Sept 1996.
[8] E. Ąyanoglu, "Signal flow graphs for path enumeration and deflection routing analysis in multihop networks," in *Proc. IEEE GLOBECOM '89*, pp. 1022–1029, Nov. 1989.
[9] K. Sivarajan and R. Ramaswami, "Multihop Lightwave networks based on de Bruijn graphs," in *Proc. IEEE INFOCOM '91*, pp. 1001–1011, March 1991.
[10] I. Chlamtac, A. Ganz, and G. Karmi, "Lightpath communication: An approach to high bandwidth optical WAN's," *IEEE trans. Commun.*, vol. COM-40, pp. 1171–1182, July 1992.
[11] K.-C. Lee and V. O. K. Li, "A wavelength-convertible optical network," *J. Lightwave Technol.*, vol. JLT-11, pp. 962–970, May/June 1993.
[12] R. A. Barry and P. A. Humblet, "On the number of wavelengths and switches in all-optical networks," *IEEE trans. Commun.*, vol. COM-42, pp. 583–591, Feb./March/May 1994.
[13] D. Huang and K. Kiasaleh, "Multiple-ring ShuffleNet: A novel topology for WDMA optical networks," in *Porc. International Conference on Communications (ICC'95)*, (Seattle, WA), pp. 1187–1190, June 1995.
[14] Z. Chen and T. Berger, "Reliability and availability analysis of Manhattan street networks," *IEEE trans. Commun.*, vol. COM-42, pp. 511–522, Feb./March/May 1994.
[15] K. Kiasaleh and D. Huang, "A survivability analysis of WDM optical fiber systems with regular topologies," in *Proc. OFC '97*, Feb. 1997.
[16] G. Panchapakesan and A. Sengupa, "On multi-hop optical network topology using Kautz digraphs," in *Proc. IEEE INFOCOM'95*, (Boston, MA), pp. 675–682, April 1995.
[17] S. Jiang and T. E. Stern, "Regular multicast multi-hop Lightwave networks," in *Proc. IEEE INFOCOM'95*, (Boston, MA), pp. 692–700, April 1995.
[18] A. K. Choudhury, "A comparative study of architecture for deflection routing," in *Proc. IEEE GLOBECOM '92*, pp. 1911–1920, 1992.
[19] T. H. Cormen, C. E. Leiserson, and R. L. Rivest, *Introduction to algorithms.* Cambridge, Massachusetts: The MIT Press, 1994.
[20] B. Sanso, F. Soumis, and M. Gendreau, "On the Evaluation of telecommunications network reliability using routing models," *IEEE trans. Commun.*, vol. COM-39, pp. 1494–1501, Oct. 1991.
[21] J. L. Kennington and R. V. Helgason, *Algorithms for network programming.* New York: John Wiley & Sons, Inc., 1980.

Erik Jonsson School of Engineering and Computer science The University of Texas at Dallas P.O. Box 830688, EC33 Richardson, TX 75083
E-mail address: `dongjie@utdallas.edu`

Erik Jonsson School of Engineering and Computer science The University of Texas at Dallas P.O. Box 830688, EC33 Richardson, TX 75083
E-mail address: `kamran@utdallas.edu`

DIMACS Series in Discrete Mathematics
and Theoretical Computer Science
Volume 42, 1998

Multispace Search for Quorumcast Routing

Jun Gu, Bin Du, Danny H.K. Tsang, and Wei Wang

ABSTRACT. The quorumcast routing problem is a generalization of multicast routing. For a given quorum size, quorumcasting sends a message from a source node to any subset of nodes whose size equals to the quorum size. The goal of quorumcast routing is to find a minimum cost tree spanning the subset. In this paper, we present a multispace search algorithm for the quorumcast routing problem. In the algorithm, we first prune the candidate set to reduce the original search space. A multispace search algorithm is then used to solve the quorumcast routing problem. By reformulating the original objective function, the original problem instance is transformed into a series of gradually simplified problem instances with smoother terrain surfaces. A local search algorithm is used to solve each problem instance, from the simplest structure to the original structure, and the solutions of the more simplified problem instances are used to guide the search of more complicated ones. Experimental results showed that this new method improved the performance of the existing heuristic search algorithms for the quorumcast routing problem.

1. Introduction

Many networking applications require multicast connections in addition to conventional unicast connections. Multicasting delivers a message from a single source node to a subset of nodes (the *multicast destinations*) in the network. Such could be the cases of multipoint video conferencing, the distribution of a document to a selected number of persons via a computer network, or the request of certain information from a distributed database. The problem of connecting a subset of nodes in an undirected graph using the minimal cost subgraph, such that there is a path between each pair of nodes, is known to be NP-complete [**GP68, Kar72, Win87**]. A variety of connection schemes, such as steiner tree and spanning tree methods, can be used for multicast connections.

1991 *Mathematics Subject Classification.* Primary 68M10, 90B12, 90C27, 90B40, 05C85, 68R10.

Key words and phrases. Computer networks, communication networks, combinatorial optimization, graph theory, search.

This work was supported in part by NSERC Research Grant OGP0046423, NSERC Strategic Grant MEF0045793, NSERC Strategic Grant STR0167029, and is presently supported in part by RGC Research Grant HKUST 756/96E and Hong Kong Telecom Institute of Information Technology Grant HKTIIT 93/94.EG01.

Quorumcasting delivers a message from a single source node to any subset of $|Q|$ nodes in an n-node network ($|Q| < n$). The problem of quorumcast routing is important in a number of distributed applications and the management of replicated data [**CK94b**]. Processes in the distributed synchronization problem need to communicate with a subset of the multicast destinations. Generally, a set of processes in a distributed application form a *distributed process group.* The processes in a distributed process group want to coordinate their execution with others to ensure correctness. For example, when two processes want to update the shared data in a distributed database system, both update operations are not allowed to proceed at the same time. *Quorum consensus* gives a general class of synchronization protocols in distributed systems [**GMB85**]. In many cases, these distributed operations proceed only after it coordinates with nodes that constitute a *quorum set* [**BGM86**]. The collection of quorum groups used by an operation is called a *quorum set.* A well-known method for defining quorum sets is *voting* [**JM90, AAC91**]. In voting, each member in the process group is assigned one vote. Critical operations that require synchronization must obtain a *quorum* before further proceeding.

Although the multicasting problems have been researched actively in the past, the quorumcasting problem has only been addressed recently. In this paper, we propose a multispace search algorithm to solve the quorumcast routing problem. The algorithm dynamically reformulates the objective function in conjunction with a local search optimization algorithm for quorumcast routing. Experimental results show that multispace search based on the objective function reformulation is effective in handling the pathological behavior of local minima encountered in the quorumcast routing problem.

The rest of the paper is organized as follows: In the next section, we will give an overview of some existing multicast and quorumcast routing methods. Section 3 describes the basic ideas of multispace search. In Section 4, we describe scrambling operation in the objective function space for quorumcast routing. A multispace search algorithm for the quorumcast routing problem is introduced in Section 5. The experimental results of the multispace search algorithm for quorumcast routing are shown in Section 6. We will compare our algorithm with other existing algorithms for the quorumcast routing problem. Section 7 concludes the paper.

2. Previous Work

Much research work has been conducted to solve the multicast problem. We will briefly discuss a number of them here.

Chow [**Cho91**] proposed for networks with directed links three multicast path finding algorithms: an optimal algorithm based on dynamic programming, a heuristic algorithm with the assumption that all vertices have the multicast capability, and a heuristic algorithm for networks where some vertices do not have the multicast capacity. Computational results show that the heuristic algorithms can find multicast paths whose costs are close to optimal.

Kompella, Pasquale, and Polyzos [**KPP92**] investigated multicast routing for high-bandwidth, delay-sensitive applications. Their routing algorithm ensures that the delay between the source and any destination is bounded by a specified value. The algorithm operates on a condensed graph and builds a tree by selecting edges based on a metric combining cost and residual delay.

In packet-switched networks, copies of a packet for multicast communication should be made available at each intermediate node. However, the concentration of packet copy operation at a particular node causes performance degradation of other calls which go through this node. Tode, Sakai, Yamamoto, Okada, and Tezuka [**TSY$^+$92**] proposed two multicast routing algorithms, a link added algorithm and a loop constructed algorithm, which distribute copy operation of packets over all nodes along the multicast path. Both algorithms derive an approximate solution for the minimum cost path and then improve the solution to prevent concentration of packet copy operation at one switching node. The solution produced is slightly poorer than optimal.

Some applications may allow destinations to join and leave the multicast group dynamically. Under these conditions, Doar and Leslie [**DL93**] did not pursue an optimal spanning tree at each change. Instead, they used the minimal cost source to destination routing for each destination, effectively ignoring the existing multicast tree.

Waxman [**Wax93**] investigated the weighted greedy algorithm in the multicast routing for ATM networks. Ammar, Cheung, and Scoglio [**ACS93**] studied the same problem in a virtual-path based ATM network.

Chen and Kumar [**CK94a**] used the cube concept which consists of a group of outlets reachable through the network for the self-routing multistage network. Later, they gave an improved two-phase multicast routing algorithm [**RCK95**].

Noronha and Tobagi [**JT94**] formulated the problem of optimal routing multicast streams using integer programming. The decomposition principle was extended to speed up the linear relaxation of the problem. The enhanced value-fixing rules were used to prune the search space for the integer problem.

Zhu, Parsa, and Garcia-Luna-Aceves [**ZPGLA95**] presented a heuristic algorithm for constructing minimum-cost trees with delay constraints. The algorithm sets variable delay bounds on destinations and minimizes the total bandwidth utilization of the tree and the maximal link cost of the most congested link. The algorithm starts with the minimum-delay tree and monotonically decreases the cost by iterative improvement of the delay-bounded tree.

After studying multicast trees, Bauer and Varma [**BV95**] showed that efficient multicast trees could be found in large, sparse networks with small multicast groups even with limited multicast capability in the individual switches.

Quorumcast routing is a generalization of multicast routing. It can be stated as follows:

DEFINITION 1. In an n-node communication network, a quorum set Q consists of any $|Q|$ members ($|Q| < n$). A message is to be sent from a source node in the network to the quorum set, such that the routing tree spanning the source node and the quorum set has the minimum communication cost, determined as the sum of costs of the links in the routing tree.

For an n-node network, there are $\binom{n}{|Q|}$ ways of selecting $|Q|$ nodes from the group of n nodes. The number of routing trees that span the source node and the quorum set can be very large. The optimum routing tree is the one that has the minimum communication cost. As the multicast routing problem is NP-complete, efficient optimal solutions to the quorumcast routing problem are unlikely to be found [**CK94b**].

Although the multicasting problems have been researched actively in the past, the quorumcasting problem has only been addressed recently. Cheung and Kumar [**CK94b**] presented three heuristics, the minimum cost path (MPH) heuristic, the improved minimum path (IMP) heuristic, and the modified average distance (MAD) heuristic, to find low cost solutions for the quorumcast routing problem.

In the MPH heuristic [**CK94b**], initially, only the source node is put in the quorum set, and the size of the quorum set is 1. In each step, the algorithm examines all the nodes that do not belong to the quorum set and selects the nearest one to the quorum set until the quorum is built. The IMP and MAD heuristics are two improved heuristic methods [**CK94b**]. In the IMP heuristic, the first iteration is identical to the MPH heuristic. In each subsequent iteration, among the nodes which have not been included in any solution, the node nearest to the source node is selected. Then the algorithm is restarted with this initial solution, proceeding to add more nodes to the quorum set. Extensive experimental results showed that, among the three heuristics, the IMP heuristic gave the best solution.

Recently *multispace search* approach has been developed for solving general optimization problems [**Gu94, Gu97**]. In this paper, we apply the multispace search to solve the quorumcast routing problem. The technique is capable of smoothing the rugged terrain surface of the search space and finding a much better solution for the quorumcast routing problem.

In the next section, we will briefly describe multispace search techniques for optimization problems.

3. Optimization by Multispace Search

Many search and optimization methods have been developed in combinatorial optimization, operations research, artificial intelligence, neural networks, genetic algorithms, and evolution programming. An optimization algorithm seeks a value assignment to variables such that all the constraints are satisfied and the performance objective is optimized. The algorithm operates by changing values to the variables in the value space. Because value changing does not affect the problem structure and the search space, it is difficult for a value search algorithm to handle the pathological behavior when encountering local minima.

Multispace search is a new optimization approach developed in recent years [**Gu94, Gu97, Gu98**]. The idea of multispace search was derived from principles of nonequilibrium thermodynamic evolution that structural changes are more fundamental than quantitative changes and evolution depends on the growth of new structure in biological system rather than just information transmission. A search process resembles the evolution process and the structural operations are important to improve the performance of traditional value search methods.

In multispace search, any active component related to the given problem structure can be manipulated, and thus, be formulated as an independent search space. For a given optimization problem, for its variables, values, constraints, objective functions, and key parameters (that affect the problem structure),[1] we define the variable space, the value space (i.e., the traditional search space), the constraint space, the objective function space, the parameter space, and other search spaces, respectively. The totality of all the search spaces constitutes a *multispace*.

[1] See an example in [**GH94**].

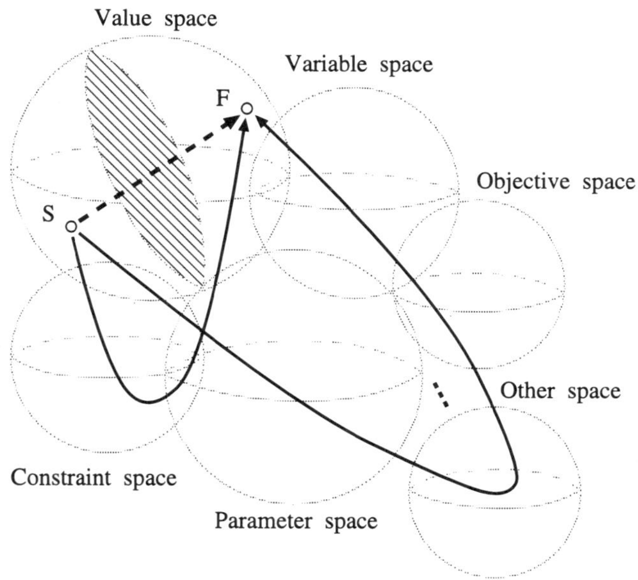

FIGURE 1. In the value space, a traditional search process (dashed line) cannot pass a "wall" of high cost search states (hatched region). It fails to reach the final solution state, F. A multispace search process (solid lines) scrambles across different search spaces. It could bypass this "wall" through the other search spaces.

The basic idea of multispace search is simple. Instead of being restricted in the value space, the multispace is taken as the search space. In the multispace, components other than value can be manipulated and optimized as well. During the search, a multispace search algorithm not only alters values in the value space; as shown in Figure 1, it also walks across the variable space and other active spaces, dynamically changes the problem structure in terms of variables, parameters, and other components, and systematically *constructs* a sequence of structured, intermediate problem instances. Each intermediate problem instance is solved by an optimization algorithm and the solution found is used as the initial solution to the next intermediate problem instance. By interplaying value optimization with structured operations, multispace search incrementally constructs the final solution to the search problem through a sequence of structured intermediate problem instances. Only at the *last* moment of the search, the reconstructed problem structure approaches the original problem structure, and thus the final value assignment represents the solution to the given search problem. Because structural operations (related to variable, constraint, objective function, and parameter changes) introduce new structural information, a multispace search method is more effective to handle difficult optimization problems.

Multispace search algorithm combines traditional optimization algorithms with structural multispace operations. A typical multispace search algorithm [**Gu94, Gu97, Gu98**] consists of the following three stages:

Initialization

At the beginning of search, for the given problem instance, an initial problem instance — which is usually different from the given problem instance — is designed. This initial problem instance will approach the given problem instance through a number of the altered, intermediate problem instances in a finite number of search phases.

An important initial state for multispace search is the *uniform* state where the initial structure of the search problem is completely symmetrical [**Gu94, Gu97, Gu98**]. In a *uniform* search state, there is no local minimum in the search space. Any solution point in the search space gives the same performance figure.

Search

This major search step involves two fundamental operations: a traditional value search and the structured reconfiguration of the intermediate problem instance during each individual search phase. A *scrambling schedule*, or *scrambling scheme*, specifies a sequence of events as *when* to make *what* structural changes to the intermediate problem instances in *which* active search space. During each search phase, the intermediate problem instance will be solved by a value search algorithm and the solution obtained will be passed to the next search phase as determined by the scrambling schedule. The last event in the schedule returns the intermediate problem instance to the original, given problem instance.

Final state

At the *last* moment of multispace search, a dynamically constructed, intermediate problem structure must be replaced by the original problem structure. This results in the *final* search state. At this state, any value assignment to the variables gives the *final solution* to the given search problem.

In the next section, we will introduce multispace search by reformulating the objective function for quorumcast routing.

4. Reformulating the Objective Function in Multispace

A network is modeled as a graph $G(V, E)$, where V represents the set of nodes (or vertices) and E the set of links (or edges). Links are assumed to be undirected. In the following discussion, let:

- n be the number of nodes in the network, $n = |V|$,
- m be the number of links, $m = |E|$,
- s be the source node,
- $i(a, b)$ be a link ($i \in E$) connecting a pair of nodes a and b,
- c_i be the cost of link i,
- Q be the quorum set ($|Q| < n$),
- R be a subset of Q,
- A and B be two subsets of the link set E, such that $A \cup B = E$ and $A \cap B = \phi$, where A represents the *quorum* link set and B the *candidate* link set,
- C be the objective function value, i.e., the total communication cost of the routing tree spanning the quorum set and the source node:

$$C = \sum_{j=1}^{|Q|} c_{l_j}, \text{ where } l_j \in A. \tag{4.1}$$

- $T(Q \cup s, A)$ be a minimum cost tree spanning the quorum set and the source node.

To search for an optimal solution efficiently, an optimization procedure must handle local minima effectively. Traditional search algorithms have difficulty in handling local minima because they optimize only in the value space. Multispace search not only alters values in the value space but also scrambles across the variable space and other active spaces. It has been proven to be efficient to overcome the pathological local minima in combinatorial optimization problems. In this work, we use a multispace search algorithm that performs structured operations in the objective function space.

For a given network, the costs of different links may be different, which makes the search space rugged. Using multispace search for quorumcast routing, we start with an uniform initial search state and proceed to reach a complicated, final state (with the original objective function) in a few, say k, iterations (Figure 2).

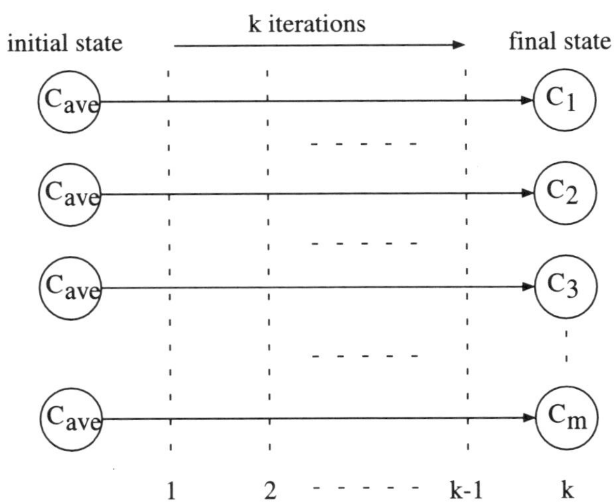

FIGURE 2. The scrambling process on the objective function space for multispace search.

For the quorumcast routing problem, one uniform initial state is the state that all the links have the same *average link cost*, that is:

$$c_{ave} = \frac{1}{m} \sum_{i=1}^{m} c_i. \tag{4.2}$$

As the costs of all links are the same, we can directly connect any $|Q|$ nodes with the source node forming a quorum. The objective value C (i.e., total cost of the routing tree) for any quorum is the same:

$$C = c_{ave} \times |Q|. \tag{4.3}$$

We use a *smoothing factor*, α, to characterize the degree of smoothing operation on the objective function and the smoothness of the resulting search space. The cost of the ith link can be determined by α as:

$$c_i(\alpha) = c_{ave} + (c_i - c_{ave}) \times \alpha, \tag{4.4}$$

where $0 \leq \alpha \leq 1$. When α increases gradually from 0 to 1, a series of gradually more complicated problem instances for quorumcast routing are generated from the simplest problem instance until it reaches the original problem instance. The series of problem instances generated as α increases corresponds to the sequence of search spaces 0 to k in Figure 3 (here we use a 1-dimensional search space as an example). A search space generated from a smaller α exhibits a smoother terrain surface, and

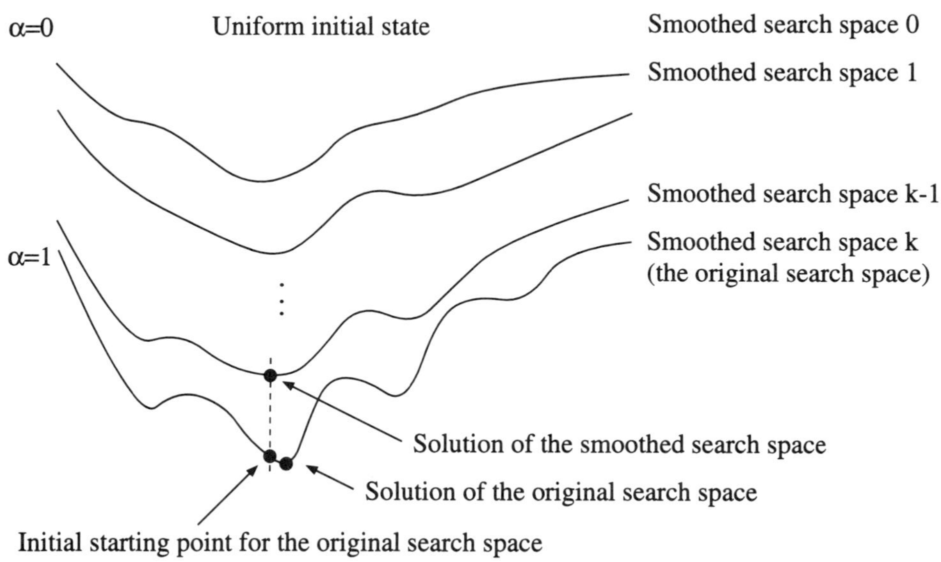

FIGURE 3. A series of smoothed search spaces of the scrambling process on the objective function space.

a search space generated from a larger α exhibits a more rugged terrain surface. Two extreme cases in the series of the problem instances are:

- If $\alpha = 0$, then $c_i(\alpha) \rightarrow c_{ave}$, which is the *simplest case*, corresponding to smoothed search space 0 in Figure 3;
- If $\alpha = 1$, then $c_i(\alpha) \rightarrow c_i$, which is the original problem instance, corresponding to search space k in Figure 3.

Once we have a series of smoothed problem instances, we can use any existing local search algorithm to solve them. We start from the simplest problem instance ($\alpha = 0$), i.e., the case with a *uniform* state. A solution to the simplest problem can easily be found by some local search algorithm. The solution of this problem instance is then taken as the initial solution to the next problem instance that has a slightly more complicated search space. The problem is again solved using the same local search algorithm. The above procedure is repeated until the final problem instance having the original search space ($\alpha = 1$) is solved. In the search process, the solutions of the more simplified problem instances are used to guide the search of the more complicated ones.

```
procedure MSAQ()
begin
   /* pruning search space */
   C_0 = heuristic_search(G);
   G' = remove_bad_nodes(G, C_0);
   R = choose_critical_links(G');
   if the optimal routing tree is found
      return the optimal routing tree;

   /* multispace search */
   A_opt = create_an_initial_route();
   α = 0;
   δ = set_threshold(G);
   while α < 1 do
   begin
      α = increment(α);
      c = set_link_cost(α);
      C_0 = Σ_{j=1}^{|Q|} c_{l_j}, where l_j ∈ A_opt;
      /* local search */
      A_opt = local_search(A_opt, C_0, δ);
   end;
   return the optimal routing tree A_opt;
end;
```

FIGURE 4. **MSAQ**: A multispace search algorithm for quorumcast routing.

In the next section, based on objective function reformulation, we will describe a multispace search algorithm for the quorumcast routing problem.

5. A Multispace Search Algorithm for Quorumcast Routing

The *M*ultispace *S*earch *A*lgorithm for *Q*uorumcast routing, *MSAQ*, is shown in Figure 4. It consists of two stages, a search space pruning stage and a multispace search stage.

5.1. Pruning Search Space. In the following, we give two procedures, *remove_bad_nodes* and *choose_critical_links*, to reduce the search space, i.e., the candidate set of nodes and links. The two procedures reduce the search space while maintaining the optimality. They quickly examine all the nodes in the network and remove nodes giving poorer solutions.

To start, procedure *heuristic_search*() uses some heuristic algorithm to obtain an approximate solution C_0. A simple heuristic algorithm, for example, can start with the source node. Each time, it selects the source node or a node nearest to the quorum set until the quorum set is found.

The first procedure to reduce the candidate set, *remove_bad_nodes*(), makes use of the obtained approximate solution C_0. We explain its operation through the following example.

Example. A simplified NSFNet T3 backbone [**JT94**] is shown in Figure 5. There are 12 nodes and 15 links in the network. Suppose the source node s is node *Argonne* and $|Q|$ is 5. We need to find a minimum cost tree spanning a set of 5 nodes and the source node *Argonne*.

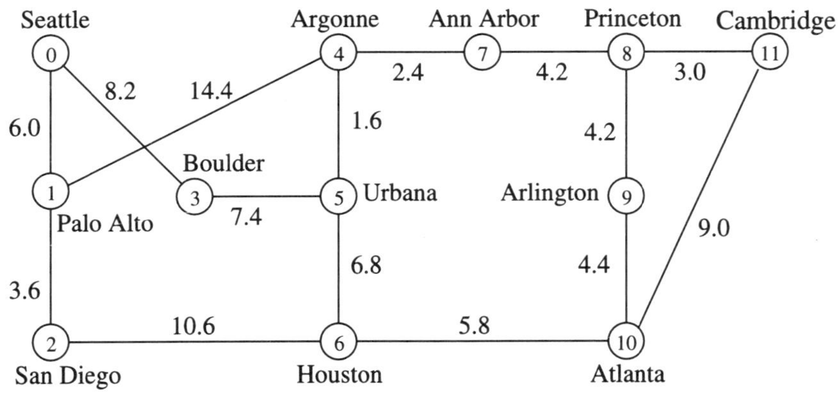

FIGURE 5. An example of the 12-node communication network NSFNet T3.

Initially, a simple heuristic algorithm is used to find an approximate solution. Each time, it selects the nearest node to the quorum set or the source node. In the example, the nearest node to node *Argonne* is *Urbana* and the cost is 1.6. So node *Urbana* is included in the quorum set. The process continues until we find 5 nodes: *Urbana, Ann Arbor, Princeton, Cambridge,* and *Arlington.* The approximate solution obtained by using this simple heuristic algorithm is $C_0 = 15.4$.

To reduce the candidate set, we first sort the link set in an ascending order of the link cost c_i. In this example, $c_1 = 1.6$ (*Argonne* to *Urbana*) is the minimum link cost and $c_{15} = 14.4$ (*Argonne* to *Palo Alto*) is the maximum link cost. Each time when we add a link to the routing tree, the link must connect a new node to the quorum set. For the first $(|Q| - 1)$ links with minimal costs, the total cost

$$sum = \sum_{j=1}^{4} c_j = 1.6 + 2.4 + 3.0 + 3.6 = 10.6$$

Since C_0 is 15.4, any link whose cost is larger than $C_0 - sum$ (i.e., 4.8) can not be chosen as a link in the candidate set. Because if any link whose cost is larger than 4.8 is selected in conjunction with the first four minimal links, their total costs will exceed C_0. The obtained cost tree will not be a minimum cost quorumcast tree. Thus, eight links whose costs are larger than 4.8 must be excluded. This produces a simplified network (Figure 6).

The simplified network may be a disconnected graph. Among all the disconnected trees, we only keep the one connecting with the source node for later search. Since nodes *Palo Alto* and *San Diego* have no connection to the tree containing the source node *Argonne*, the link between *Palo Alto* and *San Diego* can be removed. Originally, there were 15 links in the network. We had to search five links among the 15 links and there were 3,003 choices. Now the size of the candidate link set is reduced to six. We only need to select five links among the six links. This

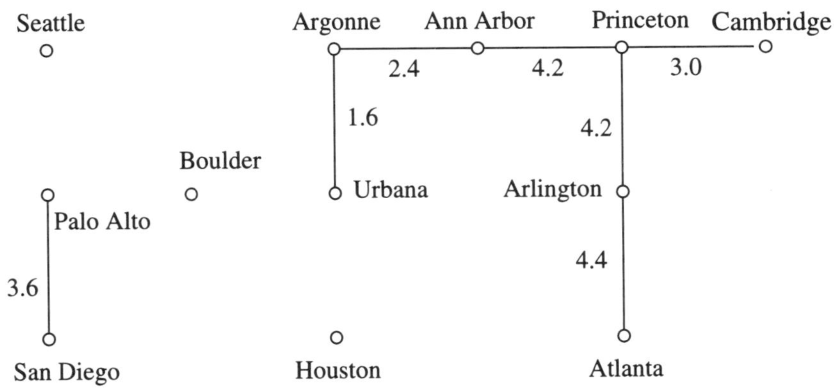

FIGURE 6. The simplified NSFNet after removing 8 links whose costs are larger than 4.8.

has significantly increased the computing efficiency of the later multispace search procedure.

The second procedure to reduce the candidate set, *choose_critical_links*(), chooses some nodes on a critical path and adds them to the quorum set, building a partial optimal tree. This path must be added to the quorum set in order to obtain a minimum cost tree. Two methods below are proposed for this procedure.

Method 1: Initially, only source node s exists. The current quorum set is empty. We examine nodes which are directly connected to s or to the current quorum set. Assume link $i(a, b)$, for which node a is in the candidate node set and node b is the source node or is in the current quorum set, has the minimum cost in the candidate link set. Then node a must be included in the optimal quorum set and link i is removed from the candidate link set B and added to the quorum link set A. We continue this process until either the final quorum set is built (in this case, an optimal solution is found) or none of the links in the candidate link set (which are directly connected to the source node or the current quorum set) have the minimum cost.

Proof of Optimality: We show, by disproof, that a routing tree without link i is not optimal (this proof can be extended to every links of the routing tree). Suppose that link i is not included in the quorum link set of the minimum cost tree spanning the source node s and the final quorum set. There are two cases, i.e., node a is in the final quorum set and node a is not in the final quorum set.

Case 1: When node a is in the final quorum set, we denote the remaining $|Q|-1$ nodes in the quorum set and the source node s as set Q', as shown in Figure 7. Since there is no cycle in the quorumcast routing tree, Q' does not contain any cycle. As a part of the minimum cost tree, link j connects node a and set Q'. Since link i can also connect node a to set Q' via node b, the connectivity of node a and set Q' remains if we replace link j with link i. As the cost of link i is smaller than that of link j, replacing link j with link i will reduce the cost of constructing the spanning tree. This is in contradiction with the assumption that the original spanning tree without link i is minimum.

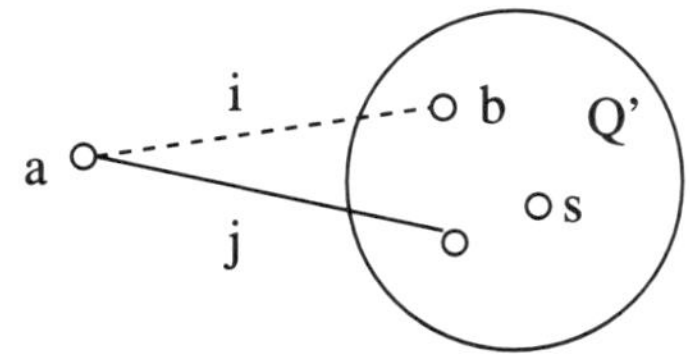

FIGURE 7. The first illustration for proving the optimality of method 1.

Case 2: When node a is not in the final quorum set (Figure 8), a similar proof follows. Assume that link i has the minimum cost in the candidate link set and

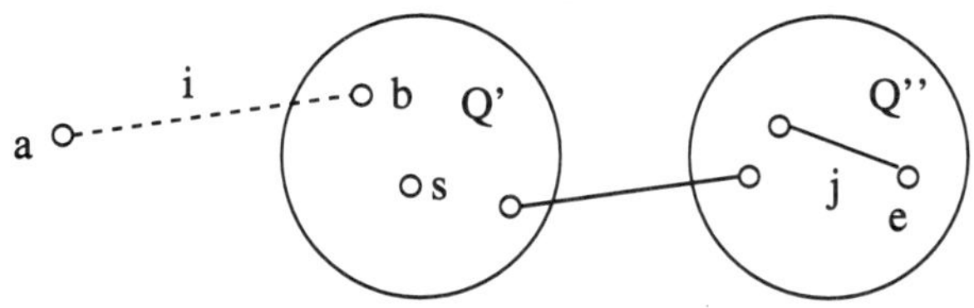

FIGURE 8. The second illustration for proving the optimality of method 1.

other links with smaller costs than link i, if any, have already been added to the quorum set. Denote the nodes to be added to Q' as set Q''. The costs for links in Q'' are larger than the cost of link i. Since set Q'' is a routing tree and does not contain any cycle, we can find a leaf node e connecting the routing tree via link j. If we replace link j with link i, i.e., remove node e from the quorum set and add node a to the quorum set, we reduce the cost of constructing the spanning tree. This is in contradiction with the assumption that the original spanning tree without link i is the minimum.

Method 2: If only one link (a, b) exists, where node a is in the candidate node set and node b is the source node or is in the quorum set, then node a must be included in the quorum set and the size of the quorum set is increased by 1. Accordingly, link (a, b) is removed from the candidate link set B and is added to the quorum link set A.

Proof of Optimality: Since source node s must be connected to the quorum set, the proof for Method 2 is straightforward.

The above two methods can be used to construct an initial, partial optimal quorumcast routing tree. If procedure *choose_critical_links*() produces the complete quorum set, an optimal quorum set with the minimum cost is found; otherwise, a multispace search algorithm will be used to search for the complete quorum set. We explain these two methods using the above example.

Example (continued). Initially, only the source node s is on the routing tree. The other nodes are in the candidate node set. The quorum set is empty. There are two links connected to the source node s (Figure 6), and their costs are 1.6 and

2.4, respectively. The minimum link cost in the network is 1.6. Using Method 1, node *Urbana* is added to the current quorum set. The link between *Argonne* and *Urbana* is removed from the candidate link set and is added to the quorum link set.

There is only one link, i.e., the link between *Argonne* and *Ann Arbor*, connecting the current quorum set and the source node to other candidate nodes. Using Method 2, node *Ann Arbor* should be added to the current quorum set. The link between *Argonne* and *Ann Arbor* is removed from the candidate link set. In the same way, nodes *Princeton, Cambridge*, and *Arlington* are added to the quorum set, giving the size of the quorum set $|Q| = 5$. Since the methods for reducing the candidate set do not compromise the optimality, the solution obtained is optimal.

In the above example we can see that, for some simple problems, the two-step algorithm for reducing candidate set is sufficiently efficient to find an optimal solutions. In this case, the algorithm would simply terminate and it will not use the succeeding multispace search. For more complicated problems, although the algorithm may not find an optimal solution, it can efficiently reduce the original search space. Accordingly, it can help the subsequent multispace search to find a good solution efficiently.

In the following, we describe the multispace search stage for the quorumcast routing problem.

5.2. Multispace Search. During search space pruning stage, nodes and links contributing to poor solutions are removed from the original graph $G(V, E)$. The remaining nodes and links compose a simplified graph $G'(V', E')$. Usually, V' and E' are smaller than the original sets V and E. In the multispace search stage, we apply the multispace search to the simplified graph, finding a good solution to the quorumcast routing problem. Nodes that have been included in the quorum set by procedure *choose_critical_links*() are put in set R. We only need to choose $|Q| - |R|$ nodes in the simplified graph $G'(V', E')$ to obtain a complete quorum set.

Procedure *create_an_initial_route*() in Figure 4 gives an initial quorumcast routing assignment. It chooses $|Q|$ nodes which are connected together with the source node s. Three heuristic methods can be used to produce the initial routing assignment. The first method, a random method, randomly chooses $|Q| - |R|$ nodes which are directly connected to the source node and the current quorum set R as an initial routing assignment. The second and third methods use the quorum set Q produced by the minimum cost path (MPH) heuristic and the improved minimum path (IMP) heuristic, respectively. If set Q does not contain the partial tree R, we add the partial tree R to the quorum set Q, and then remove redundant nodes from $Q \cup R$ making it a connected quorum set containing R. Procedure *set_threshold*() decides the threshold parameter δ to be used in the local search procedure (Figure 9).

The initial value of the smoothing factor α is set to 0. In each iteration, α is increased gradually from 0 to 1 in procedure *increment*(). There are many ways to increase α. In this work, α is increased by 0.2 in each iteration. Procedure *set_link_cost*() reevaluates the costs of all candidate links by Equation (4.4). Accordingly a quorumcast routing instance, which is more complicated than the instance in the last iteration, is generated.

Procedure *local_search*() (Figure 9) is used to find a routing tree having a locally minimum cost. The minimum cost tree spanning quorum set Q and source node s is calculated by procedure *build_a_routing_tree*() using a Kruskal's algorithm. The

```
procedure local_search(A_opt, C, δ)
begin
  C = C_0; A = A_opt;
  while there are improvements do
    for each node a in the candidate set
      for each node b in the quorum set
      begin
        C = build_a_routing_tree(A, a, b);
        if cost (C_0 + δ) does not increase after swapping nodes (a, b)
        begin
          A = swap_nodes(a, b);
          if C < C_0 then C_0 = C; A_opt = A;
        end;
      end
  return the optimal routing tree A_opt;
end;
```

FIGURE 9. A local search algorithm to find the quorum set.

quorumcast routing cost C is obtained by summing costs of all the links in A. For node a in the candidate set $(V' - Q - s)$ and node b in the quorum set $(Q - R)$, if the cost of the routing tree obtained by swapping node a and node b does not exceed the sum of C_0 (Eq. (4.1)) and δ, node a and node b are swapped. Thus, procedure *swap_nodes*(a, b) removes node b from the quorum set and adds it to the candidate set. At the same time, it removes node a from the candidate set and adds it to the quorum set. The local search procedure repeatedly swaps nodes in the two sets in a sequential order until no further improvement can be found, i.e., a routing tree with locally minimum cost is found.

In procedure *local_search*(), we use a simple and efficient local handler, called *climbing* heuristic, to avoid being stuck at local minima. The condition that "swapping two nodes (a, b) does not increase the current routing cost by δ" includes two cases: the objective function *reduces* its value or the objective function *increases* its value (i.e., *climbing*) but does not exceed its best value plus δ. In the climbing heuristic, two nodes are swapped even if the objective function increases but does not exceed $(C_0+\delta)$. The climbing heuristic have effectively improved the convergent performance of the *MSAQ* algorithm.

Threshold δ is a parameter controlling the height of local climbing. It is related to the link costs (i.e., c_i for $i = 1, 2, ...m$) of the network. In this research, we empirically selected the following formula for δ:

$$\delta = \frac{1}{3} \times \sqrt{\frac{\sum_{i=1}^{m} c_i^2}{m}}.$$

Instead of making random moves in the vertical direction of the search space as in simulated annealing, whenever a local minimum is encountered, we climb through the rugged terrain structure of the search space, moving from one local minimum to another in an attempt to locate a better locally optimal solution. As illustrated in Figure 10, a climbing can be thought of as a short-cut passing through

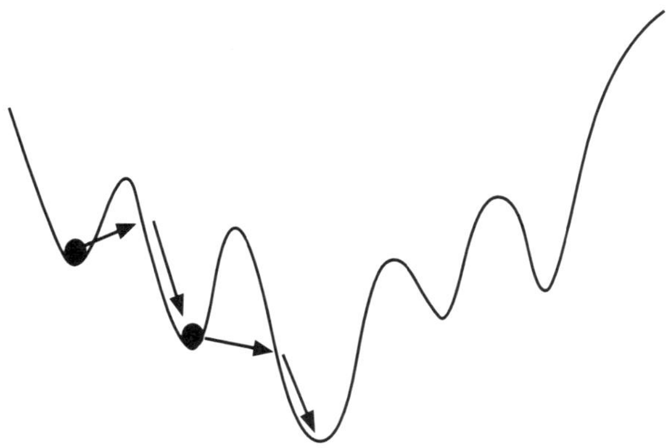

FIGURE 10. Climbing heuristic.

TABLE 1. Quorumcast routing cost by *MSAQ* algorithm for a 12-node NSFNet network.

		MSAQ algorithm										*Cheung et al.* [**CK94b**]	
		Random initialization								Initialized by			
s	$\vert Q\vert$	100 Trials		Execution Time (s)			Iterations			MPH	IMP	MPH	IMP
		Best	Ave	Min	Mean	Max	Min	Mean	Max				
8	9	39.8	39.80	<0.001	0.010	0.020	18	19.29	21	39.8	39.8	41.2	41.2
0	8	35.4	35.40	0.010	0.013	0.020	19	19.30	22	35.4	35.4	36.4	36.4
2	5	25.0	25.02	0.010	0.011	0.020	18	19.24	21	25.0	25.0	26.8	26.2
2	7	32.2	32.20	0.010	0.014	0.030	19	19.72	22	32.2	32.2	33.4	33.4

a mountain separating points whose elevations do not exceed the threshold value δ. Thus, a climbing operation can be effectively used to search a region with many local minima. Whenever a local minimum is encountered, a climbing is made through a mountain to a neighboring valley as long as this does not exceed the current value of the objective function by δ. Note that the best solution found is always kept.

6. Numerical Results

The multispace search algorithm for the quorumcast routing problem was written in C language. We tested our algorithm with various network configurations, different source nodes, and different quorum sizes. Our experiments were performed on a SUN SPARC 20 workstation. All CPU times recorded are in seconds.

The first network we tested is a 12-node NSFNet network (Figure 5) [**JT94**]. Table 1 gives performance comparison between the multispace search algorithm (*MSAQ*) and the MPH and IMP heuristics [**CK94b**] for the NSFNet network. The first two columns give the source node s and the size of the quorum set $|Q|$. From the third to the tenth columns, we give the experimental results of multispace search obtained by the random initial assignment. To reduce the effect of random initial assignments, we ran the program 100 times and obtained the probabilistic behavior of the algorithm. For each routing case, the minimum cost (*Best*) and the average cost (*Ave*) from 100 runs are given in Columns 3 and 4. Columns 5, 6, and 7 give the minimum, mean, and maximum CPU time from 100 runs (*Min*, *Mean*, and *Max*), respectively. In Columns 8, 9, and 10, the minimum, mean, and maximum iterations from 100 runs (*Min*, *Mean*, and *Max*) are given respectively. From the 100 trials, the *MSAQ* algorithm with random initial assignment was able to find the optimal solutions for almost 100 cases (note that each optimal solution in the table was obtained and verified by an exhaustive search algorithm).

In our algorithm, we also used the solutions obtained by the MPH and IMP heuristics as the initial assignments to test the *MSAQ* algorithm. The results are shown in Columns 11 and 12, respectively. The *MSAQ* algorithm performed equally well with these heuristic initial assignments. They were able to locate solutions with the minimum cost. *MSAQ* has improved the solutions produced by the MPH and IMP heuristics. It is not sensitive to a special type of initial assignments.

The last two columns in Table 1 show the experimental results, in terms of the quorumcast routing cost, for the MPH and IMP heuristics by Cheung et al. [**CK94b**]. Apparently, it was difficult for algorithms based on MPH and IMP heuristics to find an optimal solution.

The second network as shown in Figure 11 is a 28-node network [**WW95**]. The

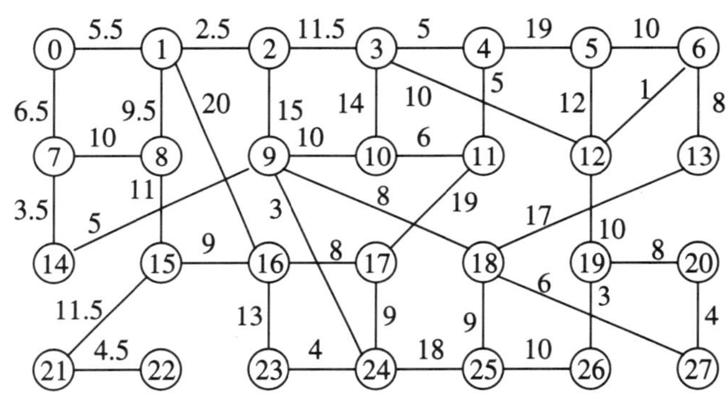

FIGURE 11. A 28-node communication network.

original network is represented as a directed graph. We modify it to an undirected graph in which the cost of a link is the average of the two corresponding, directed links. The performance comparison between the *MSAQ* and the MPH and IMP heuristics for this network is shown in Table 2. We simulated the network with different source nodes and different quorum sizes. Again, for each case, we ran the algorithm using three different initial assignments: random ones, the solutions given by the MPH heuristic, and the solutions given by the IMP heuristic. Similar performance as in the 12-node NSFNet for the *MSAQ* algorithm was observed.

The third network is an 100-node network [**WBE94**] (Figure 12). Table 3 shows the performance comparison of the multispace search algorithm (*MSAQ*) and the MPH and IMP heuristics for the 100-node network. For each case with 100 executions, the average cost obtained by the multispace search algorithm is very close to the minimum cost obtained from 100 runs.

For the first case with source node $s = 3$ and quorum size $|Q| = 12$ (Table 3), for example, *MSAQ* was able to find cost 133.2 in most cases. For the 100-node network, we are unable to find out the optimal solution of the network by

TABLE 2. Quorumcast routing cost by *MSAQ* algorithm for a 28-node network.

		MSAQ algorithm										*Cheung et al.*	
		Random initialization								Initialized by		**[CK94b]**	
s	$\|Q\|$	100 Trials		Execution Time (s)			Iterations			MPH	IMP	MPH	IMP
		Best	Ave	Min	Mean	Max	Min	Mean	Max				
3	12	62.5	62.52	0.140	0.186	0.250	19	20.56	24	62.5	62.5	64.0	64.0
10	10	51.0	51.15	0.110	0.145	0.200	19	20.48	23	51.0	51.0	53.5	53.5
19	13	66.0	66.08	0.160	0.209	0.300	19	20.93	23	66.0	66.0	68.0	67.5
25	14	75.0	75.26	0.170	0.234	0.330	19	20.63	25	75.0	75.0	77.0	77.0

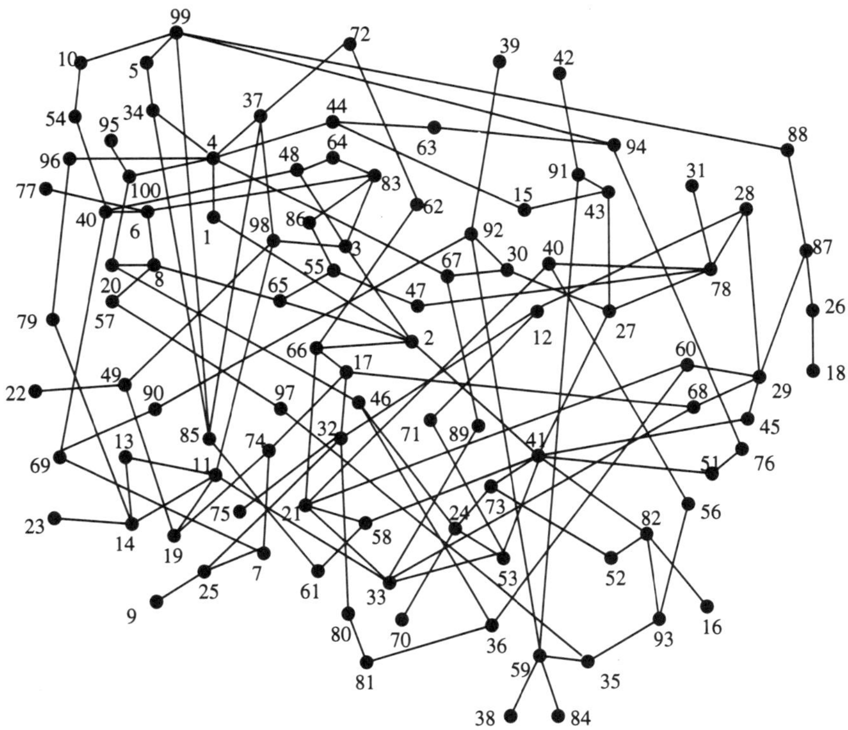

FIGURE 12. A 100-node communication network.

exhaustive search. We would use the minimum cost produced from 100 runs (with random initial assignments) as a reference point. We found that the MPH and IMP algorithms located, at best, 142.0 and 137.6, compared to the minimum cost 133.2 obtained by the *MSAQ* algorithm.

7. Conclusion

The quorumcast routing problem is a generalization of multicast routing. It has many practical application in networking communications. There are many possible routing trees which span the source node and a quorum set. The optimal routing

TABLE 3. Quorumcast routing cost by *MSAQ* algorithm for an 100-node network.

s	$\|Q\|$	*MSAQ* algorithm										*Cheung et al.* [CK94b]	
		Random initialization								Initialized by			
		100 Trials		Execution Time (s)			Iterations			MPH	IMP	MPH	IMP
		Best	Ave	Min	Mean	Max	Min	Mean	Max				
6	10	133.2	133.64	1.150	1.335	1.670	19	20.91	23	133.2	133.2	142.0	137.6
9	12	194.2	195.30	1.570	1.938	2.420	19	21.45	26	194.2	194.2	213.1	206.7
25	10	162.0	164.59	1.170	1.428	1.820	19	21.11	25	162.0	162.0	185.8	171.4
59	16	227.0	229.21	2.340	3.151	4.300	19	22.39	29	227.0	227.0	232.1	232.1

tree is the one which has the minimum quorumcast routing cost, determined as the total cost of the links in the routing tree. In this paper, we give a multispace search algorithm to minimize the quorumcast routing cost of the network. Experimental results indicate that when compared to existing heuristic approaches, the multispace search algorithm gives significant performance improvements by achieving a lower routing cost. In addition, the algorithm is insensitive to the initial solution point and is computationally efficient. This performance is achieved by an effective search space pruning method, an efficient multispace search algorithm, and a novel climbing heuristic able to handle local minima effectively. The algorithm is scalable and can be used for multicast and quorumcast routing for large size networks.

References

[AAC91] M. Ahamad, M.H. Ammar, and S.Y. Cheung, *Multi-dimentional voting*, ACM Transactions on Computer Systems **9** (1991), no. 4, 399–431.

[ACS93] M.H. Ammar, S.Y. Cheung, and C.M. Scoglio, *Routing multipoint connections using virtual paths in an ATM network*, Proc. INFOCOM (San Francisco, CA), Mar. 30-Apr. 1 1993, pp. 98–105.

[BGM86] D. Barbara and H. Garcia-Molina, *Mutual exclusion in partitioned distributed systems*, Distrib. Comput. **1** (1986), 119–132.

[BV95] F. Bauer and A. Varma, *Degree-constrained multicasting in point-to-point networks*, Proc. IEEE INFOCOM (Boston, Massachusetts), Apr. 1995, pp. 369–376.

[Cho91] C.H. Chow, *On multicast path finding algorithms*, Proc. INFOCOM (Bal Harbour, FL), Apr. 1991, pp. 1274–1283.

[CK94a] X. Chen and V. Kumar, *Multicast routing in self-routing multistage networks*, Proc. INFOCOM (Toronto, Ontario, Canada), Jun. 1994, pp. 306–314.

[CK94b] S.Y. Cheung and A. Kumar, *Efficient quorumcast routing algorithms*, Proc. INFOCOM (Toronto, Ontario, Canada), Jun. 1994, pp. 840–847.

[DL93] M. Doar and I. Leslie, *How bad is naïve multicast routing*, Proc. INFOCOM (San Francisco, CA), Mar. 30-Apr. 1 1993, pp. 82–88.

[GMB85] H. Garcia-Molina and D. Barbara, *How to assign votes in a distributed system*, Journal of ACM **32** (1985), no. 4, 841–860.

[GP68] E.N. Gilbert and H.O. Pollak, *Steiner minimal trees*, SIAM Journal of Applied Mathematics **16** (1968), no. 1, 1–29.

[Gu94] J. Gu, *Multispace search: A new optimization approach (Summary). in lecture notes in computer science, vol. 834*, pp. 252–260, 1994.

[Gu97] J. Gu, *Multispace search for satisfiability and NP-Hard problems*, DIMACS Volume Series on Discrete Mathematics and Theoretical Computer Science, Vol. 35: The Satisfiability (SAT) Problem (American Mathematical Society, 1997), 407–517.

[Gu98] J. Gu, *Optimization by multispace search*, Kluwer Academic Publishers, 1998.

[GH94] J. Gu and X. Huang, *Efficient local search with search space smoothing*, IEEE Trans. on Systems, Man, and Cybernetics **24** (1994), no. 5, 728–735.

[JM90] S. Jajodia and D. Mutchler, *Dynamic voting algorithms for maintaining the consistency of a replicated database*, ACM Transactions on Database Systems **15** (1990), no. 2, 230–280.

[JT94] C.A. Noronha Jr. and F.A. Tobagi, *Optimum routing of multicast streams*, Proc. INFOCOM (Toronto, Ontario, Canada), Jun. 1994, pp. 865–873.

[Kar72] R.M. Karp, *Reducibility among combinatorial problems. complexity of computer computations*, pp. 85–103, Plenum Press, New York, 1972.

[KPP92] V.P. Kompella, J.C. Pasquale, and G.C. Polyzos, *Multicasting for multimedia applications*, Proc. INFOCOM (Florence, Italy), May 1992, pp. 2078–2085.

[RCK95] C.S. Raghavendra, X. Chen, and V.P. Kumar, *A two phase multicast routing algorithm in self-routing multistage networks*, Proc. IEEE ICC (Seattle), Jun. 1995, pp. 1612–1618.

[TSY+92] H. Tode, Y. Sakai, M. Yamamoto, H. Okada, and Y. Tezuka, *Multicast routing algorithm for nodal load balancing*, Proc. INFOCOM (Florence, Italy), May 1992, pp. 2086–2095.

[Wax93] B.M. Waxman, *Performance evaluation of multipoint routing algorithms*, Proc. INFOCOM (San Francisco, CA), Mar. 30-Apr. 1 1993, pp. 980–986.

[WBE94] J.E. Wieselthier, C.M. Barnhart, and A. Ephremides, *A neural network approach to routing without interference in multihop radio networks*, IEEE Trans. on Communications **42** (1994), no. 1, 166–176.

[Win87] P. Winter, *Steiner problem in networks: A survey*, Networks **17** (1987), 129–167.

[WW95] C. Wang and P.N. Weissler, *The use of artificial neural networks for optimal message routing*, IEEE Network (1995), 16–24.

[ZPGLA95] Q. Zhu, M. Parsa, and J.J. Garcia-Luna-Aceves, *A source-based algorithm for delay-constrained minimum-cost multicasting*, Proc. IEEE INFOCOM (Boston, Massachusetts), Apr. 1995, pp. 377–385.

Dept. of Electrical and Computer Engineering, Univ. of Calgary, Calgary, Alberta T2N 1N4, Canada
E-mail address: j.gu@computer.org, jungu@dimacs.rutgers.edu

Dept. of Electrical and Computer Engineering, Univ. of Calgary, Calgary, Alberta T2N 1N4, Canada
E-mail address: du@enel.ucalgary.edu

Dept. of Electrical and Electronic Engineering, Hong Kong Univ. of Science and Technology, Clear Water Bay, Kowloon, Hong Kong
E-mail address: eetsang@ee.ust.hk

Dept. of Electrical Engineering, University of Utah, Salt Lake City, Utah 84112, U.S.A.
E-mail address: wang@comgroup.elen.utah.edu

DIMACS Series in Discrete Mathemati
and Theoretical Computer Science
Volume **42**, 1998

On 3-rate Rearrangeability of Clos Networks

Guo-Hui Lin, Ding-Zhu Du, Weili Wu, and Kyeongah Yoo

ABSTRACT. In this paper, we study multirate symmetric three-stage Clos network $C(n,m,r)$. We will show that $C(n, \lceil 9n/4 \rceil, r)$ is rearrangeable when every connection has a weight chosen from $\{p_1, p_2, p_3\}$ where $1 \geq p_1 > p_2 > p_3 > 0$, and that $C(n, 2n, r)$ is rearrangeable when every connection has a weight chosen from $\{p_1, p_2, p_3\}$ where $1 \geq p_1 > p_2 > p_3 > 1/5$.

1. Introduction

The multirate interconnection network is one of research topics in ATM networks with strong applications in computer networks, telecommunications, and Internet. The symmetric three-stage Clos network $C(n,m,r)$ has been widely used in the design of telecommunication networks [**1**]. $C(n,m,r)$ consists of r $n \times m$ crossbars (switches) in the first (or input) stage, m $r \times r$ crossbars in the second (or central) stage, and r $m \times n$ crossbars in the third (or output) stage. Every crossbar in the first stage has an outlet connected to an inlet of every crossbar in the second stage and every crossbar in the second stage has an outlet connected to an inlet of every crossbar in the third stage. There are totally rn inlets in the first stage, called *inputs*, and totally rn outlets in the third stage, called *outputs*. A $C(2,3,4)$ is illustrated in Figure 1.

1991 *Mathematics Subject Classification.* Primary 94A05; Secondary 05C40.
Supported in part by the National Natural Science Foundation of China.
Supported in part by the National Science Foundation under grant CCR-9530306 and Grant-in-Aid of Research in University of Minnesota.

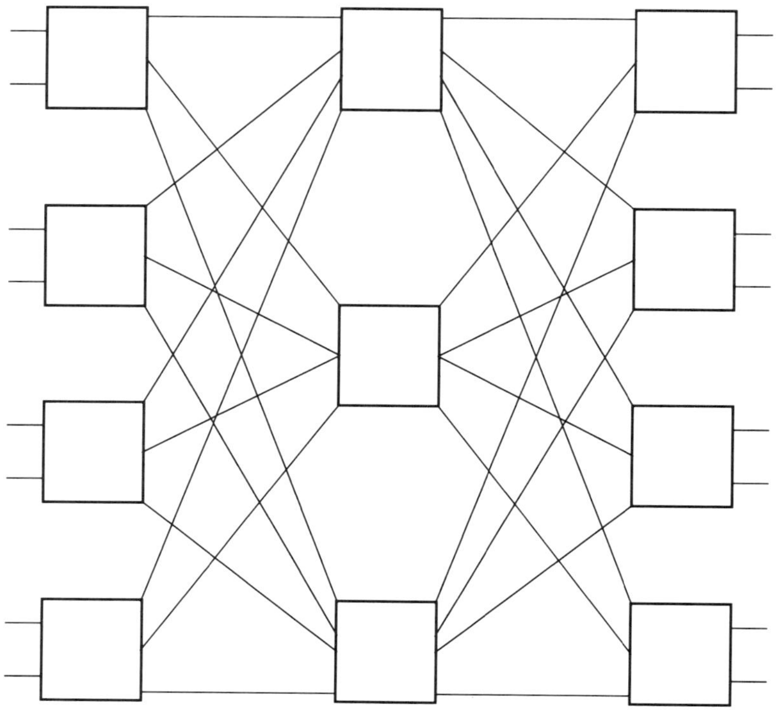

Figure 1: $C(2,3,4)$

A *connection* (*request*, or *call*) in the network is a triple (i, j, w) where i is an input, j is an output, while w is the *weight* of this connection and it represents the bandwidth required by the connection. A *route* is a path in the network joining an input crossbar (i.e., a crossbar in the first stage) to an output crossbar (i.e., a crossbar in the third stage) and a route r *realizes* a connection (i, j, w) if i and j belong to the input crossbar and the output crossbar joined by r, respectively, and the weight of r equals w.

Usually, it is assumed that links have uniform capacity which is normalized to be unity. Therefore, the range of the weights is the closed interval $[0, 1]$. A set of connections is *compatible* if for every input or output, the sum of the weights of all connections involving it is at most one. A *request frame* is a compatible set of connections. A *configuration* is a set of routes and it is *compatible* if for every link in the network the total weight of routes passing through this link is at most one. A request frame is said to be *realizable* if there exists a compatible configuration which contains routes realizing all connections in the request frame. A multirate network is said (*multirate*) *rearrangeable* if every request frame is realizable.

A connection c is said to be *compatible* with a request frame $\mathcal{F}$ if $\mathcal{F} \cup \{c\}$ is still compatible. A route r is said to be *compatible* with a compatible configuration $\mathcal{C}$ if $\mathcal{C} \cup \{r\}$ is still compatible. A network is said to be *strictly nonblocking* if for every compatible configuration $\mathcal{C}$ realizing a request frame $\mathcal{F}$ and every connection c compatible with $\mathcal{F}$, there exists a route r such that r realizes c and r is compatible with $\mathcal{C}$.

In classical circuit switching, all connections are assumed to have the same rate one. Namely, a network is said to be rearrangeable if every compatible set of connections with weight one is realizable and, it is well known that the symmetric three-stage Clos network $C(n, m, r)$ is rearrangeable in classical circuit switching if and only if $m \geq n$ [**1**]. Now, since multirate is involved, we need more crossbars in the central stage to reach the rearrangeability. Chung and Ross[**2**] conjectured that if a symmetric three-stage Clos network $C(n, m, r)$ is strictly nonblocking in classical circuit switching, then it is multirate rearrangeable in the discrete bandwidth case. It is also well known that Clos network $C(n, m, r)$ is strictly nonblocking in classical circuit switching if and only if $m \geq 2n - 1$ [**1**]. Then this conjecture states that $C(n, 2n-1, r)$ is multirate rearrangeable if every connection has a weight belong to a given finite set $\{p_1, p_2, \cdots, p_k\}$ where $1 \geq p_1 > p_2 > \cdots > p_k > 0$ and $p_1, p_2, \cdots, p_{k-1}$ are integral multiples of p_k, denoted by $p_k \mid p_i$ for $1 \leq i \leq k-1$. They verified their conjecture when $k = 2$ and $\{p_1, p_2\} = \{1, p\}$.

Specifically, if each connection is restricted to have a weight chosen from a given finite set $\{p_1, p_2, \cdots, p_k\}$ where $1 \geq p_1 > p_2 > \cdots > p_k > 0$, then it is denoted as *k-rate environment*. A network $C(n, m, r)$ which is multirate rearrangeable in k-rate environment is said to be *k-rate rearrangeable*. As it is obvious that the divisibility in the conjecture contributes nothing, the conjecture is equivalent to the proposition that $C(n, 2n-1, r)$ is multirate in general case.

One way to prove the conjecture is to show that $C(n, 2n-1, r)$ is k-rate rearrangeable for any k. Along this approach, it has been shown previously that $C(n, n, r)$ is 1-rate rearrangeable ([**2**]) and that $C(n, 2n-1, r)$ is multirate rearrangeable if every connection has a weight chosen from a given finite set $\{p_1, p_2, \cdots, p_k\}$ where $1 \geq p_1 > p_2 > \cdots > p_k > 0$ and $p_3 \mid p_2, p_4 \mid p_3, \cdots, p_k \mid p_{k-1}$ ([**4**]). A corollary of the second result is that $C(n, 2n-1, r)$ is 2-rate rearrangeable.

In this paper, we study 3-rate rearrangeability. We show that $C(n, \lceil 9n/4 \rceil, r)$ is 3-rate rearrangeable and that $C(n, 2n, r)$ is 3-rate rearrangeable if those three rates satisfy $1 \geq p_1 > p_2 > p_3 > 1/5$.

2. Preliminaries

It has been mentioned in Section 1 that $C(n, n, r)$ is 1-rate rearrangeable. Du *et al.* [**3**] generalize this result under a more flexible weight condition, that is, a network which is rearrangeable for the classical circuit switching is multirate rearrangeable if all weights are in the interval $[p, 1/\lfloor 1/p \rfloor]$ for some p, $1 \geq p > 0$. This implies the following.

LEMMA 2.1. *$C(n, n, r)$ is multirate rearrangeable when every connection has weight in $(\frac{1}{k+1}, \frac{1}{k}]$ for a fixed positive integer k.*

For any compatible request frame $\mathcal{F}$ of $C(n, m, r)$, construct a weighted connection bipartite graph G with the input switches as one part, the output switches as the other part, and an edge of weight w between vertices I and J for each request (i, j, w) in $\mathcal{F}$, where i (j) is an input (output) on input switch I (output switch J) and w stands the weight of this request. We call that this G is *induced* by $\mathcal{F}$. G can be *m-colored* if there exists an edge-coloring of G using m distinct colors such that the weights of like-colored edges incident to each vertex sum up to a value of at most 1. Given such an edge-coloring of G, let all connections corresponding like-colored edges pass through one central switch. Then we get a compatible configuration which realizes $\mathcal{F}$.

We can also construct another weighted connection bipartite graph, denoted by H, for $\mathcal{F}$ as follows: H takes the rn inputs as one part, the rn outputs as the other part, and an edge of weight w between vertices i and j for each request (i, j, w) in

$\mathcal{F}$, where i (j) is an input (output) and w stands the weight of this request. For a spanning subgraph T of H, contracting the n inputs/outputs at each input/output switch into one vertex, we obtain a spanning subgraph S of G. We call that S is the *translation* of T.

Employing edge-coloring, Du *et al.* [**3**] proved the following two lemmas.

LEMMA 2.2. *Given a connection graph G (induced by some $\mathcal{F}$), suppose that all edges with a weight $> 1/f$, f an integer, are colored by a set C of c colors, $c \geq 2n$. Then at most $\lceil (c-2)/f - c + 2n \rceil$ additional colors are needed which, together with C, color all edges with a weight $\leq 1/f$.*

Let $M_{(b,B)}(n,r)$ denote the minimum value of m such that $C(n,m,r)$ is rearrangeable when b and B are given. Then,

LEMMA 2.3. *If $b > \frac{1}{4}$, then $M_{(b,1)}(n,r) \leq \lceil \frac{9}{4} n \rceil$.*

The proof of lemma 2.3 involves a useful lemma on partition of bipartite graphs as follows.

LEMMA 2.4 (de Werra [**6**], 1971). *Let G be an arbitrary bipartite graph. Then for any $k \geq 1$, G can be partitioned into k edge-disjoint subgraphs $G_1, G_2, \cdots, G_k$ such that for every vertex x in G,*

$$\lfloor \frac{d_G(x)}{k} \rfloor \leq d_{G_i}(x) \leq \lceil \frac{d_G(x)}{k} \rceil$$

where $d_G(x)$ denote the degree of the vertex x in G.

Melen and Turner [**5**] gave an elegant routing algorithm CAP which is able to do the following: Divide all requested connections at each input/output switch into groups of size at most m. Then algorithm CAP can obtain a configuration with m central switches such that each central switch takes care of at most one connection in each group of input/output switch. When we organize the groups consecutively distributed in the linear ordering of the weights of connections at each input/output switch, this implies that the load of an internal link of $C(n,m,r)$ never exceeds

$$B + \frac{n-B}{m},$$

where b and B denote the lower and upper bound of the weight of a connection. Setting $B + \frac{n-B}{m} \leq 1$, we have

$$m \geq \frac{n-B}{1-B}.$$

LEMMA 2.5. $M_{(0,B)}(n,r) \leq \frac{n-B}{1-B}$.

Let $\{1 \geq p_1 > p_2 > \cdots > p_k > 0\}$ and $\{1 \leq q_1 > q_2 > \cdots > q_k > 0\}$ be two sets of rates. They are said to be *isomorphic* if for every set of k natural numbers $n_1, n_2, \cdots, n_k$, $\sum_{i=1}^k n_i p_i \leq 1$ if and only if $\sum_{i=1}^k n_i q_i \leq 1$. Clearly, if two sets of rates are isomorphic, then $C(n,m,r)$ is multirate rearrangeable with one set if and only if it is rearrangeable with another set. The concept of isomorphism would make more applications of the following results [**4**].

LEMMA 2.6. *Symmetric three stage Clos network $C(n, 2n-1, r)$ is multirate rearrangeable when each connection has a weight chosen from a given finite set $\{p_1, p_2, \cdots, p_k\}$ where $p_3 \mid p_2$, $p_4 \mid p_3$, $\cdots$, $p_k \mid p_{k-1}$, and $1 \geq p_1 > p_2 > \cdots > p_k > 0$.*

For example, consider three rates $1 \geq p_1 > \frac{1}{2} \geq p_2 > \frac{1}{3}, \frac{1}{4} p_3 > \frac{1}{5}$. If $p_1 + p_3 > 1$ and $2p_2 + p_3 > 1$, then $\{p_1, p_2, p_3\}$ is isomorphic to $\{1, 0.42, 0.21\}$. Thus, $C(n, 2n - 1, r)$ is rearrangeable with such three rates.

3. 3-rate Rearrangeability

In this section, we presents two results on 3-rate rearrangebility. For simplicity, an edge is called a p-edge if it corresponds to a connection with weight p.

THEOREM 3.1. *Symmetric three stage Clos network $C(n, 2n, r)$ is multirate rearrangeable when every connection has a weight chosen from $\{p_1, p_2, p_3\}$ where $1 \geq p_1 > p_2 > p_3 > 1/5$.*

PROOF. Notice that $C(n, 2n, r)$ is three-rate rearrangeable if one of the following two conditions is satisfied:

(a): $\frac{1}{2} \geq p_1$;
(b): At least two of the three rates lie in the same interval $(\frac{1}{k+1}, \frac{1}{k}]$, for some positive integer k.

Therefore, we need only to consider the following three cases.

Case 1. $1 \geq p_1 > \frac{1}{2} \geq p_2 > \frac{1}{3} \geq p_3 > \frac{1}{4}$. Given any compatible request frame $\mathcal{F}$, construct the induced graph G. Notice that an input/output switch can generate at most $2n$ calls associated with a weight larger than $\frac{1}{3}$ and at most $3n$ calls associated with a weight larger than $\frac{1}{4}$. Applying CAP algorithm, when using $2n$ colors to color the edges in G, at each vertex, p_1-edge and p_2-edge can never have a same color. Thus, if $p_1 + p_3 \leq 1$, then the sum of the weights of like-color edges at any vertex never exceeds unity. That means G can be $2n$-colored. Hence $2n$ central switches suffice.

If $p_1 + p_3 > 1$, then

$$m_I \leq 3n - 3l_I.$$

where l_I denotes the number of p_1-edges incident to vertex I and m_I denotes the number of p_2- or p_3-edges incident to vertex I. By Lemma 2.4, we can partition the subgraph with only p_2- or p_3-edges into three subgraphs such that in each of which the degree of I is at most $n - l_I$. Now, one of three such subgraph together with all p_1-edges can be n-colored. The remaining two together can be n-colored. Thus, $2n$ central switches suffice.

Case 2. $1 \geq p_1 > \frac{1}{2} \geq p_2 > \frac{1}{3}, \frac{1}{4} \geq p_3 > \frac{1}{5}$. Similarly, given any compatible request frame $\mathcal{F}$, construct the induced graph G. Note that at each vertex, there are at most $2n$ edges with a weight of p_1 or p_2 and totally at most $4n$ edges. If $p_1 + p_3 \leq 1$, by the same argument as in Case 1, $2n$ central switches suffice to route $\mathcal{F}$.

If $p_1 + p_3 > 1$, let L denote the subgraph of G consisting of all p_1-edges, M consisting of all p_2- and p_3-edges. Then, the degree of vertex I in L is $l_I \leq n$, and the degree of I in M is $m_I \leq 4n - 4l_I$. By Lemma 2.4, partition M into four subgraphs such that in each of which the degree of I is at most $n - l_I$. It is clear that the union of one such subgraph of M and L can be n-colored. Since an input/output generates at most two p_2-calls, the union of the other three subgraph can be n-colored if $2p_2 + p_3 \leq 1$.

If furthermore $2p_2 + p_3 > 1$, then $\{p_1, p_2, p_3\}$ is isomorphic to $\{1, 0.42, 0.21\}$. Therefore, $2n - 1$ central switches suffice to route $\mathcal{F}$ by 2.6.

Case 3. $1 \geq p_1 > \frac{1}{2}, \frac{1}{3} \geq p_2 > \frac{1}{4} \geq p_3 > \frac{1}{5}$. If $p_1 + p_2 > 1$, then we have $p_1 + 2p_3 > 1$. For any request frame $\mathcal{F}$, construct the induced graph G and let l_I denote the number of p_1-edges incident to vertex I, and m_I the number of p_2- and p_3-edges incident to vertex I. If $p_1 + p_3 > 1$, we have

$$m_I \leq 4n - 4l_I.$$

Let L denote the subgraph of G consisting of all p_1-edges and M the subgraph of G consisting of all p_2- and p_3-edges By Lemma werra, M can be partitioned into four edge-disjoint spanning subgraphs, such that the degree of vertex I in each subgraph is less than or equal to $n - l_I$. Therefore, the union of one such subgraph and L can be n-colored and the union of the other three subgraphs can be n-colored too. Hence, G can be $2n$-colored. If $p_1 + p_3 \leq 1$, we construct the weighted connection bipartite graph H and let l_i denote the number of p_1-edges incident to vertex i, and m_i the number of p_2- and p_3-edges incident to vertex i. It follows that

$$l_i \leq 1, \ m_i \leq 4; \text{ and if } l_i = 1, \text{ then } m_i \leq 1.$$

Let L' denote the subgraph of H consisting of edges with a weight of p_1 and M' the subgraph of H consisting of edges with a weight of p_2 or p_3. By Lemma 2.4, M' can be partitioned into four matchings. Let us first consider the union of one such matching and L'. The translation of this union is denoted by L^*. Clearly, in L^*, there are l_I p_1-edges, at most $n - l_I$ p_2-edges and at most n p_3-edges incident to vertex I. Thus, this translation can be n-colored by CAP algorithm, and it is obvious that M^*, the complement of L^* in G, can also be n-colored since $p_2 \leq 1/3$. Hence, the induced graph G can be $2n$-colored.

Finally, we may assume $p_1 + p_2 \leq 1$. Since at each vertex I of G there are at most n edges with a weight greater than $\frac{1}{2}$ and totally at most $4n$ edges (with a weight greater than $\frac{1}{5}$), $2n$ colors suffice to color the edges such that the sum of the weights of like-colored edges incident to every vertex I never exceeds unity. Therefore, $C(n, 2n, r)$ is showed, in any case, to be rearrangeable. □

THEOREM 3.2. *Symmetric three stage Clos network $C(n, \lceil\frac{9}{4}n\rceil, r)$ is 3-rate rearrangeable, i.e., $C(n, \lceil\frac{9}{4}n\rceil, r)$ is multirate rearrangeable when every connection has a weight chosen from $\{p_1, p_2, p_3\}$ where $1 \geq p_1 > p_2 > p_3 > 0$.*

We prove this theorem through several lemmas each of which deals with a special case of the theorem.

LEMMA 3.1. *Suppose $\frac{1}{5} \geq p_2$. Then $C(n, \lceil\frac{9}{4}n\rceil, r)$ is 3-rate rearrangeable.*

PROOF. Setting $B = \frac{1}{5}$ in Lemma 2.5, it is clear that all p_2- and p_3-calls in any compatible request frame can be routed through $\lceil\frac{5}{4}n\rceil$ central switches without blocking. From the 1-rate rearrangeability, all p_1-calls can be routed through n central switches. Thus, the lemma follows. □

LEMMA 3.2. *Suppose $\frac{1}{8} \geq p_3$. Then $C(n, \lceil\frac{9}{4}n\rceil, r)$ is 3-rate rearrangeable.*

PROOF. From the 1-rate rearrangeability, all p_1-calls and p_2-calls in any compatible request frame can be routed through $2n$ central switches without blocking. By Lemma 2.2, the lemma follows immediately. □

From the above two lemmas, we may assume $p_2 > \frac{1}{5}$ and $p_3 > \frac{1}{8}$. Moreover, note that $C(n, 2n, r)$ is three-rate rearrangeable if one of the following two conditions is satisfied:

(a): $\frac{1}{2} \geq p_1$;
(b): At least two of the three rates lie in the same interval $(\frac{1}{k+1}, \frac{1}{k}]$, for some positive integer k;
(c): $p_3 > \frac{1}{5}$.

Therefore, we may assume $1 \geq p_1 > \frac{1}{2} \geq p_2 > \frac{1}{5} \geq p_3 > \frac{1}{8}$.

LEMMA 3.3. *Suppose* $1 \geq p_1 > \frac{1}{2} \geq p_2 > \frac{1}{3}$, $\frac{1}{k} \geq p_3 > \frac{1}{k+1}$, $7 \geq k \geq 5$. *Then* $C(n, \lceil (2+\frac{1}{k})n \rceil, r)$ *is 3-rate rearrangeable.*

PROOF. Without loss of generality, assume that k divides n. If $p_1+p_2 \leq 1$, it is clear that all p_1-calls and p_2-calls in a compatible request frame can be successfully routed through n central switches by CAP algorithm. It follows that $2n$ central switches suffice. Thus, we may assume $p_1+p_2 > 1$. We also assume that k divides n (the case that k does not divide n can be similarly discussed).

From $p_1 + p_2 > 1$, we learn that an input/output is able to generate calls, at its best, in one of the following four forms:

$$\begin{array}{rl} \text{one } p_1\text{-call and } k_1\ p_3\text{-calls} & (k_1 = \lfloor \frac{1-p_1}{p_3} \rfloor), \\ \text{one } p_2\text{-call and } k_2\ p_3\text{-calls} & (k_2 = \lfloor \frac{1-p_2}{p_3} \rfloor), \\ \text{two } p_2\text{-calls and } k_3\ p_3\text{-calls} & (k_3 = \lfloor \frac{1-2p_2}{p_3} \rfloor), \\ k\ p_3\text{-calls.} & \end{array}$$

Given any frame $\mathcal{F}$, let the first n central switches route all p_1-calls, the second n central switches route all p_2-calls. This is possible according to 1-rate rearrangeability. And then route every p_3-call (in an arbitrary ordering) through any central switch on which there is enough space. We claim that no blocking will occur. Suppose to the contrary an (i, j, p_3)-call is blocked, where i is an input on input switch I and j an output on output switch J, define the I-load (the J-load) of a central switch as the sum of weights of all calls from input switch I to the central switch (from the central switch to output switch J). Then each of the $(2+\frac{1}{k})n$ central switches must have its I-load or its J-load greater than $1-p_3$. Note that every load which is greater than $1-p_3$ must be in the following four forms:

$$\begin{array}{l} p_1 + k_1 p_3, \\ p_2 + k_2 p_3, \\ 2p_2 + k_3 p_3, \\ k p_3. \end{array}$$

Among the first $2n$ central switches we assume that there are x_1 (y_1) ones each of which has its I (J)-load of $p_1 + k_1 p_3$; $x_2(y_2)$ ones each of which has its $I(J)$-load of $p_2 + k_2 p_3$; $x_3(y_3)$ ones each of which has its $I(J)$-load of $2p_2 + k_3 p_3$; $x_4(y_4)$ ones each of which has its $I(J)$-load of kp_3; and they are all distinct (i.e., for example, if a central switch has its I-load of $2p_2 + k_3 p_3$ and its J-load of $p_2 + k_2 p_3$, then this switch is counted either into x_3 or into y_2, but not both.) It follows that

$$(x_1 + y_1) + (x_2 + y_2) + (x_3 + y_3) + (x_4 + y_4) = 2n$$

and

$$x_1 + y_1 \leq n, \quad (x_2 + y_2) + (x_3 + y_3) \leq n.$$

Let $N(X)$ be the number of p_3-calls generated by (input or output) switch X in $\mathcal{F}$. As the last n/k central switches are left for p_3-calls only, we have

$$N(I) + N(J) \geq k_1(x_1 + y_1) + k_2(x_2 + y_2) + k_3(x_3 + y_3) + k(x_4 + y_4) + n + 2. \quad (1)$$

Suppose there are two inputs on input switch I which are asked to generate totally two p_2-calls and as many p_3-calls as possible. If each of them generates one p_2-calls, then each is still able to generate k_2 p_3-calls, or equivalently, one p_2-call in the call form containing exactly a p_2-call is accompanied with k_2 p_3-calls; If one of the two inputs generates two p_2-calls, while the other generates no p_2-call, then they are still able to generate $k_3 + k$ p_3-calls, or equivalently, one p_2-call in the call form containing exactly two p_2-calls is accompanied with $(k_3 + k)/2$ p_3-calls. We define a *contribution* $f(l)$ of the call form containing l p_2-call(s) as the number of p_3-calls accompanied with a p_2-call in that form, $l = 1, 2$. Thus, $f(1) = k_2$ and $f(2) = (k_3 + k)/2$.

Obviously, $2 - 2p_3 < 2p_2 - 2k_2p_3 \leq 2$, $2 - 2p_3 < 2p_2 - (k_3 + k)p_3 \leq 2$. It follows that $|2k_2 - (k_3 + k)| \leq 1$, and henceforth,

$$|f(1) - f(2)| \leq \frac{1}{2}.$$

Since switch I generates in $\mathcal{F}$ at least x_1 p_1-calls and at least $s = x_2 + 2x_3 (\leq 2(n - x_1))$ p_2-calls, the number of p_3-calls generated by switch I is ($s = x_2 + 2x_3$)

$$\begin{aligned} N(I) \leq\ & k_1x_1 + k(n - x_1 - s/2) + k_3 \cdot s/2 + s/2 \\ =\ & k_1x_1 + k(n - x_1) + (k_3 + 1 - k)s/2 \\ =\ & k_1x_1 + k(n - x_1 - x_2 - x_3) + (k + k_3 + 1)x_2/2 + (k_3 + 1)x_3 \\ \leq\ & k_1x_1 + k(n - x_1 - x_2 - x_3) + (k_2 + 1)x_2 + (k_3 + 1)x_3. \end{aligned}$$

Therefore, the sum of the number of p_3-calls generated by switch I and the number of p_3-calls generated by switch J is

$$\begin{aligned} & N(I) + N(J) \\ \leq\ & k_1(x_1 + y_1) + k_2(x_2 + y_2) + k_3(x_3 + y_3) + k(x_4 + y_4) + (x_2 + y_2) + (x_3 + y_3) \\ \leq\ & k_1(x_1 + y_1) + k_2(x_2 + y_2) + k_3(x_3 + y_3) + k(x_4 + y_4) + n \\ \leq\ & N(I) + N(J) - 2. \end{aligned}$$

This is a contradiction. The proof is finished. □

LEMMA 3.4. *Suppose* $1 \geq p_1 > \frac{1}{2}, \frac{1}{3} \geq p_2 > \frac{1}{4}, \frac{1}{k} \geq p_3 > \frac{1}{k+1}$, $7 \geq k \geq 5$. *If* $k = 5$, *then* $C(n, \lceil \frac{9}{4}n \rceil, r)$ *is 3-rate rearrangeable. If* $k = 6$, *then* $C(n, \lceil \frac{20}{9}n \rceil, r)$ *is 3-rate rearrangeable. If* $k = 7$, *then* $C(n, \lceil \frac{15}{7}n \rceil, r)$ *is 3-rate rearrangeable.*

PROOF. Consider three cases.

Case 1. $k = 5$. For any request frame $\mathcal{F}$, construct the induced graph G and let L denote the subgraph of G consisting of edges with a weight of p_1, M denote the subgraph consisting of edges with a weight of p_2 and S denote the subgraph consisting of edges with a weight of p_3. Let l_I (m_I, s_I) be the number of p_1- (p_2-, p_3-) edges incident to vertex I.

If $p_1 + p_2 + p_3 \leq 1$, then by CAP algorithm G can be $5n/3$-colored since $l_I \leq n$, $l_I + m_I \leq 3n$ and $l_I + m_I + s_I \leq 5n$.

If $p_1 + p_2 + p_3 > 1 \geq p_1 + 2p_3$, then $m_I + s_I \leq 5n - 3l_I$. By Lemma 2.4, $M \cup S$ can be partitioned into three edge-disjoint spanning subgraphs, such that the degree of vertex I in each subgraph is less than or equal to $5n/3 - l_I$. Note that the union of one such subgraph and L can be n-colored by CAP algorithm and the union of the other two subgraphs can be $10n/9$-colored. Hence, G can be $19n/9$-colored.

If $p_1 + 2p_3 > 1 \geq p_1 + p_2$, then $m_I + s_I \leq 5n - 4l_I$. Similarly, $M \cup S$ can be partitioned into four edge-disjoint spanning subgraphs, such that the degree of vertex I in each subgraph is less than or equal to $5n/4 - l_I$. Therefore, the union of one such subgraph and L can be n-colored by CAP algorithm and the union of the other three subgraphs can be $5n/4$-colored. Hence, G can be $9n/4$-colored.

If $p_1+p_2 > 1 \geq p_1+p_3$, then $m_I = m_{I1}+2m_{I2}+3m_{I3}$, where m_{Ij} denotes the number of inputs/outputs on switch I each of which generates exactly j p_2-calls, $j = 1,2,3$. We also have $s_I \leq 5n - 4l_I - m_{I1} - 3m_{I2} - 4m_{I3}$. Using Lemma 2.4, together with CAP algorithm, the union of one third of M, one fifth of S and L can be n-colored; the union of two thirds of M and four ninths of S can be $22n/27$-colored; the union of the other sixteen forty-fifths of S can be $16n/45$-colored. Therefore, G can be $(2 + 23/135)n$-colored.

If $p_1 + p_3 > 1$, then $m_I = m_{I1} + 2m_{I2} + 3m_{I3}$, where m_{Ij} denotes the number of inputs/outputs on switch I each of which generates exactly j p_2-calls, $j = 1,2,3$. We also have $s_I \leq 5n - 5l_I - m_{I1} - 3m_{I2} - 4m_{I3}$. Using Lemma 2.4, let M^* denote the union of M and two thirds of S. Then, by CAP algorithm, the union of L and three elevenths of M^* can be n-colored; the union of eight elevenths of M^* can be $8n/9$-colored; the remaining one third of S can be $n/3$-colored. Therefore, G can be $20n/9$-colored.

Cases 2 and 3. $k = 6$ and $k = 7$. Let us first discuss these two cases together. Without loss of generality, assume that 9 divides n for $k = 6$ (7 divides n for $k = 7$). Note that when $k = 6$, if $p_1 + p_2 + p_3 \leq 1$, then $C(n, 2n, r)$ is 3-rate rearrangeable by CAP algorithm; when $k = 7$, if $p_1+p_2+p_3 \leq 1$, then in H, let L', M', S' denote the subgraph consisting of all p_1-, p_2-, p_3-edges respectively. Partition S' into seven matchings using Lemma 2.4, it follows that the translation of one such matching can be $n/7$-colored and its complement can be $2n$-colored. That is, $C(n, 15n/7, r)$ is 3-rate rearrangeable. Thus, we may assume in the following that $p_1+p_2+p_3 > 1$.

For any request frame $\mathcal{F}$, let the first n central switches route all p_1-calls, the second n central switches route all p_2-calls. This is possible by 1-rate rearrangeability. Then route the p_3-calls in an arbitrary ordering. We claim that for any p_3-call, there is a space on some central switch which is large enough to place this p_3-call. Suppose to the contrary, there is an (i, j, p_3)-call blocked, where i is an input on input switch I and j an output on output switch J, define the I-load (the J-load) of a central switch as the sum of weights of all calls from input switch I to the central switch (from the central switch to output switch J).

Clearly, every central switch should have its I-load or its J-load greater than $1 - p_3$. Note that every load which is greater than $1 - p_3$ must be in the following five forms according to the above routing rule:

$$\begin{aligned}
p_1 + k_1p_3, &\quad (k_1 = \lfloor \tfrac{1-p_1}{p_3} \rfloor),\\
p_2 + k_2p_3, &\quad (k_2 = \lfloor \tfrac{1-p_2}{p_3} \rfloor),\\
2p_2 + k_3p_3, &\quad (k_3 = \lfloor \tfrac{1-2p_2}{p_3} \rfloor),\\
3p_2 + k_4p_3, &\quad (k_4 = \lfloor \tfrac{1-3p_2}{p_3} \rfloor),\\
kp_3. &
\end{aligned}$$

And these parameters, k_2, k_3, k_4, satisfy:

$$\begin{aligned}
|2k_2 - (k_3 + k)| &\leq 1,\\
|3k_2 - (k_4 + 2k)| &\leq 2,\\
|3k_3 - (2k_4 + k)| &\leq 2.
\end{aligned} \tag{2}$$

If $p_1 + p_2 > 1$, then an input/output is able to generate calls, at its best, in one of the following five forms:

one p_1-call and k_1 p_3-calls,
one p_2-call and k_2 p_3-calls,
two p_2-calls and k_3 p_3-calls,
three p_2-calls and k_4 p_3-calls,
k p_3-calls.

If $p_1 + p_2 \leq 1$, then an input/output is able to generate calls, at its best, in an additional form:

one p_1-call, one p_2-call and h p_3-calls.

Similarly as in the proof of Lemma 3.3, define a contribution $f(l)$ of the call form containing l p_2-call(s) as the number of p_3-calls accompanied with a p_2-call in that call form, $l = 1, 2, 3$. Then $f(1) = k_2$, $f(2) = (k_3 + k)/2$ and $f(3) = (k_4 + 2k)/3$, and

$$|f(1) - f(2)| \leq \tfrac{1}{2},$$
$$|f(1) - f(3)| \leq \tfrac{2}{3},$$
$$|f(2) - f(3)| \leq \tfrac{1}{3}.$$

Among the first $2n$ central switches we assume that there are x_1 (y_1) ones each of which has its I (J)-load of $p_1 + k_1p_3$; $x_2(y_2)$ ones each of which has its $I(J)$-load of $p_2 + k_2p_3$; $x_3(y_3)$ ones each of which has its $I(J)$-load of $2p_2 + k_3p_3$; $x_4(y_4)$ ones each of which has its $I(J)$-load of $3p_2 + k_4p_3$; $x_5(y_5)$ ones each of which has its $I(J)$-load of kp_3; and they are all distinct (i.e., for example, if a center switch has its I-load of $p_2 + k_2p_3$ and its J-load of $2p_2 + k_3p_3$, then this switch is counted either into x_2 or into y_3, but not both). It follows that

$$(x_1 + y_1) + (x_2 + y_2) + (x_3 + y_3) + (x_4 + y_4) + (x_5 + y_5) = 2n$$

and

$$x_1 + y_1 \leq n, \quad (x_2 + y_2) + (x_3 + y_3) + (x_4 + y_4) \leq n.$$

Let $N(X)$ be the number of p_3-calls generated by (input or output) switch X in $\mathcal{F}$. Note that the last $2n/9$ ($n/7$ for $k = 7$) central switches are left for p_3-calls only, we have

$$\begin{aligned} & N(I) + N(J) \\ \geq \; & k_1(x_1 + y_1) + k_2(x_2 + y_2) + k_3(x_3 + y_3) + k_4(x_4 + y_4) + k(x_5 + y_5) + N_1 + 2, \end{aligned} \tag{3}$$

where $N_1 = 4n/3$ when $k = 6$, $N_1 = n$ when $k = 7$. Since switch I generates in $\mathcal{F}$ at least x_1 p_1-calls and $s = x_2 + 2x_3 + 3x_4$ p_2-calls, in order to generate the most number of p_3-calls, it generates the more number of call form(s) with a larger contribution the better. Notice that $f(1)$ is an integer, and $f(2), f(3)$ is an integral multiple of $\frac{1}{2}$, $\frac{1}{3}$, respectively. Moreover, if $f(1) > f(2) > f(3)$, then it is clear by (2) that $2k_2 - (k_3 + k) = 1$, $3k_2 - (k_4 + 2k) = 2$ and $3k_3 - (2k_4 + k) = 1$. It follows from these that $k_2 + k_4 = 2k_3$. Next, we discuss $k = 6, 7$ respectively.

Case 2. $k = 6$. In this case, $4 \leq k_2 \leq 5$, $2 \leq k_3 \leq 3$, $0 \leq k_4 \leq 1$. First, let us assume that $2p_3 < p_2 < 3p_3$. From (2), we derive that $(k_2, k_3, k_4) = (4, 2, 0)$.

Clearly, $f(3) = f(2) = f(1) = 4$. If $p_1 + p_2 > 1$ or $p_2 < k_1 p_3$, then $k_1 = 3$ and $(s = x_2 + 2x_3 + 3x_4)$

$$\begin{aligned} N(I) \leq & \begin{cases} k_1 x_1 + 6(n - x_1 - s/3), & s \leq 3(n - x_1), \\ k_1(3(n - x_1) + x_1 - s), & 3(n - x_1) < s, \end{cases} \\ \leq & \ k_1 x_1 + 6(n - x_1) - 2s. \end{aligned}$$

If $p_1 + p_2 \leq 1$ and $p_2 > k_1 p_3$, then $k_1 = 2$ and $(s = x_2 + 2x_3 + 3x_4 - x_1)$

$$\begin{aligned} N(I) \leq & \begin{cases} 6(n - x_1) - k_1 s, & s < 0, \\ 6(n - x_1) - 2s, & s \geq 0, \end{cases} \\ \leq & \ 6(n - x_1) - 2s, \end{aligned}$$

Therefore, the sum of the number of p_3-calls generated by switch I and the number of p_3-calls generated by switch J is

$$N(I) + N(J) \leq k_1(x_1 + y_1) + 4(x_2 + y_2) + 2(x_3 + y_3) + 6(x_5 + y_5),$$

which contradicts (3).

If $p_3 < p_2 < 2p_3$, from (2), we derive that the vector (k_2, k_3, k_4) takes a value of $(5, 3, 1)$, $(4, 3, 1)$, $(4, 2, 1)$ or $(4, 2, 0)$.

Subcase 2.1. $(k_2, k_3, k_4) = (5, 3, 1)$. Clearly, $f(1) = 5 > f(2) = 4.5 > f(3) = 13/3$. If $p_1 + p_2 > 1$ or $p_2 < k_1 p_3$, then $k_1 = 2$ and the number of p_3-calls generated by switch I is $(s = x_2 + 2x_3 + 3x_4)$

$$\begin{aligned} N(I) \leq & \begin{cases} k_1 x_1 + 6(n - x_1 - s) + 5s, & s \leq n - x_1, \\ k_1 x_1 + 5((n - x_1) - \frac{s-(n-x_1)}{2}) + \frac{s-(n-x_1)}{2}, & n - x_1 < s \leq 3(n - x_1), \\ (n - x_1) + k_1(3(n - x_1) + x_1 - s), & 3(n - x_1) < s, \end{cases} \\ \leq & \ k_1 x_1 + \min\{6(n - x_1) - s, 7(n - x_1) - 2s\}. \end{aligned}$$

If $p_1 + p_2 \leq 1$ and $p_2 > k_1 p_3$, then $k_1 = 1$ and the number of p_3-calls generated by switch I is $(s = x_2 + 2x_3 + 3x_4 - x_1)$

$$\begin{aligned} N(I) \leq & \begin{cases} 6(n - x_1) - k_1 s, & s < 0, \\ 6(n - x_1) - s, & 0 \leq s \leq n - x_1, \\ 7(n - x_1) - 2s, & n - x_1 < s, \end{cases} \\ \leq & \ \min\{6(n - x_1) - s, 7(n - x_1) - 2s\}, \end{aligned}$$

Therefore, the sum of the number of p_3-calls generated by switch I and the number of p_3-calls generated by switch J is

$$\begin{aligned} & N(I) + N(J) \\ \leq & \ k_1(x_1 + y_1) + 5(x_2 + y_2) + 3(x_3 + y_3) + (x_4 + y_4) + 6(x_5 + y_5) \\ & \quad + \min\{(x_3 + y_3) + 2(x_4 + y_4), (x_1 + y_1) + (x_5 + y_5)\} \\ \leq & \ k_1(x_1 + y_1) + 5(x_2 + y_2) + 3(x_3 + y_3) + (x_4 + y_4) + 6(x_5 + y_5) + 4n/3, \end{aligned}$$

which contradicts (3).

Subcase 2.2. $(k_2, k_3, k_4) = (4, 3, 1)$. Clearly, $f(2) = 4.5 > f(3) = 13/3 > f(1) = 4$. If $p_1 + p_2 > 1$ or $p_2 < k_1 p_3$, then $k_1 = 2$ and $(s = x_2 + 2x_3 + 3x_4)$

$$\begin{aligned} N(I) \leq & \begin{cases} k_1 x_1 + 6(n - x_1 - s/2) + 3 \cdot s/2, & s \leq 2(n - x_1), \\ k_1 x_1 + 3(3(n - x_1) - s) + (s - 2(n - x_1)), & 2(n - x_1) < s \leq 3(n - x_1), \\ (n - x_1) + k_1(3(n - x_1) + x_1 - s), & 3(n - x_1) < s, \end{cases} \\ \leq & \ k_1 x_1 + 6(n - x_1) - 1.5s, \end{aligned}$$

If $p_1 + p_2 \le 1$ and $p_2 > k_1 p_3$, then $k_1 = 1$ and $(s = x_2 + 2x_3 + 3x_4 - x_1)$

$$\begin{aligned} N(I) \le & \begin{cases} 6(n - x_1) - k_1 s, & s < 0, \\ 6(n - x_1) - 1.5s, & s \ge 0, \end{cases} \\ \le & \ 6(n - x_1) - 1.5s, \end{aligned}$$

Therefore, the sum of the number of p_3-calls generated by switch I and the number of p_3-calls generated by switch J is

$$\begin{aligned} & N(I) + N(J) \\ \le & \ k_1(x_1 + y_1) + 4.5(x_2 + y_2) + 3(x_3 + y_3) + 1.5(x_4 + y_4) + 6(x_5 + y_5) + 0.5(x_1 + y_1) \\ \le & \ k_1(x_1 + y_1) + 4(x_2 + y_2) + 3(x_3 + y_3) + (x_4 + y_4) + 6(x_5 + y_5) + n, \end{aligned}$$

which contradicts (3).

Subcase 2.3. $(k_2, k_3, k_4) = (4, 2, k_4)$ with $0 \le k_4 \le 1$. Clearly, $f(3) = 4 + k_4/3 \ge f(1) = f(2) = 4$. If $p_1 + p_2 > 1$ or $p_2 < k_1 p_3$, then $k_1 = 2$ and $(s = x_2 + 2x_3 + 3x_4)$

$$\begin{aligned} N(I) \le & \begin{cases} k_1 x_1 + 6(n - x_1 - s/3) + k_4 \cdot s/3, & s \le 3(n - x_1), \\ k_4(n - x_1) + k_1(3(n - x_1) + x_1 - s), & 3(n - x_1) < s, \end{cases} \\ \le & \ k_1 x_1 + 6(n - x_1 - s/3) + k_4 \cdot s/3, \end{aligned}$$

If $p_1 + p_2 \le 1$ and $p_2 > k_1 p_3$, then $k_1 = 1$ and $(s = x_2 + 2x_3 + 3x_4 - x_1)$

$$\begin{aligned} N(I) \le & \begin{cases} 6(n - x_1) - k_1 s, & s < 0, \\ 6(n - x_1) - 2s + k_4 \cdot s/3, & s \ge 0, \end{cases} \\ \le & \ 6(n - x_1) - 2s + k_4 \cdot s/3, \end{aligned}$$

Therefore, the sum of the number of p_3-calls generated by switch I and the number of p_3-calls generated by switch J is

$$\begin{aligned} & N(I) + N(J) \\ \le & \ k_1(x_1 + y_1) + 4(x_2 + y_2) + 2(x_3 + y_3) + k_4(x_4 + y_4) + 6(x_5 + y_5) \\ & \quad + k_4(x_2 + y_2)/3 + 2k_4(x_3 + y_3)/3 + (1 - k_4)(x_1 + y_1)/3 \\ \le & \ k_1(x_1 + y_1) + 4(x_2 + y_2) + 2(x_3 + y_3) + k_4(x_4 + y_4) + 6(x_5 + y_5) + 4n/3, \end{aligned}$$

which contradicts (3).

These contradictions, together with the previous argument, indicate that for $k = 6$, $\lceil \frac{20}{9} n \rceil$ central switches suffice to route all compatible request frame.

Case 3. $k = 7$. In this case, $4 \le k_2 \le 5$, $2 \le k_3 \le 3$, $0 \le k_4 \le 1$. Let us first assume $2p_3 < p_2 < 3p_3$. From (2), we derive that the vector (k_2, k_3, k_4) takes a value of $(5, 3, 1)$, $(5, 3, 0)$, $(5, 2, 0)$ or $(4, 2, 0)$.

Subcase 3.1. $(k_2, k_3, k_4) = (5, 3, 1)$. Clearly, $f(1) = f(2) = f(3) = 5$. If $p_1 + p_2 > 1$, then

$$\begin{aligned} N(I) \le & \ k_1 x_1 + 7(n - x_1 - s/3) + s/3, \\ = & \ k_1 x_1 + 7(n - x_1) - 2s, \end{aligned}$$

where $s = x_2 + 2x_3 + 3x_4$.

If $p_1 + p_2 \le 1$ and $k_1 p_3 > p_2$, then $k_1 = 3$ and

$$\begin{aligned} N(I) \le & \begin{cases} k_1 x_1 + 7(n - x_1) - 2s, & s \le 3(n - x_1), \\ (n - x_1) + k_1(x_1 + 3(n - x_1) - s), & 3(n - x_1) < s, \end{cases} \\ \le & \ k_1 x_1 + 7(n - x_1) - 2s, \end{aligned}$$

where $s = x_2 + 2x_3 + 3x_4$.

If $p_1 + p_2 \le 1$ and $k_1 p_3 < p_2$, then $k_1 = 2$ and

$$N(I) \le 7(n - x_1) - 2s,$$

where $s = x_2 + 2x_3 + 3x_4 - x_1$.

Therefore, the sum of the number of p_3-calls generated by switch I and the number of p_3-calls generated by switch J is

$$\begin{aligned} & N(I) + N(J) \\ \leq \ & k_1(x_1 + y_1) + 5(x_2 + y_2) + 3(x_3 + y_3) + (x_4 + y_4) + 7(x_5 + y_5), \end{aligned}$$

which contradicts (3).

Subcase 3.2. $(k_2, k_3, k_4) = (5, 3, 0)$. Clearly, $f(2) = f(1) = 5 > f(3) = 14/3$. If $p_1 + p_2 > 1$, then $k_1 \leq 2$ and

$$\begin{aligned} N(I) \leq \ & k_1 x_1 + \begin{cases} 7(n - x_1 - s/2) + 3 \cdot s/2, & s \leq 2(n - x_1), \\ 3(3(n - x_1) - s), & 2(n - x_1) < s, \end{cases} \\ \leq \ & k_1 x_1 + 7(n - x_1) - 2s, \end{aligned}$$

where $s = x_2 + 2x_3 + 3x_4$.

If $p_1 + p_2 \leq 1$ and $k_1 p_3 > p_2$, then $k_1 = 3$ and

$$\begin{aligned} N(I) \leq \ & \begin{cases} k_1 x_1 + 7(n - x_1) - 2s, & s \leq 3(n - x_1), \\ k_1(x_1 + 3(n - x_1) - s), & 3(n - x_1) < s, \end{cases} \\ \leq \ & k_1 x_1 + 7(n - x_1) - 2s, \end{aligned}$$

where $s = x_2 + 2x_3 + 3x_4$.

If $p_1 + p_2 \leq 1$ and $k_1 p_3 < p_2$, then $k_1 = 2$ and

$$N(I) \leq 7(n - x_1) - 2s,$$

where $s = x_2 + 2x_3 + 3x_4 - x_1$.

Therefore, the sum of the number of p_3-calls generated by switch I and the number of p_3-calls generated by switch J is

$$\begin{aligned} & N(I) + N(J) \\ \leq \ & k_1(x_1 + y_1) + 5(x_2 + y_2) + 3(x_3 + y_3) + (x_4 + y_4) + 7(x_5 + y_5), \\ \leq \ & k_1(x_1 + y_1) + 5(x_2 + y_2) + 3(x_3 + y_3) + 7(x_5 + y_5) + n, \end{aligned}$$

which contradicts (3).

Subcase 3.3. $(k_2, k_3, k_4) = (5, 2, 0)$. Clearly, $f(1) = 5 > f(3) = 14/3 > f(2) = 4.5$. If $p_1 + p_2 > 1$, then $k_1 \leq 2$ and

$$\begin{aligned} N(I) \leq \ & k_1 x_1 + \begin{cases} 7(n - x_1 - s) + 5s, & s \leq n - x_1, \\ 5(n - x_1 - (s - (n - x_1))/2), & n - x_1 < s, \end{cases} \\ \leq \ & k_1 x_1 + 7(n - x_1) - 2s, \end{aligned}$$

where $s = x_2 + 2x_3 + 3x_4$.

If $p_1 + p_2 \leq 1$ and $k_1 p_3 > p_2$, then $k_1 = 3$ and

$$\begin{aligned} N(I) \leq \ & \begin{cases} k_1 x_1 + 7(n - x_1) - 2s, & s \leq 3(n - x_1), \\ k_1(x_1 + 3(n - x_1) - s), & 3(n - x_1) < s, \end{cases} \\ \leq \ & k_1 x_1 + 7(n - x_1) - 2s, \end{aligned}$$

where $s = x_2 + 2x_3 + 3x_4$.

If $p_1 + p_2 \leq 1$ and $k_1 p_3 < p_2$, then $k_1 = 2$ and

$$N(I) \leq 7(n - x_1) - 2s,$$

where $s = x_2 + 2x_3 + 3x_4 - x_1$.

Therefore, the sum of the number of p_3-calls generated by switch I and the number of p_3-calls generated by switch J is

$$\begin{aligned} & N(I)+N(J) \\ \leq\ & k_1(x_1+y_1)+5(x_2+y_2)+3(x_3+y_3)+(x_4+y_4)+7(x_5+y_5), \\ \leq\ & k_1(x_1+y_1)+5(x_2+y_2)+2(x_3+y_3)+7(x_5+y_5)+n, \end{aligned}$$

which contradicts (3).

Subcase 3.4. $(k_2,k_3,k_4)=(4,2,0)$. Clearly, $f(3)=14/3>f(2)=4.5>f(1)=4$. If $p_1+p_2>1$, then $k_1\leq 2$ and

$$N(I)\leq k_1x_1+7(n-x_1-s/3),$$

where $s=x_2+2x_3+3x_4$.

If $p_1+p_2\leq 1$ and $k_1p_3>p_2$, then $k_1=3$ and

$$\begin{aligned} N(I) &\leq \begin{cases} k_1x_1+7(n-x_1-s/3), & s\leq 3(n-x_1), \\ k_1(x_1+3(n-x_1)-s/3), & 3(n-x_1)<s, \end{cases} \\ &\leq k_1x_1+7(n-x_1-s), \end{aligned}$$

where $s=x_2+2x_3+3x_4$.

If $p_1+p_2\leq 1$ and $k_1p_3<p_2$, then $k_1=2$ and

$$N(I)\leq 7(n-x_1)-7s/3,$$

where $s=x_2+2x_3+3x_4-x_1$.

Therefore, the sum of the number of p_3-calls generated by switch I and the number of p_3-calls generated by switch J is

$$\begin{aligned} & N(I)+N(J) \\ \leq\ & k_1(x_1+y_1)+14(x_2+y_2)/3+7(x_3+y_3)/3+7(x_5+y_5)+(x_1+y_1)/3, \\ \leq\ & k_1(x_1+y_1)+4(x_2+y_2)+2(x_3+y_3)+7(x_5+y_5)+n, \end{aligned}$$

which contradicts (3).

If $p_3<p_2<2p_3$, we derive from (2) that $(k_2,k_3,k_4)=(5,3,1)$. Furthermore, since $f(1)=f(2)=f(3)=5$, if $p_1+p_2>1$, then

$$N(I)\leq k_1x_1+7(n-x_1-s/3)+s/3,$$

where $s=x_2+2x_3+3x_4$.

If $p_1+p_2\leq 1$ and $k_1p_3>p_2$, then $k_1=2$ and

$$\begin{aligned} N(I) &\leq \begin{cases} k_1x_1+7(n-x_1)-2s, & s\leq 3(n-x_1), \\ (n-x_1)+k_1(x_1+3(n-x_1)-s), & 3(n-x_1)<s, \end{cases} \\ &\leq k_1x_1+7(n-x_1)-2s, \end{aligned}$$

where $s=x_2+2x_3+3x_4$.

If $p_1+p_2\leq 1$ and $k_1p_3<p_2$, then $k_1=1$ and

$$N(I)\leq 7(n-x_1)-2s,$$

where $s=x_2+2x_3+3x_4-x_1$.

Therefore, the sum of the number of p_3-calls generated by switch I and the number of p_3-calls generated by switch J is

$$\begin{aligned} & N(I)+N(J) \\ \leq\ & k_1(x_1+y_1)+5(x_2+y_2)+3(x_3+y_3)+(x_4+y_4)+7(x_5+y_5)+(x_1+y_1), \\ \leq\ & k_1(x_1+y_1)+5(x_2+y_2)+3(x_3+y_3)+(x_4+y_4)+7(x_5+y_5)+n, \end{aligned}$$

which contradicts (3).

These contradictions, together with previous argument, show that for $k = 7$, $\lceil \frac{15}{7}n \rceil$ central switches suffice to route all compatible request frame.

The proof of the lemma is finished. □

LEMMA 3.5. *Suppose* $1 \geq p_1 > \frac{1}{2}$, $\frac{1}{4} \geq p_2 > \frac{1}{5}$, $\frac{1}{k} \geq p_3 > \frac{1}{k+1}$, $k = 5, 6, 7$. *Then* $C(n, \lceil \frac{9}{4}n \rceil, r)$ *is 3-rate rearrangeable.*

PROOF. Consider three cases.

Case 1. $k = 5$. The proof is rather easy in this case. Since any input/output switch generates at most $5n$ p_2- and p_3-calls, together with $\frac{1}{4} \geq p_2$, all p_2- and p_3-calls in any compatible request frame can be routed through $\lceil \frac{5}{4}n \rceil$ central switches by the edge-coloring technique [**3**]. It follows from 1-rate rearrangeability that $C(n, \lceil \frac{9}{4}n \rceil, r)$ is 3-rate rearrangeable.

Case 2. $k = 6$. Note that if $p_1 + p_2 + p_3 \leq 1$, then clearly $C(n, 2n, r)$ is rearrangeable by employing CAP algorithm. Thus, we assume in the following that $p_1 + p_2 + p_3 > 1$.

Given any compatible request frame $\mathcal{F}$, construct the induced graph G. Let L denote the subgraph of G consisting of all p_1-edges, M consisting of all p_2- and p_3-edges. If $p_1 + p_3 > 1$, then the degree of vertex I in L is $l_I \leq n$, and the degree of I in M is $m_I \leq 6n - 6l_I$. Partition M into six subgraphs by Lemma 2.4 such that in each of which the degree of I is at most $n - l_I$. It is clear that the union of one such subgraph of M and L can be n-colored and the union of the other five subgraphs can be $5n/4$-colored since $p_2 \leq 1/4$. If $p_1 + p_2 \leq 1$, then the degree of vertex I in L is $l_I \leq n$, the degree of I in M is $m_I \leq 6n - 4l_I$. Partition M into three subgraphs such that in each of which the degree of I is at most $2n - 4l_I/3$. It is clear that the union of one such subgraph of M and L can be n-colored by CAP algorithm and the union of the other two subgraphs can be n-colored since $p_2 \leq 1/4$.

Now, we may assume $p_1 + p_2 > 1 \geq p_1 + p_3$. Consider the weighted connection bipartite graph H constructed from the request frame $\mathcal{F}$. Let $L'(M', S')$ denote the subgraph of H consisting of all $p_1(p_2, p_3)$-edges. Partition S' into three subgraphs S'_1, S'_2, S'_3 using Lemma 2.4, then the degree of vertex i in $M' \cup S'_1 \cup S'_2$ is not greater than 5, while in S'_3 is not greater than 2. Now translate the subgraphs $L' \cup S'_3$, $M' \cup S'_1 \cup S'_2$ of H into subgraphs, denoted by L^*, M^*, of the induced graph G, respectively. It is obvious that the degree of vertex I in L^* is not greater than $2n$, while in M^* is not greater than $5n$. Therefore, L^* can be n-colored by CAP algorithm and M^* can be $5n/4$-colored since $p_2 \leq 1/4$. In short, $C(n, \lceil \frac{9}{4}n \rceil, r)$ is rearrangeable for $k = 6$.

Case 3. $k = 7$. Given any compatible request frame $\mathcal{F}$, construct the induced graph G. Let L denote the subgraph of G consisting of all p_1-edges, M consisting of all p_2- and p_3-edges. If $p_1 + p_2 \leq 1$, then partition M into seven subgraphs using Lemma 2.4. It is clear that the union of five such subgraphs can be $\frac{5}{4}n$-colored since $\frac{1}{4} \geq p_2$ and its complement can be n-colored by CAP algorithm. Thus, we may assume in the following that $p_1 + p_2 > 1$ and turn to the technique developed in the above.

Note that an input/output is able to generate calls, at its best, in one of the following six forms:

$$\begin{array}{rl} \text{one } p_1\text{-call and } k_1\ p_3\text{-calls} & (k_1 = \lfloor \frac{1-p_1}{p_3} \rfloor), \\ \text{one } p_2\text{-call and } k_2\ p_3\text{-calls} & (k_2 = \lfloor \frac{1-p_2}{p_3} \rfloor), \\ \text{two } p_2\text{-calls and } k_3\ p_3\text{-calls} & (k_3 = \lfloor \frac{1-2p_2}{p_3} \rfloor), \\ \text{three } p_2\text{-calls and } k_4\ p_3\text{-calls} & (k_4 = \lfloor \frac{1-3p_2}{p_3} \rfloor), \\ \text{four } p_2\text{-calls and } k_5\ p_3\text{-calls} & (k_5 = \lfloor \frac{1-4p_2}{p_3} \rfloor), \\ \text{seven } p_3\text{-calls.} & \end{array}$$

These parameters, k_1, k_2, k_3, k_4, k_5, satisfy:

$$\begin{array}{c} 0 \le k_1 \le 1,\ 5 \le k_2 \le 6,\ 3 \le k_3 \le 4,\ 1 \le k_4 \le 3,\ 0 \le k_5 \le 1 \text{ and} \\ |2k_2 - (k_3 + k)| \le 1, \quad |3k_2 - (k_4 + 2k)| \le 2, \\ |4k_2 - (k_5 + 3k)| \le 3, \quad |3k_3 - (2k_4 + k)| \le 2, \\ |2k_3 - (k_5 + k)| \le 1, \quad |4k_4 - (3k_5 + k)| \le 3, \end{array}$$

where $k = 7$. It follows that the vector (k_2, k_3, k_4, k_5) takes a value of $(6,4,3,1)$, $(6,4,2,1)$, $(6,4,2,0)$, $(5,4,3,1)$, $(5,4,2,1)$, $(5,4,2,0)$, $(5,3,2,0)$, or $(5,3,1,0)$.

Given any compatible frame $\mathcal{F}$, let the first n central switches route all p_1-calls, the second n central switches route all p_2-calls. This is possible according to 1-rate rearrangeability. And then route every p_3-call in an arbitrary ordering through any center switch on which there is enough space. We claim that no blocking will occur. Suppose to the contrary an (i, j, p_3)-call is blocked, where i is an input on switch I and j an output on switch J, define the I-load (the J-load) of a central switch as the sum of weights of all calls from input switch I to the central switch (from the center switch to output switch J). Then each of the $\frac{9}{4}n$ center switches must have its I-load or its J-load greater than $1 - p_3$. Note that every load which is greater than $1 - p_3$ must be in the following six forms:

$$\begin{array}{l} p_1 + k_1 p_3, \\ p_2 + k_2 p_3, \\ 2p_2 + k_3 p_3, \\ 3p_2 + k_4 p_3, \\ 4p_2 + k_5 p_3, \\ 7p_3. \end{array}$$

Among the first $2n$ central switches we assume that there are x_1 (y_1) central switches each of which has its I (J)-load of $p_1 + k_1 p_3$; $x_2(y_2)$ central switches each of which has its $I(J)$-load of $p_2 + k_2 p_3$; $x_3(y_3)$ central switches each of which has its $I(J)$-load of $2p_2 + k_3 p_3$; $x_4(y_4)$ central switches each of which has its $I(J)$-load of $3p_2 + k_4 p_3$; $x_5(y_5)$ center switches each of which has its $I(J)$-load of $4p_2 + k_5 p_3$; $x_6(y_6)$ central switches each of which has its $I(J)$-load of $7p_3$; and they are all distinct (i.e., for example, if a central switch has its I-load of $p_2 + k_2 p_3$ and its J-load of $2p_2 + k_3 p_3$, then this switch is counted either into x_2 or into y_3, but not both.) It follows that

$$(x_1 + y_1) + (x_2 + y_2) + (x_3 + y_3) + (x_4 + y_4) + (x_5 + y_5) + (x_6 + y_6) = 2n,$$

and

$$x_1 + y_1 \le n, \quad (x_2 + y_2) + (x_3 + y_3) + (x_4 + y_4) + (x_5 + y_5) \le n.$$

Let $N(X)$ be the number of p_3-calls generated by (input or output) switch X in $\mathcal{F}$. Note that the last $\frac{1}{4}n$ central switches are left to route p_3-calls only. We have

$$N(I)+N(J) \geq \begin{array}{l} k_1(x_1+y_1)+k_2(x_2+y_2)+k_3(x_3+y_3)+k_4(x_4+y_4) \\ +k_5(x_5+y_5)+7(x_6+y_6)+7n/4+2. \end{array} \quad (4)$$

Clearly, switch I generates in $\mathcal{F}$ at least x_1 p_1-calls and $x_2+2x_3+3x_4+4x_5(\leq 4(n-x_1))$ p_2-calls. Similarly as in the proofs of the previous lemmas, we define a contribution $f(l)$ of the call-form with l p_2-call(s) as the number of p_3-calls accompanied with a p_2-call in that form, $l=1,2,3,4$. That is, $f(1)=k_2$, $f(2)=(k_3+7)/2$, $f(3)=(k_4+14)/3$ and $f(4)=(k_5+21)/4$.

Subcase 3.1. $(k_2,k_3,k_4,k_5)=(6,4,3,1)$. Since $f(1)>f(3)>f(4)=f(2)$, the number of p_3-calls generated by switch I is

$$\begin{aligned} N(I) \leq\ & k_1x_1+\begin{cases} 7(n-x_1-s)+6s, & s\leq n-x_1, \\ 6(n-x_1-\frac{s-(n-x_1)}{2})+3\cdot\frac{s-(n-x_1)}{2}, & n-x_1<s\leq 3(n-x_1), \\ 3(4(n-x_1)-s)+(s-3(n-x_1)), & 3(n-x_1)<s, \end{cases} \\ \leq\ & k_1x_1+7.5(n-x_1)-1.5s, \end{aligned}$$

where $s=x_2+2x_3+3x_4+4x_5$. Therefore, the sum of the number of p_3-calls generated by switch I and the number of p_3-calls generated by switch J is

$$\begin{aligned} & N(I)+N(J) \\ \leq\ & k_1(x_1+y_1)+6(x_2+y_2)+4.5(x_3+y_3)+3(x_4+y_4)+1.5(x_5+y_5)+7.5(x_6+y_6) \\ \leq\ & k_1(x_1+y_1)+6(x_2+y_2)+4(x_3+y_3)+3(x_4+y_4)+(x_5+y_5)+7(x_6+y_6)+n, \end{aligned}$$

which contradicts (4).

Subcase 3.2. $(k_2,k_3,k_4,k_5)=(6,4,2,1)$. Clearly, $f(1)>f(4)=f(2)>f(3)$, thus, the number of p_3-calls generated by switch I is

$$\begin{aligned} N(I) \leq\ & k_1x_1+\begin{cases} 7(n-x_1-s)+6s, & s\leq n-x_1, \\ \frac{s-(n-x_1)}{3}+6((n-x_1)-\frac{s-(n-x_1)}{3}), & (n-x_1)<s, \end{cases} \\ \leq\ & k_1x_1+23(n-x_1)/3-5s/3, \end{aligned}$$

where $s=x_2+2x_3+3x_4+4x_5$. Therefore, the sum of the number of p_3-calls generated by switch I and the number of p_3-calls generated by switch J is

$$\begin{aligned} & N(I)+N(J) \\ \leq\ & k_1(x_1+y_1)+6(x_2+y_2)+13(x_3+y_3)/3+8(x_4+y_4)/3+(x_5+y_5)+23(x_6+y_6)/3 \\ \leq\ & k_1(x_1+y_1)+6(x_2+y_2)+4(x_3+y_3)+2(x_4+y_4)+(x_5+y_5)+7(x_6+y_6)+4n/3, \end{aligned}$$

which contradicts (4).

Subcase 3.3. $(k_2,k_3,k_4,k_5)=(6,4,2,0)$. Since $f(1)>f(2)>f(3)>f(4)$, and notice that $k_2+k_4=2k_3$, the number of p_3-calls generated by switch I is

$$\begin{aligned} N(I) \leq\ & k_1x_1+\begin{cases} 7(n-x_1)-s, & s\leq n-x_1, \\ (s-(n-x_1))+4((n-x_1)-\frac{s-(n-x_1)}{2}), & n-x_1<s\leq 3(n-x_1), \\ 2(4(n-x_1)-s), & 3(n-x_1)<s, \end{cases} \\ \leq\ & k_1x_1+\min\{7(n-x_1)-s,\ 8(n-x_1)-2s\}, \end{aligned}$$

where $s=x_2+2x_3+3x_4+4x_5$. Therefore, the sum of the number of p_3-calls generated by switch I and the number of p_3-calls generated by switch J is

$$\begin{aligned} & N(I)+N(J) \\ \leq\ & k_1(x_1+y_1)+6(x_2+y_2)+4(x_3+y_3)+2(x_4+y_4)+7(x_6+y_6) \\ & +\min\{(x_3+y_3)+2(x_4+y_4)+3(x_5+y_5),(x_6+y_6)\} \\ \leq\ & k_1(x_1+y_1)+6(x_2+y_2)+4(x_3+y_3)+2(x_4+y_4)+7(x_6+y_6)+3n/2, \end{aligned}$$

which contradicts (4).

Subcase 3.4. $(k_2, k_3, k_4, k_5) = (5, 4, 3, 1)$. Since $f(3) > f(4) = f(2) > f(1)$, the number of p_3-calls generated by switch I is

$$\begin{aligned} N(I) \leq & \ k_1x_1 + \begin{cases} 7(n - x_1 - s/3) + s, & s \leq 3(n - x_1), \\ 3(4(n - x_1) - s) + (s - 3(n - x_1)), & 3(n - x_1) < s, \end{cases} \\ \leq & \ k_1x_1 + 7(n - x_1) - 4s/3, \end{aligned}$$

where $s = x_2 + 2x_3 + 3x_4 + 4x_5$. Therefore, the sum of the number of p_3-calls generated by switch I and the number of p_3-calls generated by switch J is

$$\begin{aligned} & N(I) + N(J) \\ \leq & \ k_1(x_1 + y_1) + 17(x_2 + y_2)/3 + 13(x_3 + y_3)/3 + 3(x_4 + y_4) + 5(x_5 + y_5) + 7(x_6 + y_6) \\ \leq & \ k_1(x_1 + y_1) + 5(x_2 + y_2) + 4(x_3 + y_3) + 3(x_4 + y_4) + (x_5 + y_5) + 7(x_6 + y_6) + 2n/3, \end{aligned}$$

which contradicts (4).

Subcase 3.5. $(k_2, k_3, k_4, k_5) = (5, 4, 2, 1)$. Since $f(4) = f(2) > f(3) > f(1)$, the number of p_3-calls generated by switch I is

$$\begin{aligned} N(I) \leq & \ k_1x_1 + 7(n - x_1 - s/4) + s/4, \\ \leq & \ k_1x_1 + 7(n - x_1) - 1.5s, \end{aligned}$$

where $s = x_2 + 2x_3 + 3x_4 + 4x_5$. Therefore, the sum of the number of p_3-calls generated by switch I and the number of p_3-calls generated by switch J is

$$\begin{aligned} & N(I) + N(J) \\ \leq & \ k_1(x_1 + y_1) + 5.5(x_2 + y_2) + 4(x_3 + y_3) + 2.5(x_4 + y_4) + (x_5 + y_5) + 7(x_6 + y_6) \\ \leq & \ k_1(x_1 + y_1) + 5(x_2 + y_2) + 4(x_3 + y_3) + 2(x_4 + y_4) + (x_5 + y_5) + 7(x_6 + y_6) + 0.5n, \end{aligned}$$

which contradicts (4).

Subcase 3.6. $(k_2, k_3, k_4, k_5) = (5, 4, 2, 0)$. Clearly, we have $f(2) > f(3) > f(4) > f(1)$. Notice that $k_3 + k_5 = 2k_4$, the number of p_3-calls generated by switch I is

$$\begin{aligned} N(I) \leq & \ k_1x_1 + \begin{cases} 7(n - x_1 - s/2) + 4 \cdot s/2, & s \leq 2(n - x_1), \\ 4(2(n - x_1) - s/2), & 2(n - x_1) < s, \end{cases} \\ \leq & \ k_1x_1 + 7(n - x_1) - 1.5s, \end{aligned}$$

where $s = x_2 + 2x_3 + 3x_4 + 4x_5$. Therefore, the sum of the number of p_3-calls generated by switch I and the number of p_3-calls generated by switch J is

$$\begin{aligned} & N(I) + N(J) \\ \leq & \ k_1(x_1 + y_1) + 5.5(x_2 + y_2) + 4(x_3 + y_3) + 2.5(x_4 + y_4) + (x_5 + y_5) + 7(x_6 + y_6) \\ \leq & \ k_1(x_1 + y_1) + 5(x_2 + y_2) + 4(x_3 + y_3) + 2(x_4 + y_4) + 7(x_6 + y_6) + n, \end{aligned}$$

which contradicts (4).

Subcase 3.7. $(k_2, k_3, k_4, k_5) = (5, 3, 2, 0)$. Since we have $f(3) > f(4) > f(2) = f(1)$, the number of p_3-calls generated by switch I is

$$\begin{aligned} N(I) \leq & \ k_1x_1 + \begin{cases} 7(n - x_1 - s/3) + 2 \cdot s/3, & s \leq 3(n - x_1), \\ 2(4(n - x_1) - s), & 3(n - x_1) < s, \end{cases} \\ \leq & \ k_1x_1 + 7(n - x_1) - 5s/3, \end{aligned}$$

where $s = x_2 + 2x_3 + 3x_4 + 4x_5$. Therefore, the sum of the number of p_3-calls generated by switch I and the number of p_3-calls generated by switch J is

$$\begin{aligned} & N(I) + N(J) \\ \leq & \ k_1(x_1 + y_1) + 16(x_2 + y_2)/3 + 11(x_3 + y_3)/3 + 2(x_4 + y_4) + (x_5 + y_5)/3 + 7(x_6 + y_6) \\ \leq & \ k_1(x_1 + y_1) + 5(x_2 + y_2) + 3(x_3 + y_3) + 2(x_4 + y_4) + 7(x_6 + y_6) + 2n/3, \end{aligned}$$

which contradicts (4).

Subcase 3.8. $(k_2, k_3, k_4, k_5) = (5, 3, 1, 0)$. Since we have $f(4) > f(3) = f(2) = f(1)$, the number of p_3-calls generated by switch I is

$$\begin{aligned} N(I) \leq & \ k_1 x_1 + 7(n - x_1 - s/4), \\ \leq & \ k_1 x_1 + 7(n - x_1) - 7s/4, \end{aligned}$$

where $s = x_2 + 2x_3 + 3x_4 + 4x_5$. Therefore, the sum of the number of p_3-calls generated by switch I and the number of p_3-calls generated by switch J is

$$\begin{aligned} & N(I) + N(J) \\ \leq & \ k_1(x_1 + y_1) + 21(x_2 + y_2)/4 + 14(x_3 + y_3)/4 + 7(x_4 + y_4)/4 + 7(x_6 + y_6) \\ \leq & \ k_1(x_1 + y_1) + 5(x_2 + y_2) + 3(x_3 + y_3) + (x_4 + y_4) + 7(x_6 + y_6) + 3n/4, \end{aligned}$$

which contradicts (4).

These contradictions show that for $p_1 + p_2 > 1$, $\lceil \frac{9}{4} n \rceil$ (factually, $\lceil (2 + \frac{3}{14}) n \rceil$) central switches suffice to route $\mathcal{F}$. Thus ends the proof of the lemma. □

References

[1] V.E. Beneš, Mathematical theory of connecting networks and telephone traffic, Academic Press, New York, 1965.

[2] S.-P. Chung and K.W. Ross, On nonblocking multirate interconnection networks, *SIAM J. Comput.*, 20(1991) 726–736.

[3] D.Z. Du, B. Gao, F.K. Hwang and J.H. Kim, On multirate rearrangeable Clos networks, accepted by *SIAM J. Comput.*

[4] G.-H. Lin, D.-Z. Du, X.-D. Hu and G. Xue, On Rearrangeability of Multirate Clos Networks, Dept. of Computer Science, University of Minnesota, TR 96-072.

[5] R. Melen and J.S. Turner, Nonblocking multirate networks, *SIAM J. Comput.*, 18(1989) 301–313.

[6] D. de Werra, Balanced schedules, *Infor. J.*, 9(1971) 453–465.

Institute of Applied Mathematics, Chinese Academy of Sciences, P.O. Box 2734, Beijing 100080, People's Republic of China

E-mail address: `ghlin@AMath11.amt.ac.cn`

Department of Computer Science and Engineering, University of Minnesota, Minneapolis, MN 55455

E-mail address: `dzd@cs.umn.edu`

Department of Computer Science and Engineering, University of Minnesota, Minneapolis, MN 55455

E-mail address: `wuw@cs.umn.edu`

Department of Computer Science and Engineering, University of Minnesota, Minneapolis, MN 55455

E-mail address: `yoo@cs.umn.edu`

Selected Titles in This Series

(Continued from the front of this publication)

13 **Jin-Yi Cai, Editor,** Advances in Computational Complexity Theory

12 **David S. Johnson and Catherine C. McGeoch, Editors,** Network Flows and Matching: First DIMACS Implementation Challenge

11 **Larry Finkelstein and William M. Kantor, Editors,** Groups and Computation

10 **Joel Friedman, Editor,** Expanding Graphs

9 **William T. Trotter, Editor,** Planar Graphs

8 **Simon Gindikin, Editor,** Mathematical Methods of Analysis of Biopolymer Sequences

7 **Lyle A. McGeoch and Daniel D. Sleator, Editors,** On-Line Algorithms

6 **Jacob E. Goodman, Richard Pollack, and William Steiger, Editors,** Discrete and Computational Geometry: Papers from the DIMACS Special Year

5 **Frank Hwang, Fred Roberts, and Clyde Monma, Editors,** Reliability of Computer and Communication Networks

4 **Peter Gritzmann and Bernd Sturmfels, Editors,** Applied Geometry and Discrete Mathematics, The Victor Klee Festschrift

3 **E. M. Clarke and R. P. Kurshan, Editors,** Computer-Aided Verification '90

2 **Joan Feigenbaum and Michael Merritt, Editors,** Distributed Computing and Cryptography

1 **William Cook and Paul D. Seymour, Editors,** Polyhedral Combinatorics